European Commission

The Agricultural Situation in the European Union

1994 Report

Report published in conjunction with the
'General Report on the Activities
of the European Union — 1994'

BRUSSELS ● LUXEMBOURG ● 1995

Cataloguing data can be found at the end of this publication

Luxembourg: Office for Official Publications of the European Communities, 1995

ISBN 92-826-8676-0

Foreword

This report, which is the 20th annual agricultural situation report, is published in conjunction with the 1994 *General Report on the Activities of the European Union*. It is presented in accordance with the procedure laid down in the Declaration on the system for fixing Community farm prices contained in the Accession Documents of 22 January 1972. It was completed in December 1994.

As in previous years, the report falls into two parts. The first presents the agricultural situation and the year's out-turn and opens with a review of the implementation of the 1992 CAP reform. The main policy developments during the year are outlined, together with a broad description of weather and production patterns across the Community.

The economic situation, the trends on principal markets, issues affecting rural development, financing of the common agricultural policy and external trade relations are then discussed.

The second part of the report provides the main statistics on European Union agriculture and gives, in a form that has now become standard, updates of the tables produced in previous reports.

Unfortunately, this year it has still not been possible to include all of the statistics concerning the former German Democratic Republic. However, some of the tables include data covering all of Germany. Production and consumption data have been drawn up covering the whole Community, i.e. including the five new *Länder*.

The statistics are based mainly on data supplied by the Statistical Office of the European Communities (Eurostat). The Directorate-General for Agriculture has updated some of the figures and has sometimes added estimates when final figures were unavailable because of the report's publication date.

As in earlier years, certain subjects covered by the report have also been dealt with in other Commission documents. The reader will thus find references to various publications available from either the Commission, the Office for Official Publications of the European Communities or Eurostat.

Contents

Principal abbreviations used

ACP	=	African, Caribbean and Pacific countries
AWU	=	Annual work unit
CAP	=	Common agricultural policy
COM	=	Common organization of the market
EAGGF	=	European Agricultural Guidance and Guarantee Fund
EFTA	=	European Free Trade Association
EMS	=	European Monetary System
FADN	=	Farm accountancy data network
GATT	=	General Agreement on Tariffs and Trade
MCA	=	Monetary compensatory amount
MGQ	=	Maximum guaranteed quantity
NVA	=	Net value-added
UAA	=	Utilized agricultural area
USDA	=	United States Department of Agriculture

I — The agricultural year

CAP reform working well

1. For European Union agriculture, 1994 was, more than anything else, the year which provided indisputable evidence that reform of the common agricultural policy (CAP) was working and, what was more, working better than had originally been predicted. Production and surpluses were down for the second year running whilst internal consumption and farm incomes were up. It had provided the Union with the means to match its domestic policy needs with its external trade policy objectives and was showing itself to be generally beneficial in environmental terms. This was in striking contrast to much of the ill-informed and outdated criticism of the CAP that continued to be repeated regardless of the facts that were clearly visible for those who cared to look.

2. The year, of course, saw other important events with significance for our farm sector. There was the favourable conclusion of intensive accession negotiations with Austria, Finland, Norway and Sweden, followed by encouragingly positive referendum results in three of these countries but with one preferring to go its own way.

3. Signature and ratification of the agreements reached in the Uruguay Round negotiations also took place during 1994 prior to their entering into force the following year and presenting our farm sector with challenges which would have been impossible to accept without a successfully reformed CAP.

4. Alongside the real and solid achievements of successful results in those sectors where reform was introduced in 1993, the Commission came forward with its ideas for future policy in major areas remaining: wine, sugar and fruit and vegetables. As a result, most product sectors have now been covered either by unanimous Council decision or by Commission proposals for reform.

5. The encouraging early results of reform already emerging towards the end of 1993 and which were noted in that year's report,[1] were unambiguously confirmed when firm figures for the programme's first full year became available during 1994. This was particularly the case in the cereals sector.

---

[1] *The Agricultural Situation in the Community — 1993 Report.*

Reform: second successful year

6.   However, and even more encouraging, the release of preliminary data for the 1994/95 marketing year showed that the second year of reform appeared to be just as successful. Not only had total cereals production been brought down two years running, from a typical pre-reform figure of 180 million tonnes to 163 million tonnes in 1993/94 and to an estimated 161 to 162 million tonnes in 1994/95, with latest figures edging even lower, but production figures for both years were below the forecasts of 164 to 165 million tonnes on which reform proposals were made and on which decisions were taken by ministers and on the basis of which the Commission made its assertion that the Uruguay Round settlement was compatible with CAP reform.

7.   One of the major contributory causes of this cut-back in production was, undoubtedly, the success of the set-aside programme, introduced as a flexible market management tool and forming an integral part of the whole reform process.

8.   The area set aside under reform programmes increased by 1.3 million hectares, or by almost 30 %, in 1994/95 to 5.9 million hectares. This expansion was hardly surprising given the generous increase in premia awarded by ministers at the end of 1993 and with the options of voluntary and non-rotational set-aside being made available. The total area of all land set aside both under reform and under the old five-year scheme has now reached 7.2 million hectares which is 970 000 hectares more than the previous year and 1 million more than forecast.

9.   Consequently, the area devoted to cereals has been reduced by more than was predicted earlier. It had stood at around 36 million hectares before reform and has now declined to 32 million compared with the 33 million hectares originally forecast. Equally significant, the first two years have not seen the increase in yields which some outside analysts had warned would offset any reduction in area. In this context it is not without interest to note reports from the agro-chemical industry showing substantially reduced purchases of fertilisers and pesticides.

10.   Another essential pillar of the cereal reform programme was, of course, a substantial reduction in support prices aimed at making indigenous grain more competitive and thus more attractive to the animal-feed sector. This was certainly the case in 1993/94 when uptake increased by some 6 million tonnes which was well above expectations for the first year. Another example of actual performance improving on forecasts.

11.   Factors other than reform were clearly at work here; not least the relatively strong world market prices for soya and maize following severe flooding in the United States during the summer of 1993. Unfavourable weather conditions in the Union — although these should not be exaggerated — also played their part in the reduction of production levels in 1993 and

1994. A timely reminder, if one were needed, that everything does not depend on political decisions taken in Brussels, Geneva or Washington.

12. This combination of two years of declining grain production, on the one hand, and of increased consumption on the other has led to a substantial reduction in intervention stocks — another of the objectives of reform. Not only was storage and disposal of surpluses in the past taking a disproportionate slice of the agricultural budget but their very existence was depressing prices to producers and their release on world markets antagonizing our trading partners. Stocks have fallen from 33 million tonnes at the end of the last pre-reform marketing year to a current 11.5 million tonnes. They could drop further to around 10 million tonnes by the end of 1994/95, and further still during the following year.

13. However, the pronounced drawing down of intervention stocks has been accompanied by a strong rise in prices to levels, in some cases, well above those for intervention. And to levels, moreover, that could have raised questions about the rates of compensation agreed as part of the reform package. If this tendency is not corrected one of the major goals of the reform, which was to make indigenous products more competitive, would be defeated. Escalating grain prices also present distinct drawbacks elsewhere — in the poultry and pig sectors, for example.

14. The quantity of cereals finding its way to the animal-feed sector in 1994/95, whilst still higher than before reform is expected to be about 1 million tonnes less than in 1993/94 despite the second programmed cut in institutional prices. At the same time non-grain feed ingredients have increased their attraction through a decline in the value of the dollar.

15. It was in the light of these circumstances, and of the possibility of their developing to the serious detriment of reform and with the clear intention of not imposing on farmers a higher set-aside obligation than strictly called for by market conditions that the Commission proposed a reduction for the 1995/96 marketing year. [1] There was general recognition by the Council that set-aside was a market management tool which had to respond to market changes and it decided to reduce rates by 3 percentage points for the 1995/96 marketing year. [2]

16. There were also satisfactory results from other reformed sectors. Beef intervention stocks, which had exceeded 1.2 million tonnes before reform, had been spectacularly reduced to about 100 000 tonnes by November 1994, despite unhelpful factors such as the unfavourable impact on beef's image of health and veterinary problems and strong competition from cheaper meats such as pork and poultry, although there is the danger that beef production could increase again in response to favourable prices to producers.

---

[1]   COM(94) 417, 5.10.1994.
[2]   Regulation (EC) No 2990/94, 5.12.1994, OJ L 316, 9.12.1994, p. 1.

17. In the dairy sector, where the Council had already decided to maintain quotas unchanged for 1994/95 and 1995/96, quantities of milk delivered remained stable as did butter production quantities. The intervention price for butter was, however, reduced by 3 %[1] as from July 1994 in order to further improve its competitiveness following earlier price cuts in 1993. These moves have been accompanied by a sharp fall in intervention stocks of butter and of skimmed-milk powder which now stand at historically low levels.

18. For those major sectors which remain unreformed, the Commission has, in two cases, forwarded proposals to the Council — for wine and sugar — and in the case of the fruit and vegetables sector has sent a communication to the European Parliament and the Council.

Reform of remaining sectors brought closer

19. The proposal for the reform of the common organization of the wine market[2] was sent to the Council in May. However, despite the fact that it follows the Commission's earlier reflection document[3] and on the basis of which the Council asked for a proposal to be made, it has run into considerable difficulty in that forum and some hard negotiation lies ahead. As was envisaged in COM(93) 380, its overriding aim is to achieve balance in a market which is currently seriously oversupplied. And whilst, because of unfavourable climatic conditions, production in 1994 fell below the target foreseen in the reform proposals, a solution remains both necessary and urgent.

20. The Commission proposal for changes in the common organization of the sugar market[4] presented in November does not involve the fundamental reforms applied in some other sectors, such as cereals, as it was felt that the present system of quotas and self-financing had functioned satisfactorily. It is proposed that production quotas be maintained for six marketing years until 2000/2001 but with the possibility of adjustment, bearing in mind the European Union's GATT commitments.

21. The Commission's communication to the Council and European Parliament on the future of the fruit and vegetable sector[5] which was designed to stimulate wide discussion, drew attention to some serious structural imbalance and stressed that policy changes should be aimed at encouraging producers to improve the quality of their often highly perishable products and at developing their ability to adapt to rapidly changing markets.

---

[1] Including a reduction of 2% already agreed by the Council in 1994.
[2] COM(94) 117.
[3] COM(94) 380.
[4] COM(94) 439.
[5] COM(94) 360.

22.   The year was thus largely one of consolidation for reformed sectors with changes kept to an absolute minimum in the interest of stability. Some modifications were introduced via the price package (see below); developments in other areas included the following.

23.   The Commission persisted in its determined efforts to counter fraud not least in its proposal [1] to introduce a procedure for reporting to all relevant national authorities the identity of operators guilty of irregular behaviour involving EAGGF.

24.   In the field of rural development, 1994 saw nearly all the programmes from the first phase of the reform of the Structural Funds finalized. For the second phase (1994-99) a significant number of changes were introduced due not only to the incorporation of five new *Länder* but also to the reclassification of other regions. As for Objective 5b initiatives, the eligible areas for the period 1994-99 will cover a population 72 % greater than during the period 1989-93.

25.   Towards the end of the year the Commission presented a report on the agrimonetary regime and made proposals for the future [2] which emphasized the absolute necessity to abolish the switch-over mechanism but to retain the possibility of compensating farmers for appreciable revaluations.

26.   A question which engendered a great deal of sound but very little light in the Council during the autumn was the question of transporting live animals. Neither the Commission's proposal [3] and its pragmatic approach of achieving progress step by step nor the Council president's compromise found sufficient support. The subject remains open.

27.   Another question which surfaced in the Council towards the end of the year was that of simplification of the CAP with four delegations presenting memoranda on the subject. This is a discussion that promises to continue for some time but is one that brings with it the danger that simplification can sometimes result in a relaxing of reform and the risk this would hold for the balanced package of May 1992 and for our GATT commitments.

Weather and production patterns

28.   The preparation and sowing of fields in the autumn of 1993 got off to a generally poor start due to above-average rain in many northern and central areas. Wet conditions continued through the winter and well into spring; there was serious and widespread flooding,

---

[1] COM(94) 498.
[2] COM(94) 498, 16.11.1994.
[3] COM(94) 380.

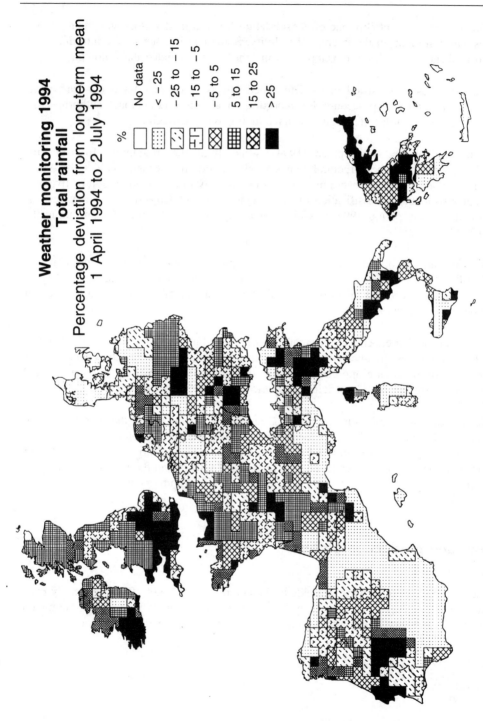

Weather monitoring 1994
Total rainfall
Percentage deviation from long-term mean
1 April 1994 to 2 July 1994

%

No data
< − 25
− 25 to − 15
− 15 to − 5
− 5 to 5
5 to 15
15 to 25
> 25

Sources: Data provided by the national meterological services of the European Union and adjacent countries, the MARS project and Corine. System and software development by the Winand Staring Centre, the Netherlands. Analysis and mapping by the MARS/IATD Computer Laboratory for Agricultural Statistics.

Weather monitoring 1994
Water balance
Percentage deviation from long-term mean
1 April 1994 to 2 July 1994

%

	No data
	< − 25
	− 25 to − 15
	− 15 to − 5
	− 5 to 5
	5 to 15
	15 to 25
	> 25

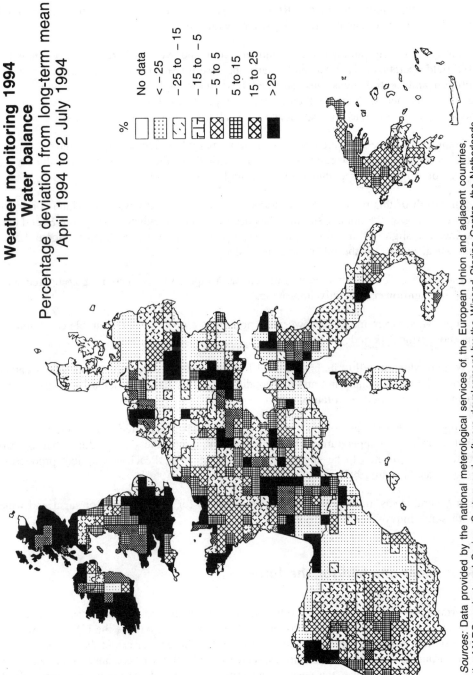

Sources: Data provided by the national meterological services of the European Union and adjacent countries, the MARS project and Corine. System and software development by the Winand Staring Centre, the Netherlands. Analysis and mapping by the MARS/IATD Computer Laboratory for Agricultural Statistics.

particularly in the Meuse, Rhine and Rhone valleys. Southern areas were, in contrast, much drier, with Spain experiencing critically low moisture levels.

29.   Spring and early summer were relatively cool and wet and excessive soil moisture delayed field operations. This was particularly the case for sugar. Common wheat, especially in northern and central areas, also suffered from these high moisture levels which were frequently combined with cooler than average temperatures. But, as was the case earlier in the season, Spain continued to experience unusually dry conditions (see maps).

30.   Summer told a very different story with abnormally high temperatures, often 5°C above average, combined with low rainfall; this was particularly the case in July and during the early part of August. Temperatures remained high in the south.

31.   The result of this not entirely favourable year, which broadly was one which combined a cold wet spring with a hot dry summer, has been to reduce yields for barley, maize and oats but to increase slightly those for wheat. Overall, the cereals yield declined by about 1.5%. Durum wheat yields, on the other hand, rose sharply by 14%.

32.   The effect on oilseeds yields was mixed, with rape and soya declining compared with 1993/94 but sunflower generally improving.

33.   Sugar yields were markedly lower than normal as a result of unfavourable conditions at all the most critical periods of the season.

34.   Wine production fell to its lowest level for many years as a result of unhelpful weather conditions; the exceptionally hot and dry conditions in July had a particularly adverse effect in Spain, Italy and southern France.

35.   Apple and pear production north of the Alps was adversely affected by some late frosts and below average temperatures which prevailed up until the end of June, but overall production was expected to be slightly higher than in 1993. Peach and apricot production also did better than in 1993, particularly in France.

36.   In general terms, therefore, 1994 saw lower yields for most cereals, for sugar and wine but with some improvement for fruit.

Solid foundations laid for the future

37.   Farming, as is the case with other sectors of economic activity, has to face the consequences of changing economic conditions both inside and outside the Union. However, it is also subject to factors largely outside its control to a far greater extent than other sectors; to flood, drought, plant and animal disease and to fickle shifts in consumer preference for example. The CAP has, over the years, gone some way towards providing a framework of stability although in the recent past a range of short-term measures frequently taken at the last minute has tended to intensify rather than diminish agriculture's feelings of uncertainty. Farmers have also felt threatened by the Uruguay Round negotiations — particularly during the later stages.

38.   But the clearly successful launching of major reform, a reform, moreover, which has been shown to work and to bring benefits to the farmer, in 1994, for example, farm incomes increased on average by 5.7 %,[1] with the prospect of the new regimes continuing largely unchanged for the foreseeable future, with a distinct move towards healthier market balance, signalled by spectacular reductions in surplus stocks and with the successful conclusion of the Uruguay Round, stable foundations are now in place for the future of European farming.

39.   Obviously 1994 does not mark the end of the reform process but it does mark the successful conclusion of the first two stages. If the results so far achieved are set against the objectives envisaged when the programme was first launched and against the discouraging background of doubts expressed in some quarters, it is clear that the Commission has put in place a mechanism that has been properly run-in, shown to be in excellent working order and capable of providing good service in the years ahead.

Prices and market organization

40.   As in 1993, traditional price proposals were rendered superfluous in many areas because of the 1992 reform which had already fixed prices in a number of key sectors. With this in mind, and taking into account the restrictions imposed by the budgetary guideline, the Commission's proposals were essentially to leave unchanged those prices which had already been decided and, where decisions were needed, to maintain existing levels. At the same time, some adjustments were needed, notably:

(i)   changing the base year for the reference quantities for the special male bovine premiums, experience having shown that 1992 was an atypical year;

(ii)   the abolition of the consumption aid for olive oil and a corresponding increase in production aid;

(iii)   a significant reduction in the basic price for pigmeat to more realistically reflect the market situation;

(iv)   reductions in sugar storage refunds and cereals and rice monthly increments reflecting lower interest rates and, in the case of cereals, the price reductions agreed as part of the reform.

---

[1]   More detailed information on changes in farm incomes in 1994 can be found in points 135-137.

41. The Commission adopted its proposals at the end of January but, despite the fact that the package proposed was relatively straightforward, final agreement by the Council did not emerge until late July. The principal features of this agreement were:

(i)   cereals and rice: the monthly increments were reduced although by a lower amount than the Commission had proposed. In the case of durum wheat, an additional aid for production in France comes into force from 1995/96, Spain and Portugal's production ceilings were increased and Umbria was included among the eligible Italian regions, although these measures were not formally adopted by the Council until December 1994;

(ii)   sugar: the monthly refunds for sugar storage costs were also reduced by a lower amount than that proposed by the Commission;

(iii)   potato starch: a new regime of reference quantities was set up which includes provision for exports above these quantities to be made without the benefit of export refunds in much the same way as for 'C' sugar;

(iv)   dried fodder: the Commission agreed to make a formal proposal for a new regime before the end of the year.[1] This was submitted to the Council in November;

(v)   olive oil: consumption aid was reduced to ECU 10/100 kgs with appropriate adjustments to related prices and aids;

(vi)   linseed: the aid was left unchanged at ECU 87/tonne (compared to the Commission proposal of ECU 78 tonne);

(vii)   fruit and vegetables and wine: prices were left unchanged pending reform of these sectors;

(viii) tobacco: prices were left unchanged but certain premiums were increased;

(ix)   dairy products: the butter intervention price was cut by 3 %,[2] intervention for cheese was abolished and the revised provisional Italian quota was made definitive;

(x)   beef: prices and premiums were already adjusted as part of the reform but the Council decided to adjust the regional ceilings to take account of the problems caused by the use of 1992 as a base year;

(xi)   pigmeat: the basic price was cut to ECU 1 300/tonne from ECU 1 872/tonne.

42. The Commission estimated that the changes agreed by the Council would cost an additional ECU 81 million in 1994 and ECU 293 million in 1995 whereas it had estimated that its own proposals would have led to savings of ECU 30 million in 1994 and ECU 1 700 million in 1995.

---

[1] COM(94) 508 final, 18.11.1994.
[2] See point 17.

Main institutional prices and aids applicable for the 1994/95 marketing year

	Amount[1]	Percentage change over 1993/94
Cereals[2]		
Target price[2]	118.45	− 7.7
Intervention price[3]	106.60	− 7.7
Compensatory payment	35.00	+ 40.0
Set-aside compensation payment[3]	57.00	+ 26.7
Rice		
Target price (husked rice)	530.60	0
Intervention price (paddy rice)	309.60	0
Oilseeds		
Projected reference price	163.00	0
Community reference amont (ECU/hectare)	359.00	0
Protein crops		
Compensatory payment (ECU/hectare)	65.00	0
Sugar		
Basic price for sugar beet	394.80	0
Intervention price for white sugar	523.30	0
Olive oil		
Production target price	3 178.20	0
Intervention price	1 624.00	− 15.4
Representative market price	1 900.60	0
Threshold price	1 864.40	0
Production aid	1 177.60	+ 33.5
Consumption aid	100.00	− 75
Dried fodder		
Guide price	176.29	0
Linseed		
Compensatory payment	87.00	+ 2.3
Fibre flax		
Flat-rate aid (ECU/ha)	774.86	0
Hemp		
Flat-rate aid (ECU/ha)	641.60	0
Silkworms		
Aid per box	110.41	0
Cotton		
Guide price	1 014.60	0
Minimum price	963.90	0
Milk products		
(a) Target price for milk	256.60	0
(b) Intervention price		
Butter[4]	2 718.00	− 3.0
Skimmed-milk powder	1 702.00	0
Beef/veal		
Guide price for adult bovine animals	1 974.20	0
Intervention price for adult bovine animals	3 047.10	− 5
Sheepmeat		
Basic price (1995 marketing year)	4 174.50	0
Pigmeat		
Basic price	1 300.00	− 30.6

	Amount[1]	Percentage change over 1993/94
Table wine		
Guide price Type RI (ECU/%/hl)	3.17	0
RII (ECU/%/hl)	3.17	0
RIII (ECU/hl)	51.47	0
AI (ECU/%/hl)	3.17	0
AII (ECU/hl)	68.58	0
AIII (ECU/hl)	78.32	0
Tobacco		
Premiums		
Groups II, III, V	1 795.00	0
Group I	2 244.00	0
Group IV	1 974.00	0
Group VI	3 109.00	+ 5
Group VII	2 638.00	+ 5
Group VIII	1 885.00	+ 5
Fruit and vegetables		
Basic and buying-in prices	5	0

[1] Unless otherwise indicated these are in ecus per tonne and include agrimonetary adjustments.
[2] Cereals prices were significantly reduced as part of CAP reform but this reduction was compensated for by the introduction of direct, per hectare compensatory payments.
[3] The payments are calculated on the basis of average regionalized yields per hectare.
[4] See point 17.
[5] A series of prices covering different periods and products are fixed each year.

Agrimonetary measures

43.   Since 2 August 1993 and the widening of the margins for fluctuation within the EMS to 15 %, all Community currencies have been considered floating currencies from the agri-monetary viewpoint. The correcting factor for the ecu therefore remains unaltered at 1.207509.

44.   To take account of the resulting greater variability of the agricultural conversion rates, the Council decided at the end of December 1993 to amend the rules for adjusting those rates by the end of 1994.[1]

45.   The new rules permit wider gaps, of up to 5 points instead of 4, between any two currencies. They delay the downward adjustment of the agricultural conversion rate for a revalued currency as long as the monetary gap does not exceed 5 points.

46.   In addition, the Council introduced stricter conditions for the granting of compensation for revaluations where they follow or precede devaluations.

---

[1] Regulation (EC) No 3528/93, OJ L 320, 22.12.1993, p. 32.

47. The Council reviewed the way the agrimonetary arrangements introduced on 1 January 1993 function, using as a basis the report and proposals submitted by the Commission in November (see point 25).[1]

48. As currency fluctuations were relatively small in 1994, the agricultural conversion rates varied little.

49. As a result of the new agrimonetary rules, no agricultural conversion rate was reduced although at times the positive monetary gaps were in excess of 4 points for the Irish pound, 3 points for the Belgian franc and the Luxembourg franc, and 2 points for the Danish crown and the Dutch guilder.

50. There was no change for currencies considered fixed from the agrimonetary viewpoint on 1 August 1993. In the case of the other currencies, namely the Greek drachma, the Portuguese escudo, the Italian lira, the Spanish peseta and the pound sterling, some increases in the agricultural conversion rates followed relatively small devaluations.

Rural development

51. The first phase of the reform of the Structural Funds (1989-93) ended in 1993. However, payments continued throughout 1994.

52. The year 1994 saw the presentation, negotiation and approval of most programming documents for the implementation of structural measures for the second period of application of the reform of the Structural Funds (1994-99). Appropriations for and territorial coverage of eligible zones were increased substantially.

53. In the negotiations on accession, the applicant States attached much importance to the Structural Funds policy. In view of the geographical position of those countries, rural development is a significant issue. Leaving aside Sweden, where agricultural prices are close to the prices decided under the CAP, in the other three countries agricultural prices are high, aid is substantial and protection at the borders is strong.

54. The Council selected only one region under Objective 1, namely Burgenland in Austria, for which budget appropriations amounting to ECU 184 million are planned over five years (1995-99). The northern regions of the Nordic countries did not strictly meet the criteria for classification under Objective 1 laid down in Council Regulation (EEC) No 2081/93 of 20 July 1993 on the tasks of the Structural Funds;[2] accordingly, the Council decided to

---

[1] COM(94) 498, 16.11.1994.
[2] OJ L 193, 31.7.1993, p. 5.

introduce a new Objective 6 for regions with a very low population density (eight inhabitants or less per km^2) with appropriations of ECU 368 million for Norway, ECU 511 million for Finland and ECU 230 million for Sweden.

55.   The demarcation of the Objective 5b areas follows the same procedures and principles as those applying to the present Member States of the European Union.

56.   As regards Objective 5a, the aim of which is to speed up the adjustment of agricultural structures under the reform of the CAP, the applicant States attached great importance to:

(i)   the Council Directive on farming in mountain and hill areas and in certain less-favoured areas, which provides for compensatory allowances to be granted to offset natural handicaps;

(ii)   the various types of aid which may be granted towards investments on agricultural holdings;

(iii)   the Council Regulation on the processing and marketing of agricultural and forestry products.

57.   The Union has also undertaken to ensure that the applicant States can quickly implement Council Regulation (EEC) No 2078/92 of 30 June 1992 on agricultural production methods compatible with the requirements of the protection of the environment and the maintenance of the countryside, which was adopted under the reform of the CAP.

58.   Lastly, it was decided that the Nordic countries could grant long-term national aid to ensure that farming continues in the region north of the 62nd parallel and in certain regions bordering on that parallel to the south which are affected by comparable climatic conditions rendering agriculture particularly arduous. Article 142 of the Act of Accession lays down the conditions governing application.

Promotional measures

59.   Measures to promote quality beef and veal, milk and milk products, nuts, flax fibres, apples, citrus fruits and grape juice were undertaken in 1994.

60.   Invitations to tender for new campaigns to promote olive oil and fibre flax were published by the Commission.

61.   The Council also decided on specific measures relating to the quality and promotion of dried grapes (Regulation (EC) No 399/94). [1]

---

[1]   OJ L 54, 25.2.1994, p. 3.

Quality policy

62. The new Community system for the protection of geographical indications and designations of origin for agricultural products and foodstuffs[1] is frequently used by Member States wishing to have their designations recognized. As to the new arrangements on certificates of specific character for agricultural products and foodstuffs,[2] the Commission has made available to those in the trade the Community symbol and endorsement which indicate products recognized as having their 'traditional speciality guaranteed'.[3]

Assistance to the needy

63. In 1994, as in previous years, the Community continued its food aid programme for the needy.[4] Since Germany, like last year, did not participate in the programme, ECU 175 million was made available from the Community budget to be allocated among the 11 other Member States to assist the distribution of foodstuffs through social and charitable organizations. The breakdown of this amount and the quantities which may be removed from intervention storage in each of the Member States are shown in the following table.

Free distribution of agricultural products (1994)

Member State	Approps. allocated (ECU million)	Quantities (tonnes)							
		Common wheat	Durum wheat	Rice	Olive oil	Skimmed-milk powder	Butter	Cheese	Beef
Belgique/België	2.85	3 100				300	330		600
Danmark	2.00						50		250
Deutschland	0.0								
Elláda	14.10								4 000
España	41.59		30 000		4 500		6 000		7 000
France	33.56	5 000	8 500	2 000		7 500	1 500		6 000
Ireland	5.41						40		1 450
Italia	28.78	5 000	15 000	1 500	3 000		1 300	2 300	7 300
Luxembourg	0.08	20				25	15		15
Nederland	3.00						150		600
Portugal	12.26	1 850	1 850	1 200	1 500	1 000	1 350		2 500
United Kingdom	29.37								7 100
Total	173.00[1]	14 970	55 350	4 700	9 000	8 825	10 735	2 300	36 815

[1] The total, which amounts to ECU 175 million, includes ECU 2 million to finance transport costs.
Source: DG VI.

---

[1] Regulation (EEC) No 2081/92, OJ L 208, 27.7.1992, p. 1.
[2] Regulation (EEC) No 2082/92, OJ L 208, 27.7.1992, p. 9.
[3] Regulation (EC) No 2515/94, OJ L 275, 26.10.1994, p. 1.
[4] Council Regulation (EEC) No 3730/87, 10.12.1987, OJ L 352, 15.12.1987, p. 1 and Commission Decision 94/177/EEC, 17.3.1994, OJ L 82, 25.3.1994, p. 37.

Harmonization of legislation

Veterinary and zootechnical legislation

64. Major decisions have been adopted this year in the veterinary and zootechnical field. With regard to zootechnical conditions governing imports, the Council adopted Directive 94/28/EC laying down the principles relating to the zootechnical and genealogical conditions applicable to imports from third countries of animals, their semen, ova, and embryos, and amending Directive 77/504/EEC on pure-bred breeding animals of the bovine species. [1] The Council also adopted Directive 94/42/EC amending Directive 64/432/EEC on animal health problems affecting intra-Community trade in bovine animals and swine. [2] In addition, it adopted Decision 94/117/EC [3] relating to certain small establishments ensuring the distribution of fishery products in Greece, Decision 94/370/EC [4] on expenditure in the veterinary field and Decision 94/371/EC [5] on the placing on the market of certain types of eggs. In addition, the EC-San Marino Cooperation Committee adopted Decision No 1/94 [6] on veterinary regulations.

65. For its part, the Commission submitted several proposals to the Council, in particular as regards equidae intended for competition, [7] detailed rules for drawing up provisional lists of establishments in third countries, [8] controls of diseases of bivalve molluscs, [9] the financing of veterinary inspections of fishery products [10] and the conditions governing the production and placing on the market of fresh meat. [11]

66. In addition, under the powers conferred on it, the Commission adopted many implementing texts to supplement the measures laid down for the completion of the single market. It also adopted many decisions on Community financing of measures to eradicate animal diseases and to curb the spread of animal diseases within the European Union. Furthermore, the accession of the new Member States means that the Commission has had to adopt much legislation in the veterinary field and to propose measures for protection against salmonella to the Council.

---

[1] OJ L 178, 12.7.1994, p. 66.
[2] OJ L 201, 4.8.1994, p. 26.
[3] OJ L 54, 25.2.1994, p. 28.
[4] OJ L 168, 2.7.1994, p. 31.
[5] OJ L 168, 2.7.1994, p. 34.
[6] OJ L 238, 13.9.1994, p. 25.
[7] OJ C 51, 19.2.1994, p. 6.
[8] OJ C 208, 28.7.1994, p. 9.
[9] OJ C 295, 13.10.1994, p. 9.
[10] Not published.
[11] OJ C 224, 12.8.1994, p. 15.

Phytosanitary and animal feedingstuffs legislation

67. The new plant health regime, which came into force on 1 June 1993, was extended to apply to the Canary Islands from 1 January 1995 by Council Directive 94/13/EC. [1] It also contained provisions to adapt the derogation provisions of the regime relating to imports from third countries to the single market concept and to facilitate trials, scientific and varietal selection work, the movement for personal use of small quantities of plant and plant products, and the production or use of such material in frontier zones between Member and non-member States.

68. During the year, the Commission adopted Community financial contributions to programmes for the control of organisms harmful to plants and plant products in the French overseas departments, Madeira and the Azores [2] and established a standardized procedure [3] for the notification of interceptions of consignments of harmful organisms from third countries and presenting an imminent phytosanitary danger.

69. The Council continued its examination of the Commission's earlier proposal aimed at introducing the principle of Community financial solidarity and Member States' liability into the plant health regime.

70. The Community regime for authorization of plant protection products, which came into force in July 1993, provides for Member States to authorize individual preparations under harmonized rules. These rules ('Uniform principles') were adopted by the Council on 27 July 1994. [4] In the first phase of the re-evaluation programme of active substances already on the market before the Community regime, the Commission designated [5] rapporteur Member States for 89 active substances and organized in June, with the cooperation of the German authorities, at the Biologische Bundesanstalt in Braunchweig a workshop which brought together for the first time staff of the registration authorities of all the Member States to examine practical questions of implementation.

71. The programme for the control of pesticide residues in agricultural products was developed further through the adoption by Council of two more measures [6] establishing maximum residue levels to ensure both the free circulation of agricultural products and a high level of consumer protection.

---

[1] OJ L 92, 9.4.1994, p. 27.
[2] OJ L 159, 28.6.1994, p. 63.
   OJ L 187, 22.7.1994, p. 14.
   OJ L 200, 3.8.1994, p. 43.
[3] OJ L 32, 5.2.1994, p. 37.
[4] OJ L 227, 1.9.1994, p. 31.
[5] OJ L 189, 28.4.1994, p. 8.
[6] OJ L 189, 29.7.1994, p. 67 and p. 70.

72. With regard to organic farming, the Commission adopted several implementing measures under the basic Regulation (EEC) No 2092/91 and the Council continued its examination of the Commission's earlier proposal to improve the regime's functioning in the light of the first year's experience. [1] The Council adopted on 20 June Regulation (EC) No 1468/94 delaying for one year the current provisions concerning the labelling of products under conversion.

73. In the seeds and propagating material sector, intensive discussions continued in the Council on the Commission's proposal, [2] made at the end of 1993, for a directive to amend and update the seven basic seeds marketing directives relating to agricultural and vegetable seeds, adopted in the late 1960s and 1970.

74. In addition to many routine Commission measures adopted during the year, a decision [3] setting up an experiment relating to the transport of seed of agricultural plants in bulk was adopted, which has the aim of substantially reducing the cost of the transport of seed.

75. In the new area of propagating material of ornamental and fruit plants, much work was carried out at the level of the relevant standing committees to develop the approximation of the Member States' application of Community legislation.

76. On 27 July the Council adopted the Commission's proposal, made originally in 1990 as part of the White Paper programme, on Community plant variety rights establishing a Community system of protection of new varieties of plants. [4] This system provides for the grant, by a Community office, of a specific form of industrial property rights with direct and uniform effect throughout the European Union.

77. In the animal nutrition sector, the Commission adopted in July two directives [5] modifying the guidelines for the evaluation of additives to cover new categories of additives and establishing a list of intended uses for dietetic feedingstuffs respectively. The Council for its part continued examination of the series of proposals made by the Commission between 1992 and 1994 to strengthen and update the legislation in this sector. These included proposals concerning a modified procedure for the admission of certain classes of

---

[1] OJ L 159, 28.6.1994, p. 11.
[2] OJ C 29, 31.1.1994, p. 1.
[3] OJ L 252, 28.9.1994, p. 15.
[4] OJ L 227, 1.9.1994, p. 1.
[5] OJ L 208, 11.8.1994, p. 15; OJ L 207, 10.8.1994, p. 20.

additives, [1] the official inspection of feedingstuffs, [2] the approval of manufacturers [3] and specific rules for raw materials used in feedingstuffs. [4]

78.   In the hygiene field, the Commission intensified its preparatory work to develop tolerances for contaminants of agricultural origin, in particular nitrates and mycotoxins, in plant products intended for human consumption.

Veterinary inspection

79.   In addition to inspections of establishments producing fresh meat and meat products in both Member States and third countries, inspection visits were organized in the Member States to ascertain the situation in establishments which produce fresh meat in small quantities or which have been granted temporary derogations.

80.   Furthermore, in view of the scale of imports of fishery products from third countries, inspection visits were made to a certain number of the latter (Brazil, Colombia, Ecuador, Indonesia, Morocco, New Zealand and Singapore), in order to verify whether health conditions laid down for imports into the Community are met by their products.

81.   In addition to routine animal-health inspections, visits were paid to several Member States on account of the existence of classical swine fever in Germany, swine vesicular disease in the Netherlands and Italy, bovine spongiform encephalopathy in the United Kingdom, African swine fever in Spain and Portugal and foot-and-mouth disease in Greece.

82.   To monitor the foot-and-mouth disease situation in Eastern Europe, inspection visits were organized in the latter. Similarly, in order to ascertain the situation regarding diseases affecting poultry, inspections were carried out in the Far East.

83.   The lists of third countries from which the Member States authorize imports of live animals and products of animal origin and the animal-health conditions and veterinary certificates required have been updated to take account of the trend in the animal-health situation in certain third countries. New lists of countries have been drawn up for a whole range of products (e.g. casings, processed animal protein, serum of equidae, pet food) and have been supplemented by animal-health conditions and the certificates required.

---

[1] OJ C 211, 5.8.1992, p. 21; OJ C 218, 12.8.1993, p. 6.
[2] OJ C 313, 19.11.1993, p. 10.
[3] OJ C 348, 28.12.1993, p. 13.
[4] OJ C 236, 24.8.1994, p. 7.

84. In addition, border inspection posts selected for veterinary checks on animals and products of animal origin were inspected to ensure they comply with the Community provisions and to amend the list drawn up at Community level.

85. Lastly, exploratory discussions were entered into with countries in South America and Eastern Europe with a view to negotiations to adapt the conditions of trade on a non-discriminatory basis and to take account of their special features.

Plant-health inspection

86. With the adoption of the new Community plant-health strategy on 1 June 1993, on-site inspection visits, investigations and checks throughout 1994 mainly related to verification of the proper application of the Community provisions on the subject by the Member States.

87. Furthermore, comprehensive surveys and inspections were carried out in the Member States concerned by 'protected zones' with regard to organisms harmful to plants and plant products.

88. Various inspection visits took place in the Member States applying protective measures and in which organisms harmful to plants and plant products occurred or recurred in 1994 (potato ring rot in Denmark, *Thrips palmi* in the Netherlands, brown spot on conifers (*Scirrhia acicola*) in the south-west of France, potato brown rot in Belgium, etc.).

89. Several consignments of plants and plant products originating in third countries were intercepted by the Member States because they did not comply with Community plant-health requirements, with the result that inspection visits and clarifications were required with the third countries concerned (Egypt, Israel, Guatemala, China and Hong Kong).

90. In connection with the enlargement of the European Union, technical visits to the official departments of the applicant States have commenced (in Finland firstly, given the plant-health situation of potatoes).

91. As regards plant-health information, the vade-mecum for inspectors is still being drafted in computer-readable form with dissemination via EUROPHYT/FIS and training sessions have been organized in several Member States (Ireland, the United Kingdom and Italy) to introduce officials responsible for plant-health inspections in the use of this means of dissemination.

State aids

92.   Commission policy concerning the assessment of State aid under Articles 92 and 93 of the EC Treaty in the agricultural sector is largely determined by the regulations which constitute the common agricultural policy. This is in particular the case in the field of the common market organizations which are considered as complete and exhaustive regulations for the products concerned. Consequently it is as a general rule Commission policy to deem incompatible with the common market aids which concern market support measures of a type provided for in the different common market organizations. This is done because normally these national aids per unit of output or input risk disturbing Community market mechanisms and, as operating aids, have no lasting effect on the development of the sector concerned.

93.   In this context the Commission took a final negative decision in 1994 against an Irish aid for the export of mushrooms which included the requirement for reimbursement of aid. The main argument used by the Commission, and which the Irish authorities could not refute, was that according to the information available, the aid, being based on the quantity of mushrooms exported, reduced exporters' costs.

94.   The Commission examined the reorganization schemes of the England and Wales Milk Marketing Board and of three milk marketing boards in Scotland, because these schemes involved State aid in the form of involvement in the transfer of ownership of board assets, which falls under Article 92(1) of the Treaty. The Commission raised no objection to these aids because it considered that the transfer and the distribution of the boards' assets to new successor cooperatives and milk producers does not unduly favour members or non-members of the new cooperatives and consequently is not contrary to the open market principle.

95.   Furthermore, the United Kingdom authorities confirmed that the valuations placed on the assets and liabilities being transferred were at the appropriate market value and that the valuations covered all the assets and liabilities transferred. The Commission had in fact opened the Article 93(2) procedure in 1993 with regard to an initial proposal for the reorganization of the England and Wales Milk Marketing Board subsequently withdrawn by the United Kingdom — because this proposal did not correspond to the criteria mentioned above.

96.   In the pigmeat sector the Commission has taken final negative decisions in respect of two French aid schemes following initiation of the Article 93(2) procedure. One aid which the Commission regards as illegal and is operated by Stabiporc, the French income stabilization fund for pig producers, consists of advances involving public money at reduced rates of interest for pig producer groups in the absence of adequate guarantees that the groups will repay to Stabiporc all or any of the advances made to them.

97.   The other aid involves the reduction of annual repayments due for 1993 on start-up loans for young farmers taken out by pig farmers between 1990 and 1992 and is regarded as

incompatible with the common market, since compliance with the ceilings set under Commission policy in such cases is not ensured in the event of overlap with other aid, such as start-up loans and grants.

98.   In both cases the Commission decided to require France to discontinue the aids and to require reimbursement of aids illegally received by beneficiaries.

99.   Furthermore, as a result of information from the French authorities, the Commission decided to initiate the procedure laid down in Article 93(2) of the EC Treaty in respect of a State guarantee on loans granted to pig producers from private funds via Stabiporc.

100.   The Commission decided to initiate the procedure laid down in Article 93(2) of the Treaty in respect of two Italian aid measures and one French aid measure aimed at promoting the use of set-aside land for the production of biofuels from oilseeds. The measures consist principally in a premium per hectare for the production of oilseed plants on land set aside and tax exemption to benefit biofuels produced from those plants. The Commission is, on the whole, in favour of the development of biofuels. It takes the view, however, that in the case in point the means used to attain that end are not compatible with the common market. The measures in question, the Commission feels, constitute a violation of various provisions of Community law and, in particular, of Community legislation on the system of set-aside, the market organizations concerned and Article 95 of the Treaty.

101.   Concerning aid to investments for improving processing and marketing conditions for agricultural products the Commission decided not to raise objections to an investment aid which the German authorities intend to grant for environmental protection in the context of modernization of two processing undertakings in the oilseed sector.

102.   In this sector State aid for productive purposes can be granted only under very restrictive conditions according to the selection criteria established for Community aids for these investments which are applied for the assessment of State aid by analogy. Therefore it was a condition for this decision that the German authorities provided detailed information that aid would be granted only for investments which are destined exclusively for environmental protection in the sense of the Community guidelines on State aid for environmental protection which were published in 1994. [1]

---

[1]   OJ C 72, 10.3.1994, p. 3.

II — Economic situation and farm incomes in 1994

General overview

103. Despite some abnormal weather conditions, which affected the harvests of certain crops (such as cereals, sugar and wine), and despite the persistence of rather critical market conditions in other sectors (such as vegetables and pigmeat), the overall economic situation of farming in the Community in 1994 was much better than in previous years. After rising slightly in 1993 (+ 0.5 %), farm incomes rose by an average of around 6 % in 1994, with increases of more than 10 % in some Member States. However, two Member States recorded falls in incomes: Luxembourg (− 0.9 %) and Italy (− 7.6 %) (see points 136 and 137). Three factors seem to have played a decisive role in this generally favourable trend:

(a) the implementation of the reform of the CAP from 1993/94, particularly the introduction of compensatory aid to farmers to offset the reduction in institutional prices for cereals and beef and veal. These aids are now an essential component of agricultural incomes and provide some security and stability of income in the sectors concerned;

(b) a gradual rebalancing of agricultural markets thanks to the massive disposal (partly within the Community, but mostly on the world market) of intervention stocks which had built up over recent years. For example, public stocks of cereals had fallen to 11.5 million tonnes at the end of November 1994, from 30 million tonnes in July 1993. Similarly, public stocks of beef amounted to only 100 000 tonnes at the end of November 1994, as against more than one million tonnes at the end of 1992;

(c) lastly, the restoration of better balance between supply and demand, thanks to measures introduced under the reform to control production (particularly set-aside of arable land) and thanks also to the effects of a downturn in production and, in some sectors, to an upturn in demand (as in the increased consumption of cereals for animal nutrition).

104. By contrast, the monetary factor, i.e. the positive impact on both institutional and market prices of the successive devaluations of different currencies, which had had a substantial impact on farm incomes in 1993, played only a minor role in 1994.

105. Unlike the previous year, 1994 saw very little fluctuation, indeed a fair measure of stability, in exchange rates, and particularly in the agricultural conversion rates, in most Member States. The impact of the agrimonetary adjustments on prices and aid expressed in national currency was virtually nonexistent, except for Greece, the United Kingdom, Italy and, to a lesser extent, Spain and Portugal. For example, the average price rise resulting from adjustments to the green rates for products such as common wheat was 0.8 % during the first nine months of 1994, as against 6.8 % in 1993.

106. Conversely, the dollar rate fell by more than 10 % during the first 10 months of 1994 against the ecu and most of the European currencies. This factor, together with the drop in Community cereals prices from the 1993/94 marketing year and the substantial reduction in international prices for imported feedingstuffs (particularly soya cake) helped bring down the costs of animal feed, particularly compound feedingstuffs, in 1994. This trend, which was already underway in the second half of 1993, somewhat cushioned the impact on incomes of the drop in producer prices for most animal products.

107. The improvement in the economic situation of farming in 1994 can also be seen in the wider context of a general recovery of economic growth after the severe recession of recent years. Economic activity in the European Union picked up to a surprising extent during the first half of 1994, not only as a result of exports and the reconstitution of stocks, but also because demand was greater than expected. This trend firmed up and accelerated slightly during the second half of the year. In particular, private consumption, which had fallen somewhat in 1993, picked up again in 1994, although remaining at a much lower level than in the 1980s. The vigour of the recovery led to a slight improvement in the outlook for the job market. After several years of steady climbing, unemployment seems to have peaked in the spring of 1994 and has been gradually falling since then. In the same way, inflation has continued to fall in most Member States and the Community average should be around 3 %, as against 4 % in 1993.

Production levels and price trends

108. Community cereal production for 1994 is estimated at approximately 162 million tonnes, representing a drop of about 2.2 % on the most recent official estimate for 1993 (165.7 million tonnes) and of 3.8 % on the 1992 harvest (168.4 million tonnes). Thus, for the third year running, Community production is well below the record level of 1991, when the cereal harvest reached 181.3 million tonnes. Moreover, while the drop in production in 1992 was mainly due to bad weather conditions, particularly the drought which afflicted several regions of the Union, the results for 1993 and 1994 must be seen in the light of the large reduction in the sown area (about 3 million hectares less) resulting from the application of the new set-aside scheme introduced under the reform. Without that constraint, there is no doubt that cereal production would again have exceeded 180 million tonnes in 1993 and 1994.

109.   The area sown to cereals in 1994 changed little from 1993 (remaining at around 32.2 million hectares), although there was a slight drop in average yields, which fell below the general trend. There was an increase in the production of common wheat, durum wheat and rye. However, production of maize and barley was down because of the reduction in sown area and/or yields.

110.   Under the CAP reform, the intervention prices in ecus were reduced for the first time (by 24.8 % for common wheat and maize, by 44.3 % for durum wheat and by 20.9 % for other cereals) on 1 July 1993 and for the second time (by 7.7 % for all cereals) on 1 July 1994. These reductions were offset by compensatory payments per hectare on the basis of the historical yield of each region. In general market prices reflected these reductions, although sometimes with a time-lag of several months, after the beginning of the marketing year. However, these reductions were partially neutralized on the different markets, either by the successive devaluations of certain green rates or by a relative improvement in market equilibrium following the drop in production and the massive disposal of intervention stocks, or by other factors (quality of the harvest, quantities withheld by producers or collecting organizations, improvements in prices on the international market, etc.). This lag between institutional prices and market prices was particularly strong in the early months of the 1994/95 marketing year.

111.   At the end of November 1994, the prices for common wheat, for example, were still 7 % to 10 % higher than the intervention price in France and Belgium, 15 % in Germany, 25 % in the United Kingdom, and 30 % to 40 % in Italy and Spain.

112.   Community production of oilseeds (including those grown on set-aside land for non-food uses) was 1.4 million tonnes, or 13.3 %, up on the previous year. About a quarter of this increase was the result of an increase in sown areas (about + 200 000 hectares) and the remaining three quarters was due to the improvement in yields after the decrease in 1993. Rape production was up around 350 000 tonnes and sunflower production around 870 000 tonnes. In both cases, it is important to note the increase in cultivation for non-food uses (+ 743 000 tonnes for rape and + 189 000 tonnes for sunflower). With regard to soya, the entry into force of the reform and the exclusion of land on which a second harvest is gathered from eligibility for aid brought about a steep decline in the soya area and soya production in Italy and in the European Union as a whole in 1993. In 1994 there was a substantial increase in areas (+ 34 %) and Community production (+ 23 %).

113.   Unlike oilseeds, the areas under, and production of, protein crops (peas and field beans) fell in relation to the previous year (by 2.7 % for areas and 9.1 % for production).

114.   Sugar production fell sharply in 1994 (– 10 % for the European Union as a whole) as a result of a slight reduction in area and, above all, a very substantial decrease in sugar yields. This was due to two main factors. Firstly, the marketing year started late because of bad weather conditions during the sowing season. Secondly, growing conditions in most regions of northern Europe were affected by a greater water shortage than usual, particularly during the summer. The extremely heavy rainfall at the time of harvest did little to change this state of affairs.

115.   The Community wine harvest, provisionally estimated at around 150 million hl for the second year running, will be considerably less in 1994 than it was in 1992 (191 million hl). Production is down in all the main producer countries because of poor weather conditions and, in certain cases, a downward trend in areas. By contrast, the quality is expected to be fairly good everywhere. The drop in production in 1993 and 1994 is reflected in the market prices which, at the beginning of December 1994, were often considerably higher than at the same time the year before (+ 5 % in France, + 37 % in Italy and + 77 % in Spain).

116.   As in 1993, the Community vegetable market was, on the whole, fairly depressed and even critical for certain products, like cauliflower. The causes of this slump in the vegetable market over recent years are not altogether clear.

117.   However, the stagnation, or even fall, in consumption due to the recession and the increasing pressures on family budgets in recent years have helped to depress the market. Experience in 1994 has also shown that this market is strongly influenced by the weather and by the fluctuation in consumption depending on the temperature. Lastly, the drop in consumption has been paired with an increase in production, which has aggravated the drop in prices for certain vegetables. The price of cauliflowers at the end of November was less than 35 % what it had been in 1993.

118.   The only exception to this serious situation has been potatoes, because of the sharp drop in production (down about 8 % according to initial estimates). Thanks to improved market equilibrium in most Member States, market prices at the end of November 1994 were considerably higher than they had been at the same time the year before (up to 100 % in some Member States).

119.   The market situation for fruit was generally less critical than for vegetables, despite some slight increases in production. This was the case, particularly, for peaches and nectarines, for which prices were relatively satisfactory. For apples, production of which was up by about 5 %, the marketing year was fairly difficult until the end of May 1994, after which the situation improved considerably and the year ended with prices about 40 % higher than at the end of 1993. The production of pears was also up by about 5 %, but prices fell only slightly in relation to the previous marketing year.

120.   Overall milk production in 1994 was estimated to be very slightly down on the previous year (− 0.7 %), although it varied considerably from one Member State to the next. The same applied to milk deliveries to dairies. This is explained by a reduction in the dairy herd (− 1.1 % in 1994) and an increase in yields. However, because weather conditions in certain Member States were not favourable for milk production, yields increased by no more than 0.8 % on average in 1994, as against more than 2.5 % per year over previous years. The available figures show, however, that milk deliveries recovered in most Member States after the summer and even overtook deliveries during the same period the year before by as much as 5 or 6 % in some countries.

121. Butter production, which fell by 142 000 tonnes (− 7.7 %) in 1992 and by only 8 000 tonnes (− 0.5 %) in 1993, fell by 50 000 tonnes (− 2.9 %) in 1994. By contrast, cheese production increased slightly (+ 0.4 %), after a much sharper increase in 1993, and above all in 1992.

122. It is not possible at present to make a sufficiently accurate assessment of the trend in beef and veal production in 1994, because the available figures for animals slaughtered only partially cover the second half of the year. However, it is already clear that 1994, like 1993, is right at the bottom of the beef production cycle and maybe even lower than the already very low level of 1993. If compared with the 1991 peak, there has been a drop in production of around 1 million tonnes, or 11.5 %, in three years. In addition to the traditional cyclical trend, two other phenomena have contributed to this spectacular nose-dive in production, particularly in 1993 (− 7.8 %): firstly, the abnormally high level of production in 1991 as a result, *inter alia*, of massive imports from the countries of Eastern Europe and secondly the tendency of some producers to keep animals on the farm with a view to increasing their reference herd for the grant of the premiums provided for under the reform.

123. Intervention prices for beef and veal were reduced by 6.2 % at the beginning of July 1993 and by 5.3 % at the beginning of July 1994. These institutional price reductions were partially reflected in the market prices. However, actual reductions were less than those in the institutional prices because of the drop in production and the drastic reduction in intervention stocks. There has also been an additional impact on prices in national currency resulting from the monetary adjustments since September 1992. At the end of November 1994, prices for adult bovine animals were on average 3.3 % lower than they had been at the same time in 1993.

124. After hitting the lowest level in recent years in the autumn of 1993, at the same time as a substantial increase in the number of animals slaughtered, pigmeat prices firmed up a little at the end of 1993, as a consequence of the reduction in market supply following the restrictive measures introduced in Germany and Belgium as a result of the outbreaks of classical swine fever in those countries. Despite the relative improvement, at the end of 1993 prices were down around 35 % on those of December 1991 (before the pigmeat crisis, the darkest period of which was around the middle of 1992).

125. 1994 started with a new setback in the price of pigmeat, followed by stagnation at very low levels. From April onwards, the fall-back in production and the different market support measures prompted a slight recovery in prices. However, prices have continued, throughout the year, to reflect the glut of supply on the Community market in pigmeat despite a reduction in the number of animals slaughtered in the second half of the year.

126.   After the spectacular increase in 1993 (+ 6 %), pigmeat production remained fairly static in 1994, at the high level of the previous year. The Community market has been able to survive the last two years thanks only to an increase in low price Community exports, particularly to Eastern Europe.

127.   In 1994 poultrymeat production exceeded 7 million tonnes (approximately 7.1 million tonnes) for the first time, representing a 2.2 % increase over the previous year, during which production was stable after a long period of steadily increasing production. The high level of production, set against a slight drop in domestic consumption in 1993 and an increase of only around 1 % in 1994, is probably the main cause of low prices in 1993 and 1994 in this sector.

128.   In the sheepmeat and goatmeat sectors, the drop in production seen in 1992 (− 3.1 %) and 1993 (− 3.1 %) slowed down and stopped in mid-1994. Production began to rise in the second half of 1994, so that production in 1994 will be around 1.14 million tonnes, or almost the same level as 1993.

Producer prices

129.   On the basis of information available on 30 November, it is estimated that the index of nominal producer prices for all agricultural products increased by an average of 3.6 % in 1994 compared with the previous year. This corresponds to a fall of 1.1 % in real prices. The downward trend which started in 1989 has therefore continued in 1994, but at a much slower rate than in 1992 (− 8.3 %) and 1993 (− 4.1 %). In addition, it should not be forgotten that in 1993 and 1994, as a result of the implementation of the CAP reform, the support prices for certain products (particularly cereals and beef and veal) were reduced and producers received instead compensatory aid which has not been taken into account in the producer price index.

130.   In relation to 1993, the producer price index for crop products increased in real terms by an average of 0.9 %. However, the situation varies considerably from one sector to another. Thus, in the cereals sector, real producer prices fell by an average of 12.6 % (this drop was partially offset by the aids introduced under the reform). In contrast, the real price of potatoes increased by about 25 % on average following a smaller harvest. Apart from these two extremes, real producer prices have generally followed more moderate trends, whether downwards (flowers: − 2.2 %; fresh vegetables: − 1.3 %) or, more often, upwards (fruit: + 6.1 %; wine: + 6.4 %; olives: + 7.9 %).

Deflated indices of producer prices for agricultural products

(1985 = 100)

	1989	1990	1991	1992	1993	1994	1994/93 %
Belgique/België	94.6	86.4	82.6	78.8	71.5	71.9	0.5
Danmark	85.8	77.4	74.1	72.7	62.9	60.3	− 4.2
Deutschland	95.8	88.6	84.9	79.9	71.0	69.7	− 1.9
Elláda	90.2	90.8	89.6	80.5	75.1	77.2	2.8
España	91.0	85.7	80.7	71.1	72.0	76.2	5.8
France	94.0	90.7	88.0	80.0	75.0	74.0	− 1.3
Ireland	105.5	90.5	85.1	84.9	89.3	88.0	− 1.5
Italia	91.2	89.7	89.2	80.0	78.7	75.1	− 4.6
Luxembourg	109.2	103.2	93.0	85.9	81.4	77.1	− 5.3
Nederland	99.1	91.3	90.5	83.6	76.3	75.0	− 1.7
Portugal	99.0	91.0	77.6	65.7	63.4	65.4	3.2
United Kingdom	91.8	85.0	79.7	77.9	80.8	78.1	− 3.4
EUR 12	93.4	88.5	85.3	78.3	75.1	74.3	− 1.1

Deflated input price indices (Staple goods and services in agriculture)

(1985 = 100)

	1989	1990	1991	1992	1993	1994	1994/93 %
Belgique/België	87.7	81.2	79.1	77.2	74.1	72.0	− 2.9
Danmark	84.0	80.4	77.8	76.0	75.3	73.4	− 2.5
Deutschland	89.4	85.5	84.5	82.7	78.1	76.1	− 2.5
Elláda	83.1	82.4	85.7	83.6	83.0	81.5	− 1.8
España	84.3	79.7	76.8	72.7	71.4	69.9	− 2.1
France	89.0	85.6	83.7	81.8	80.1	78.7	− 1.7
Ireland	87.2	84.6	82.2	79.9	79.0	77.6	− 1.8
Italia	87.5	82.9	79.1	77.0	79.2	78.2	− 1.2
Luxembourg	90.8	88.6	87.8	84.8	82.0	79.6	− 2.9
Nederland	87.8	82.9	80.2	78.3	74.6	72.6	− 2.7
Portugal	82.4	76.4	72.4	66.0	60.9	57.9	− 4.9
United Kingdom	89.8	85.2	83.4	82.8	85.7	83.0	− 3.2
EUR 12	87.6	83.4	81.3	79.1	77.8	76.1	− 2.2

131. The index of real prices for livestock products fell by 2.9 % on average. This negative trend concerned all products, particularly eggs (− 6.5 %), poultry (− 4.4 %), cattle (− 2.8 %) and pigs (− 2.5 %).

132. In 1994 the general producer price index tended, in real terms, to rise in Spain (+ 5.8 %), Portugal (+ 3.2 %), Greece (+ 2.8 %) and Belgium (+ 0.5 %) and to fall in Luxembourg (− 5.3 %), Italy (− 4.6 %), Denmark (− 4.2 %), the United Kingdom (− 3.4 %), Germany (without the five new *Länder:* − 1.9 %), the Netherlands (− 1.7 %), Ireland (− 1.5 %) and France (− 1.3 %).

Changes in nominal agricultural input prices

(%)

	Intermediate consumption (goods and services)			Investment (goods and services)			Total		
	1992/91	1993/92	1994/93	1992/93	1993/92	1994/93	1992/91	1993/92	1994/93
Belgique/België	0.0	− 1.4	− 0.7	2.1	4.6	1.6	0.3	− 0.6	− 0.4
Danmark	− 0.3	0.3	− 0.9	1.8	2.6	0.9	0.2	0.7	− 0.5
Deutschland	1.8	− 1.7	− 0.3	4.6	3.1	1.6	2.5	− 0.6	0.1
Elláda	13.1	13.6	8.4	11.3	14.2	7.9	12.6	13.8	8.3
España	0.3	2.8	2.1	1.6	3.8	4.7	0.6	3.0	2.6
France	0.1	− 0.1	− 0.5	3.3	2.4	2.1	0.8	0.4	0.1
Ireland	0.2	0.2	0.6	1.7	2.1	2.6	0.5	0.6	1.0
Italia	2.4	7.3	1.7	6.1	4.8	4.0	4.1	6.1	2.8
Luxembourg	− 0.4	0.1	− 1.3	3.9	3.4	2.0	1.1	1.3	0.0
Nederland	0.7	− 2.3	− 0.6	3.2	1.7	− 0.4	1.3	− 1.4	− 0.5
Portugal	− 0.6	− 1.9	0.5	6.3	3.9	− 0.4	0.1	− 1.3	0.4
United Kingdom	3.1	5.0	0.5	4.1	2.1	3.1	3.2	4.5	0.1
EUR 12	2.0	2.6	1.0	4.9	4.4	3.4	2.8	3.1	1.6

Source: Eurostat.

Changes in nominal producer prices of agricultural products in 1993 and 1994

(%)

	1994/93			1993/92		
	Crop products	Livestock products	Total	Crop products	Livestock products	Total
Belgique/België	9.7	− 0.1	3.1	− 5.5	− 7.2	− 6.7
Danmark	− 5.4	− 0.8	− 2.3	− 13.6	− 11.8	− 12.4
Deutschland	4.4	− 0.7	0.7	− 5.0	− 8.2	− 7.3
Elláda	13.9	12.0	13.3	6.0	8.5	6.7
España	13.8	6.7	10.8	6.7	3.6	5.4
France	1.1	0.0	0.5	− 5.5	− 3.1	− 4.4
Ireland	1.9	1.3	1.3	4.3	6.9	6.6
Italia	− 1.4	− 0.2	− 0.9	1.1	5.1	2.6
Luxembourg	− 1.7	− 2.7	− 2.6	0.6	− 2.2	− 1.8
Nederland	8.3	− 4.1	0.6	− 3.1	− 8.0	− 6.3
Portugal	13.2	4.4	9.0	3.2	2.1	2.7
United Kingdom	4.1	− 2.2	0.0	0.5	8.0	5.3
EUR 12	6.2	1.1	3.6	0.4	0.1	0.4

Source: Eurostat.

Input prices

133. The index of purchase prices for staple goods and services in agriculture in the European Union as a whole is estimated to have risen by 1 % in 1994 in nominal terms, but to have fallen 2.8 % in real terms. There was an above-average fall in the prices of feeding-stuffs (− 5.0 %), seeds (− 3.3 %) and energy (− 3.1 %). However, even for most other intermediate consumption, the price index fell in real terms (fertilizer: − 1.4 %; plant protection products: − 2.1 %; farm animals: − 0.8 %).

134. The index of real input prices (including investment) fell by an average of 2.2 %, with above-average falls in Portugal (− 4.9 %), the United Kingdom (− 3.2 %), Belgium and Luxembourg (− 2.9 %), the Netherlands (− 2.7 %), Germany and Denmark (− 2.5 %), with smaller falls in Spain (− 2.1 %), Greece and Ireland (− 1.8 %), France (− 1.7 %) and Italy (− 1.2 %).

Trends in farm incomes

Trends in farm incomes in 1994

135. The first estimates available at the end of November 1994 indicate that farm incomes (net value-added at factor cost per annual work unit (AWU)) increased by 5.7 % in real terms in the European Union as a whole in 1994. This increase follows a smaller one in 1993 (0.5 % on average).

136. Unlike 1993, where trends varied considerably from one Member State to another, in 1994 there was an upward trend in farm incomes in almost all the Member States, except for Luxembourg (− 0.9 %) and Italy (− 7.6 %). The largest increase took place in Portugal (20.0 %) after falling for three years. Farm incomes rose considerably in Spain (12.9 %), France (12.1 %), Greece (10.2 %), the Netherlands (9.7 %), Ireland (7.6 %), Germany (6.7 %), Belgium (5.4 %) and the United Kingdom (4.2 %). A smaller, but still significant increase was recorded in Denmark (2.9 %).

137. The increase in farm incomes at Community level can mostly be explained by the following factors:

(i) a slight drop (− 0.5 %) in final agricultural production volume compared with the previous year, after a larger drop in 1993 (− 2.5 %);

(ii) relatively stable real producer prices;

(iii) a slight increase (+ 0.6 %) in the volume of intermediate consumption which was, however, more than offset by the drop in input prices in real terms, particularly the price of feedingstuffs;

(iv) a substantial increase in subsidies (+ 10.3 % in real terms);

(v) a 2.5 % reduction in farm labour compared with 1993.

The decrease of 7.6 % in Italy can be explained by a number of reasons, including reduced levels of production, unfavourable trends in producer prices, a reduced level of subsidies and a slowing-down of the drop in the farming labour force (in terms of full-time labour units).

Farm incomes over the last 10 years

138. Despite the deterioration in farm incomes between 1985 and 1987, their level is currently higher than it was at the beginning of the 1980s. The net value-added at factor cost per work unit (see Figure 1) increased by 15.8 % between '1981'(average for 1980, 1981 and 1982) and '1993' (average for 1992, 1993 and 1994), i.e. at an average annual rate in real terms of 1.2 %. The same rate of growth was recorded for the net family farm income, i.e. for the income remaining after deduction of wages, rent and interest paid.

139. This is, to a great extent, the result of the considerable reduction in the farming labour force over the last 10 years (– 3 % per year in terms of full-time labour units) and, consequently, of increased labour productivity. The productivity of intermediate consumption, on the other hand, as well as the agricultural price spread, have been stable over the

Net value-added [1] per person employed in agriculture [2]
'1985' [3] = 100

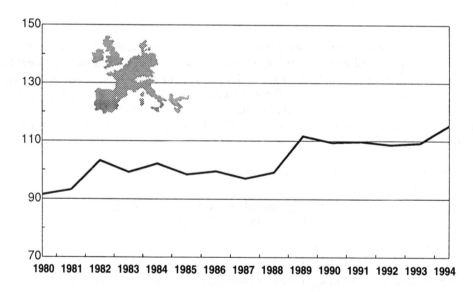

[1]   At factor cost in real terms (deflated using the GDP implicit price index).
[2]   Measured in annual work unit.
[3]   '1985' = average for 1984, 1985 and 1986.

Source: Eurostat — Sectoral income index analysis.

Figure 1

Net value-added [1] per person employed in agriculture [2]
'1985' [3] = 100

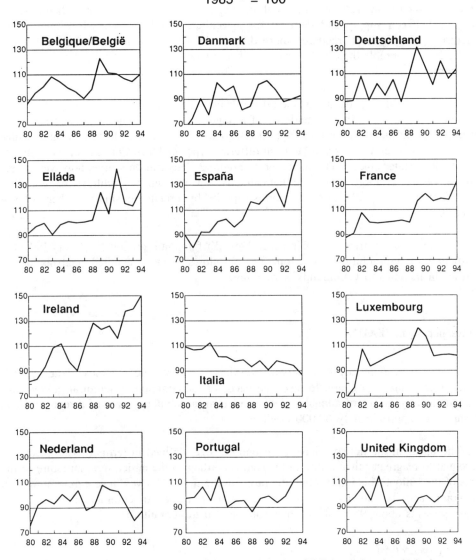

[1]   At factor cost in real terms (deflated using the GDP implicit price index).
[2]   Measured in annual work units.
[3]   '1985' = average for 1984, 1985 and 1986.

Source: Eurostat — Sectoral income index analysis.

Figure 2

long term, with a slight deterioration over the last few years. The most serious negative factor during the decade was the growing imbalance on both the Community and world markets, and the need to carry out reforms of agricultural policy in order to try to re-establish a degree of balance between supply and demand for agricultural products. The slowdown in the restructuring of agriculture, rendered more difficult by an economic environment that was generally less favourable than in the past, also had a damaging effect on the development of farm incomes.

140.   These general considerations should not cause one to lose sight of the fact that the Community agricultural sector is made up of a huge number of farms (estimated at around eight million) and the situation therefore varies greatly between Member States and, within each country, between regions and between different types of farm. Changes in farm income in the Member States since 1980 are shown in Figure 2. This can only give, however, a very rough idea of the variations in income that can exist in Community agriculture. Around 80 % of final Community agricultural production is supplied by a relatively small number of farms (around 20 %).

141.   The level and distribution of incomes between farms are obviously closely linked to the structure of agricultural production. There follows a more detailed analysis of the disparities in incomes in Community agriculture.

Incomes per farm (FADN)

142.   The farm accountancy data network (FADN) monitors the incomes of agricultural holdings in the European Union to meet the needs of the common agricultural policy. This task is attained by means of annual collection of farm accountancy data of a representative field survey of approximately 60 000 holdings.

143.   The farms in the network have an economic size, defined in terms of European size units, equal to or greater than a threshold fixed according to the agricultural structure of the regions of the Union. This implies that the farms in the field of survey have a minimum level of economic activity and, for that reason, are commonly called 'commercial farms'. The survey represents about 95 % of the total value-added in agriculture.

Incomes by type of farming

144.   The FADN user tends to ask for income by product. However, farms with only one product in the output are the exception. The rule is that most farms have several activities and produce, on average, about a dozen different products. Hence the classification of farms by type of farming is made by the proportion of standard gross margins of certain products in the total standard gross margin of production.

145.   The description of the type of farming comes from the main kind of production in terms of value in the total output of the farm. For example, a large proportion of farms produce cereals. Nevertheless, only as little as 6.7 % of all the farms are cereal specialists. The rest of the cereal producers are specialized in general cropping, mixed (crops and livestock), grazing stock and other types (see Chapter VII, Table 3.2.2).

146.   Figure 3 shows that there are important differences in income according to the type of farming. The commercial farms have an average family farm income (FFI) of about ECU 11 200 at 1990 prices. [1] The income is very high for pigs and poultry specialized farms (about ECU 30 000) and very low for cereals and permanent crops [2] (about ECU 7 000).

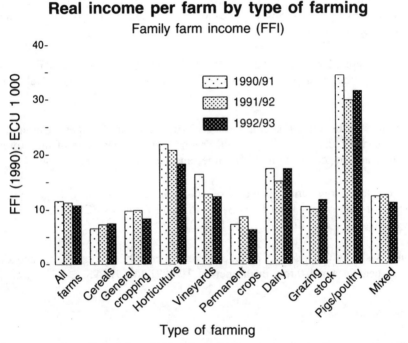

Real income per farm by type of farming

Family farm income (FFI)

Source: FADN, 1988 standaard gross margins, 1990 farm structure survey.

Figure 3

---

[1]   All FADN monetary indicators used in the sections 'Incomes by type of farming' and 'Incomes by economic size of farm' are expressed in 1990 ecu rates in real terms.

[2]   Vineyards are not included in this group and, as can be seen, have an income of ECU 13 000 to 16 000.

147. The results of the last three accounting years for which data are shown appear in Figure 4. The short-term average trend is a decrease of family farm income of a little less than 7 % for the whole group of farms for the period 1990/91 to 1992/93.

148. Year-to-year changes are difficult to evaluate because of the important variability of the factors and the results of agricultural output and, consequently, the farm income. The constant increase in income for the cereal specialists group is rather unique over this period.

Incomes by economic size of the farm

149. The variable economic size is expressed in terms of European size units (ESU), obtained by dividing the standard gross margin of the farm by a fixed number of ecus (at present ESU 1 = ECU 1 200). Several thresholds are established in order to obtain the six economic size classes shown in Figure 4.

150. As an example, large farms have between 40 and 100 ESU of standard gross margin (calculated by using regional standards of gross margin for individual crops and animals [1]).

151. According to this classification, commercial farms in the European Union are mostly of small and medium size: 27 % very small, around 20 % for each one of the groups small, lower medium, and upper medium, 10 % large, and less than 3 % very large.

152. As expected, large farms in terms of economic size also have higher family farm income. This variable ranges from ECU 3 100 in small farms to ECU 60 300 in very large farms (on average, for all farms — ECU 11 200).

153. Figure 4 shows that average income decreased mainly in large farms and there was little improvement in small farms in the period 1990/91 to 1992/93.

Incomes per family work unit [2]

154. Unpaid labour is normally provided by family members in European Union farms. The working family has 1.3 members on average, [3] ranging from one member in cereals to 1.5 members in mixed (crops and animals), dairying, and horticulture.

---

[1] For more methodological information on the subject, see the FADN publication *An A to Z of methodology*, 1988, published by the European Commission.
[2] The monetary units of this section are in current (1992) ecus and the data relate to the 1992/93 accounting year.
[3] Part-time work is computed with decimals of unit.

Real income per farm by economic size of farm

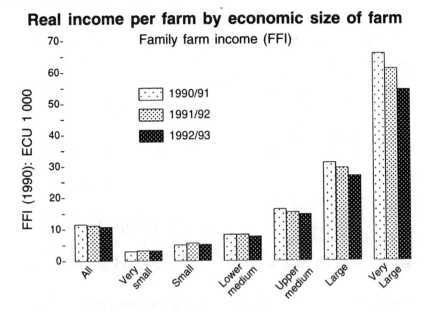

Economic size class

Source: FADN, 1988 standard gross margins, 1990 farm structure survey.

Figure 4

155.   The family work unit is particularly small in Denmark and Spain, with only 0.9 units in both cases, and exceptionally large in Greece and the Netherlands, with 1.5 units for those two Member States.

156.   The European family farm income per family work unit was about ECU 9 200 on average in 1992/93. The United Kingdom, Belgium and Luxembourg are at the top of the income ranking, with about ECU 20 000. Denmark, [1] Portugal and Greece are at the bottom of the same list.

---

[1]   The income results per family member for Denmark are very low (even negative) for the types of cereals, grazing stock, general cropping and mixed farming) compared to the high degree of modernization of Danish agriculture. The reasons for this situation have been a high level of indebtedness and the impact of a part-time agriculture in which income is not the upper priority variable for the business decision-making of the agricultural activity. There are also taxation reasons for this specific situation.

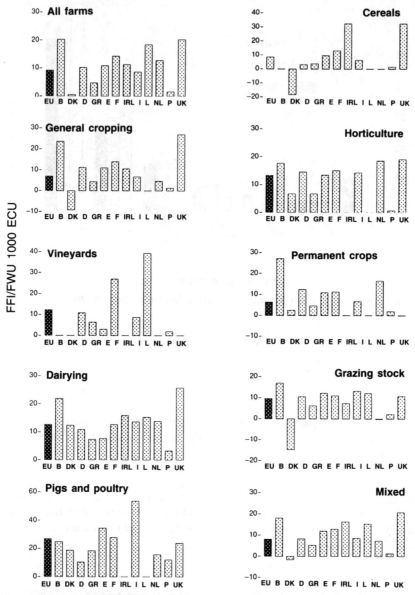

Income per family work unit — 1992/93

Family farm income per family work unit (ECU 1 000)

Source: FADN, 1988 standard gross margins, 1990 farm structure survey.

Figure 5

157.   Figure 5 illustrates the income per family member by type of farming and Member State. At the European Union level the highest incomes are on pigs and poultry farms (ECU 26 900) and the lowest on permanent crops (ECU 6 400). At national level, Denmark is the only Member State with negative incomes per unit for cereals, grazing stock and general cropping. [1] Portugal and Greece are the other two Member States with generally low results.

158.   Cereal specialists have very high income per family member in the United Kingdom and Ireland. Other types of farming have a good situation as well: general cropping and milk producers in the United Kingdom and Belgium, pigs and poultry in Italy, and vineyards in Luxembourg and France. With regard to viticulture, the particularly high result for Luxembourg is due to the fact that FADN holdings market their products directly, without going through cooperative channels, for example. For the whole of the wine sector in Luxembourg, the results should be similar to those in Germany.

---

[1]   The income results per family member for Denmark are very low (even negative) for the types of cereals, grazing stock, general cropping and mixed farming) compared to the high degree of modernization of Danish agriculture. The reasons for this situation have been a high level of indebtedness and the impact of a part-time agriculture in which income is not the upper priority variable for the business decision-making of the agricultural activity. There are also taxation reasons for this specific situation.

III — Agricultural markets

159. This chapter reviews the world and Community markets for the main agricultural products, covering price trends and the main market management measures proposed or decided at Community level. Recent developments, due to the implementation of the reform of the common agricultural policy, are explained in Chapter I.

Cereals

160. World cereal production in 1993/94 fell sharply compared with the previous year, particularly of feed grains, which were seriously affected by severe floods in the United States. Production of both wheat and coarse grains fell from the high levels reached in 1992/93 in several major producer countries such as the Union, the United States, Canada and the CIS countries as a whole, but rose in Australia and China.

161. The 1993/94 world cereals harvest amounted to 1 347 million tonnes as compared with 1 426 million tonnes in the previous marketing year, a drop of 5.5 %. Wheat production declined from 561 million tonnes in 1992/93 to 558 million tonnes in 1993/94 (− 0.4 %) and feed grains from 865 million tonnes to 789 million tonnes (− 8.8 %).

162. World wheat consumption in the 1993/94 marketing year is estimated at 565 million tonnes, i.e. 7 million tonnes more than the harvest. This represents an increase of 15 million tonnes compared with the 550 million tonnes for the preceding marketing year. The increase is due to the replacement of coarse grains by wheat. Consumption of feed grains is estimated to be 1.6 % down on 1992/93, at 824 million tonnes, but is 35 million tonnes above the figure for production.

163. International trade involved around 172 million tonnes of cereals during 1993/94, which represents a fall of 11.3 % compared with the 194 million tonnes for world trade in cereals in 1992/93. The fall can be put down mainly to lower purchases by China and the CIS countries. Those two destinations absorbed 35 % of world wheat exports in 1991/92 while their share of world trade was only 23.5 % in 1992/93. In 1993/94, their share of world wheat trade was only 11.5 million tonnes, i.e. 12.5 %. The drop to 82 million tonnes in feed grain imports was due to some extent to the sharp fall in production in the United States.

164.   Trade in cereals with the CIS continued to decline on account of the lack of credits from exporting countries. The re-establishment of a more rational prices policy also reduced consumption considerably.

165.   Forecasts for 1994/95 indicate higher world production (1 389 million tonnes against 1 347 million tonnes in 1993/94) with a stronger rise in world consumption. Consequently, world trade, which is estimated at 176 million tonnes, including 95 million tonnes of wheat (against 93 million tonnes in 1993/94), is expected to stand still.

166.   Community production for 1993/94 is assessed at 163 million tonnes, 5 million tonnes down on 1992/93, when production was affected by drought, and 17.5 million tonnes down on 1991/92.

167.   The fall in production reflects the reduction in areas sown to cereals as a result of the introduction of compulsory set-aside (32.27 million ha sown instead of 35.24 million ha in 1992/93, i.e. − 16 %).

168.   The drop in cereals production is due to a fall in the barley and durum wheat harvests. Production of common wheat, maize and rye was less affected by set-aside.

169.   The trend in cereal production varied from Member State to Member State. As a result of the exemption of small producers from the set-aside obligation, the reduction was smaller in the regions where agriculture is less highly developed.

170.   When account is taken of the resumption of production on agricultural land withdrawn under the temporary set-aside arrangements, the reduction in the area sown to cereals corresponds to set-aside of 4.6 million ha under the reform. The rate of set-aside was highest in the cereal-growing regions of the United Kingdom, France and Germany.

171.   The drop in cereal prices under the reform stimulated the use of greater quantities of cereals in animal feedingstuffs, which rose to 86 million tonnes in 1993/94, i.e. over 5 million tonnes more than in the period before the reform.

172.   At the same time, trade in cereals between the Member States has been firm.

173.   At 32.5 million tonnes (including processed products and food aid), cereal exports were below the record of nearly 35 million tonnes in 1992/93, but well above the figure for the period 1986-91. Commercial exports amounted to 19.8 million tonnes of common wheat (including flour), 8.5 million tonnes of barley (including malt) and 3.5 million tonnes of

Cereals[1]

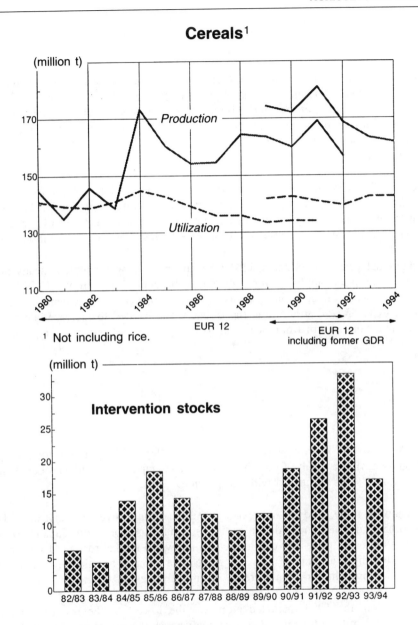

1 Not including rice.

EUR 12

EUR 12
including former GDR

NB: Intervention stocks at 31 December.

Figure 6

maize. Exports of durum wheat (1 million tonnes) and rye (0.5 million tonnes) dropped significantly.

174. With exports being maintained at a high level and animal consumption increasing, it was possible to reduce intervention stocks. From the exceptionally high figure of 33.4 million tonnes at the beginning of the 1993/94 marketing year, stocks fell to 18 million tonnes, comprising 6.5 million tonnes of common wheat, 1.1 million tonnes of durum wheat, 6.5 million tonnes of barley, 2.5 million tonnes of rye, 1.1 million tonnes of maize and 0.2 million tonnes of grain sorghum.

175. The 1993/94 marketing year was the first year of application of the reform to arable crops. The fall in prices, offset by aid per hectare, and the drop in production led to better balance on the market. Thus prices for cereals on the internal market and in particular for durum wheat and breadmaking common wheat were higher than the intervention prices.

176. The gradual price reduction decided when the reform was adopted means that the single intervention price applying for 1994/95 is brought down from ECU 115.49 to ECU 106.6 per tonne, offset by an increase in the compensatory payment of ECU 10/t bringing it to ECU 35/t.[1] Furthermore, the Council has decided to increase compensation for set-aside under the reform from ECU 45/t to ECU 57/t as set out in the regionalization plans.

Rice

177. The 1993 world harvest was lower than in 1992 (521 million tonnes against 527.2 million tonnes) as a result of poor harvests in Asia (Japan, Thailand and Laos) and the United States (– 13 % as compared with the previous year).

178. International trade rose by around 1.3 million tonnes over 1992 (from 14 million tonnes to 15.3 million tonnes), due in particular to purchases of rice by Japan to offset its poor harvest. As a result, prices for quality rice on the world market rose until January 1994 and subsequently fell on account of slackening Japanese demand and the devaluation of the CFA franc, which brought about an increase in import prices in Africa.

179. In the Union, the 1993 marketed crop was the lowest of the last four marketing years (at around 1.2 million tonnes milled rice equivalent), the severe drought in Spain and Portugal having diminished the areas sown in those two countries. In France, torrential rains

---

[1] The reduction in guaranteed prices decided under the reform is offset by an amount granted per hectare and determined on a regional basis depending on yields obtained over a reference period.

in the Camargue also affected the paddy rice harvest. However, Italian production rose by 5 % in 1993 as compared with the 1.2 million tonnes in the previous year but was lower than the 1.5 million tonnes anticipated, on account of poor weather conditions during the harvest.

180. Prices on the Italian market rose until January 1994 and subsequently fell on account of the lack of an export refund on rice in bulk and the small quantities of food aid granted by Italy. Expressed in ecus, prices on average remained above the buying-in price and only fell to about its level at the end of the marketing year.

181. No rice was offered for intervention during 1993/94.

182. As to trade with third countries, imports (mainly of Indian rice) rose by 14 % while exports fell by 34.5 % as a result of the poor Community harvest and firm demand within the Community.

183. Imports from the ACP States reached 87 267 tonnes (against 42 382 tonnes in 1992/93). Imports of processed rice from the Netherlands Antilles (OCT) amounted to 146 409 tonnes expressed as husked rice (against 111 916 tonnes in the 1992/93 marketing year). Imports of rice processed in the Netherlands Antilles are free of levy. Overall imports from the ACP States and the OCT increased by around 30 % over the 1992/93 marketing year.

184. Forecasts for the 1994/95 marketing year indicate an area sown of around 370 000 ha, of which 56 000 ha of indica rice. The self-sufficiency rate for indica rice should therefore rise from 26 % to 36 % and the shortfall will probably be made up by imports of rice from third countries.

Sugar

185. The strong decline in world stocks in 1993/94 appears to have put an end to the overall surplus on the world market since 1990. The production shortfall in relation to consumption, estimated at 3.4 million tonnes at the end of the 1993/94 marketing year, left stocks at 30 % of consumption, a level which can be considered as the point of market equilibrium.

186. The increase in consumption (+ 0.7 %) is lower than the population growth rate during this period and shows a decline in per capita consumption. The main reason for this decline is the trend in world market prices, in a context of recession which has also affected this marketing year. The downward trend which began in 1990 has therefore come to an end, with the price of raw sugar fluctuating between 10 and 12.5 cents/pound during 1993/94.

World market trends

(million tonnes of raw sugar)

	Production	Consumption	Surplus or deficit	Stock as % of consumption
	(1)	(2)	(3) = (1) − (2)	(4)
1984/85	100.4	98.2	+ 2.2	42.2
1985/86	98.6	100.5	− 1.7	38.4
1986/87	104.2	105.9	− 1.7	33.8
1987/88	104.7	107.2	− 2.5	31.1
1988/89	104.6	107.0	− 2.4	29.1
1989/90	109.2	109.4	− 0.4	28.0
1990/91	115.6	110.3	+ 5.3	31.7
1991/92	116.4	111.3	+ 5.1	35.5
1992/93	112.6	112.2	+ 0.4	34.9
Forecast 1993/94	109.6	113.0	− 3.4	30.9

Source: F.O. Licht.

Average spot price:

Paris Exchange *(white sugar)*   = ECU 21.91/100 kg in 1992/93
ECU 26.79/100 kg in 1993/94 (+ 22.3 %)

New York Exchange *(raw sugar)*   = ECU 16.95/100 kg in 1992/93
ECU 20.37/100 kg in 1993/94 (+ 20.2 %)

---

187.   The second successive year of poor harvests in Cuba, China and India have continued to affect market conditions. Cuba was unable to meet its protocol with Russia (1.5 million tonnes), thereby facilitating the flow of large quantities of Community white sugar to Russia. China has dampened the increase in its consumption in order to adjust to the smaller supply. India has had to import large quantities (1.9 million tonnes) to take account of the rapid decline in its buffer stocks resulting from increasing domestic consumption, and banned all exports from the beginning of the marketing year.

188.   Despite these events, the world supply balance for 1993/94 confirmed the downward trend in import demand since 1988/89. With a reduction of around 1.5 million tonnes in relation to the previous marketing year, the world sugar market amounted to only 27.5 % of total production, as against 33 % in 1984/85. This trend reflects the move towards self-sufficiency of importing countries in order to meet the growing demand from their population. However, white sugar took a 50 % share in this world market trend, as against 34 % in 1984/85.

189.   In this context, the first estimates for the 1994/95 harvest indicate a new production deficit in relation to world consumption. The level of stocks dropped below the point of theoretical equilibrium, situated at 30 % of consumption. The improvement in market prices

Sugar

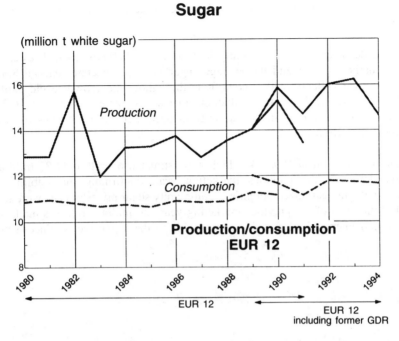

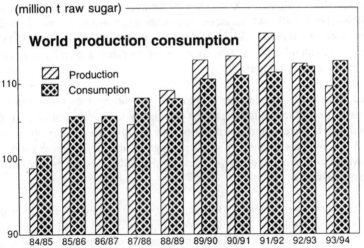

Figure 7

could therefore stabilize at the current level because of the difficult situation expected in the first quarter of 1995.

190. Areas under beet in the Community in 1993 (1 918 000 ha) fell (3.1 %) compared to the previous marketing year and the average sugar yield per hectare reached 8.32 tonnes. Early sowing and excellent weather conditions throughout the growth cycle explain these high yields, which resulted in total production, in white sugar equivalent, of 16.23 million tonnes, of which 15.96 million tonnes came from beet, 0.26 million tonnes from cane and 0.02 million tonnes from molasses.

191. Sugar consumption in 1993/94, which was estimated overall at 11.87 million tonnes, was slightly down on the previous marketing year, confirming the stability of sugar consumption in the industrialized countries. Community sugar prices in ecus were maintained in 1993/94 at the level of the previous marketing year and inulin syrup was made subject to production arrangements similar to those applicable to the sugar and isoglucose production quotas.

Olive oil

192. World production is about 1 800 000 tonnes on average, of which the European Union accounts for 80 % (about 1 450 000 tonnes). The other main producers are Tunisia (100 000 tonnes), Turkey (80 000 tonnes), Syria (55 000 tonnes) and Morocco (35 000 tonnes). Production varies considerably from one year to another but generally the world market closely mirrors that of the Community.

193. Estimated Community production in 1993/94 was about 1 280 000 tonnes as against 1 379 347 tonnes in 1992/93. The area remains practically unchanged, according to available figures, at 4.9 million hectares, equivalent to 66 % of the total area under olive cultivation in the world and 3.3 % of the Community's utilized agricultural area. Some estimates put the number of cultivated and abandoned olive trees at 450 million. Some two million farms are engaged in olive cultivation.

194. In 1992/93, Community consumption was 1 360 000 tonnes (77 % of world consumption). The most recent estimates suggest that consumption should remain at around the same level in 1993/94. Particularly as a result of the introduction of consumption aid in 1979, most of the Community uptake, 1 300 000 tonnes, is in the form of small containers. At the beginning of the 1993/94 marketing year, intervention stocks stood at 198 000 tonnes, falling to about 110 000 tonnes at the end of the year.

195. Greece and Spain are the main suppliers and, although Italy both produces and exports, it remains the main purchaser. Apart from exceptional cases, imports are restricted to the Tunisian quota of 46 000 tonnes. At 110 000 tonnes, exports in 1993/94 were down on the previous year (162 000 tonnes).

196. Developments in Community policy: a stabilizer was introduced from the 1987/88 marketing year with a maximum guaranteed quantity (MGQ) of 1.35 million tonnes. When output exceeds the MGQ plus, where applicable, the carryover from the previous year, production aid is reduced proportionately. The MGQ was exceeded in 1992/93, resulting in a 2.23 % reduction in the aid. Estimates for 1993/94 suggest that an overrun of the MGQ is not likely to occur this year.

197. Consumption seems to be less affected than in the past by price competition from other vegetable oils. It appears to be more sensitive to fluctuations in the price of olive oil, increases in consumer incomes and efforts to improve product quality and promote consumption. For these reasons, and with a view to the proper management of the consumption aid scheme, the Council decided to adjust the aid for 1993/94. This adjustment was achieved without a negative effect on consumption, which firmed throughout the Union.

Olive oil production in the European Union

(1 000 tonnes)

Member State	1986/87	1987/88	1988/89	1989/90	1990/91	1991/92	1992/93	1993/94 provisional
Elláda	246.4	321.7	334.9	171.0	237.6	430.1	314.4	260.0
España	529	770	406.5	700	702.0	610.0	636.0	560.0
France	1.5	3	1.4	2.0	2.0	3.4	1.8	3.0
Italia	383	742.5	437.1	585.0	148.0	650.0	410.0	430.0
Portugal	44.8	38	28	20.0	37.8	35.0	17.1	30.0
Total	1 204.7	1 875.2	1 143.2	1 512.3	1 041.0	1 728.5	1 379.3	1 283.0

Areas planted to olives in the Community and the number of trees

Member State	Hectares	Number of trees (millions)
Elláda	838 000	122
España	1 935 000	177
France	40 000	3
Italia	1 372 000	126
Portugal	727 000	32

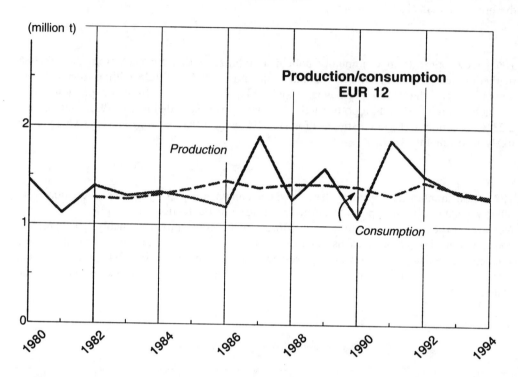

Figure 8

Oilseeds

198. Oilseeds yield cake for animal feed and oil. This means that the economic position of the sector depends on price trends for seed, oils and cake. The oils may be consumed without further processing or as prepared oils and fats such as margarine. They may be used as animal feed, for human consumption or for technical purposes.

199. The Community is a net importer of oilseeds, vegetable oils and cake, annual import volumes being largely dependent on the relative prices of seeds, cake and competing animal-feed products (cereals, corn gluten feed, etc.) and on the opportunities for exporting oils and cake from the Community.

Average oil supply balance for the Community 1991-93
(figures in brackets 1989-91)

(million t oil equivalent)

	Production	Consumption	Imports	Exports	Self-sufficiency (%)
Rapeseed	2.1 (2.2)	1.6 (1.5)	2.3 (0.9)	0.6 (0.8)	131 (146)
Sunflower	1.7 (1.6)	1.8 (1.7)	2.0 (0.8)	0.2 (0.3)	95 (94)
Soya	2.5 (2.0)	1.9 (1.9)	2.5 (1.8)	0.8 (0.6)	132 (105)
Vegetable oils[1]	5.5 (5.6)	9.4 (8.9)	9.5 (6.6)	1.8 (1.9)	58 (62)

NB: World production about 50 million tonnes.
[1] Rapeseed, sunflower, soya, olive oil, cotton, linseed, groundnut, sesame, palm, palm kernel and coconut.

Average cake supply balance for the Community 1991-93
(figures in brackets 1989-91)

(million t cake equivalent)

	Production	Consumption	Imports	Exports	Self-sufficiency (%)
Rapeseed	3.7 (3.2)	4.6 (3.9)	0.7 (0.8)	0.1 (0.1)	81 (82)
Sunflower	2.2 (2.1)	4.5 (3.6)	1.5 (1.5)	0.0 (0.0)	56 (58)
Soya	11.7 (10.9)	21.5 (20.3)	10.6 (10.6)	0.9 (0.8)	54 (53)
Cake[1]	16.2 (16.2)	33.6 (31.5)	18.6 (25.7)	1.1 (1.3)	48 (51)

NB: World production about 120 million tonnes.
[1] Rapeseed, sunflower, soya, cotton, linseed, groundnut, sesame and palm kernel.

200.   In the 1993/94 marketing year a total of some 23.6 million tonnes of oilseeds were crushed in the Community: 12.5 million tonnes of soya (down 1.3 million on 1992/93), 6.0 million tonnes of rape (up 0.6 million on 1992/93 and 4.3 million tonnes of sunflower (up 0.1 million tonnes on 1992/93).

Community soya imports

(1 000 t)

	1985	1986	1987	1988	1989	1990	1991	1992
Soya								
USA	7 532	9 591	10 256	7 821	5 883	6 383	6 186	8 260
Brazil and Argentina	5 037	2 985	3 255	3 331	3 885	5 242	5 545	5 388
Soya cake								
USA	2 112	2 792	3 109	1 380	529	253	201	369
Brazil and Argentina	8 932	7 689	6 667	7 329	8 005	9 522	9 921	10 149

Source: Eurostat.

201.   From 1993/94 the support arrangements for oilseeds growers (rape, sunflower, soya) formed part of the new support scheme for arable crops (cereals, oilseeds, protein plants) requiring a 15 % land set-aside. A specific compensatory payment is granted on oilseeds, set

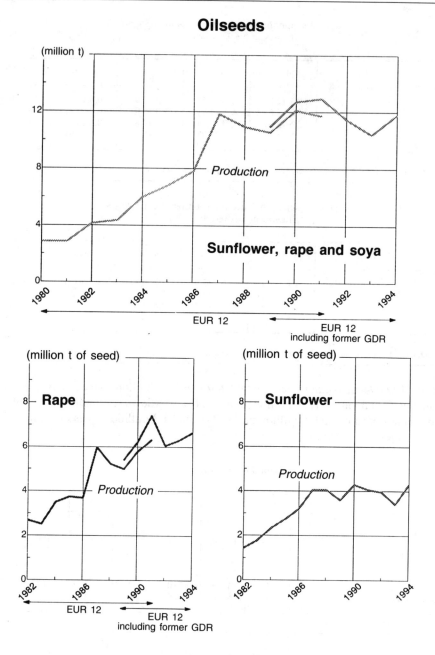

Figure 9

at ECU 359/ha (except for sunflower in Spain and Portugal and producers opting for the simplified scheme). The amount actually paid to growers is regionally differentiated according to historic yields of cereals or oilseeds and is also adjusted in line with world price fluctuations beyond a certain margin. For 1993/94 Community support was granted on some 6.0 million hectares of oilseeds, if non-food crops are included.

202.   Community oilseed production in 1993/94 was almost 10.9 million tonnes compared with 11.5 million in 1992/93. The crop estimate for 1994/95 is 11.8 million tonnes.

203.   The support arrangements for oilseed growers were radically altered with effect from 1992/93. Processing aids and intervention were abolished and growers' incomes protected by a direct payment per hectare.

Dried fodder, peas, field beans and sweet lupins

204.   These products, which go chiefly to the animal-feed industry, compete with a wide range of other raw materials.

Peas, field beans and sweet lupins

(1 000 t)

	1989/90	1990/91	1991/92	1992/93	1993/94	1994/95
Total production	4 745	5 541	4 979	4 728	5 652	5 183
Aid-eligible production	4 173	5 044	4 465	4 270	5 652	5 183

205.   From 1993/94 aid to processors and minimum prices were abolished and growers' incomes were protected under the same scheme as for cereals and oilseeds involving a land set-aside of 15%. The regionally differentiated aid is ECU 65 multiplied by the historic cereals yield.

206.   For 1993/1994 the protein crop area for which Community support was given amounted to some 1.3 million hectares.

207.   Dried fodder production in the European Union again reached a record level.

Dried fodder

(1 000 t)

	1990/91	1991/92	1992/93	1993/94	1994/95 [1]
Dehydrated	3 271	3 687	4 220	4 529	4 628
Sun-dried	568	431	440	385	400
Total	3 839	4 118	4 660	4 914	5 028

[1] Estimation September 1994.

Fruit and vegetables

Fresh fruit and vegetables

World situation

208.   Statistical information on the fresh fruit and vegetable sector worldwide is still very incomplete. Moreover, international trade is significant only in certain products, particularly apples and citrus fruit. World production trends show an increase in the production of oranges for processing following the planting of new groves in some of the main non-Community producer countries.

Trade with non-member countries

209.   The European Union is the world's largest importer of fresh fruit and vegetables, particularly citrus fruit and apples. Although imports account for a relatively small proportion of the Union's annual requirements, they play an important role in supplying the market out of season and can affect price stability in season.

210.   In the absence of specific data on world trade, import price trends in the Union give an impression of the situation. Thus, for the 1993/94 marketing year, import prices for citrus fruit remained stable, fell slightly for plums, fell substantially for pears, fell sharply for apples, rose slightly for table grapes, apricots and peaches and rose substantially for cherries. In terms of quantities, imports of apples fell (34%) after increasing for three years. Imports of vegetables and citrus fruit remained more or less stable.

211.   The surge in Community exports of fresh fruit and vegetables witnessed in recent years continued in 1993, particularly in the case of vegetables and citrus fruits. Exports of oranges increased by 33% in relation to 1992. In the case of fruit, apple exports more than doubled but exports of other fruits fell by 30% in relation to 1992.

The Community market

212.   8.55 million tonnes of apples were produced in the Community in 1993/94, 22% less than the bumper crop of 1992/93. This drop is mainly due to the consequences of biennial bearing in Germany (−50%). The reduction was also fairly sharp in the countries where much of the production is marketed (France: −15%; Italy: −10%). Community withdraw-

als (988 000 tonnes) fell by 52 %. Production of pears, which rose to more than 3 million tonnes in 1992/93, fell by 19 % to the more usual level of 2.5 million tonnes and withdrawals fell from 6.3 % to 1.6 % of the harvest. The production of table grapes fell by 6 % in Italy, the source of 64 % of total Community production, and by 20 % in Spain. In terms of volume, the production of table grapes is, however, almost on a par with that of pears.

213.   The most recent production figures for citrus fruit in 1993/94 show a drop in orange production, which was considerably lower than the 6 million tonnes produced in 1992/93 (– 25 % in Italy and – 21 % in Spain, the main Community producers). Conversely, satsuma production is up 20 % in Spain, which is responsible for almost all of the Community's production. The situation regarding clementines was the same in Spain, while production in Italy fell by 43 %. Lemon production rose by 3 % in Spain and fell by 26 % in Italy, although these two countries traditionally produce similar quantities. Mandarins, while less popular on the market than the other small citrus fruits, fell by only 14 % in Italy and increased by 20 % in Spain. Withdrawals in 1993/94 were considerably down in relation to the previous marketing year: – 50 % for oranges and lemons and – 64 % for mandarins. These withdrawals represent around 6.3 % of total Community production for oranges, 4.2 % for lemons, 1.6 % for mandarins, 0.8 % for clementines and 1.5 % for satsumas.

214.   Production of fruit with kernels fell back to normal levels in 1993/94, dropping sharply in comparison with the large crop in 1992/93 (– 30 % for peaches, – 20 % for nectarines, – 16 % for apricots and – 7 % for cherries and morello cherries). Withdrawals of peaches this year accounted for 22 % of the harvest as opposed to 25 % in 1992/93. At 17 %, withdrawals of nectarines were down considerably on the previous marketing year. Apricots were the only exception to the general drop in withdrawals in this sector, at 16 % of the harvest as opposed to 3 % in 1992/93.

215.   Tomato production, while falling in Italy (– 5 %), which is by far the largest producer, remained more or less stable in the other Mediterranean producer Member States. In Portugal, the 28 % increase in relation to 1992/93 signals a return to the levels of 1989/90, 1990/91 and 1991/92. Withdrawals shrank from 2 % to 0.4 % of the harvest, but market-support measures for tomatoes tend to concentrate on the processed product sector. Greenhouse production remained stable in the United Kingdom, Germany and Denmark, but the downward trend continued in the Netherlands (– 7 %) as did the upward trend in Belgium (5 %).

216.   Withdrawals of cauliflowers in 1993/94 were 23 % down on the previous year, representing 4.6 % of the harvest, as compared to 6.2 % in 1992/93.

The main developments in legislation and policy

217.   For the 1994/95 marketing year, the Council has kept all basic and purchase prices at their previous levels and decided to reintroduce a support scheme for grubbing apple trees

Fruit

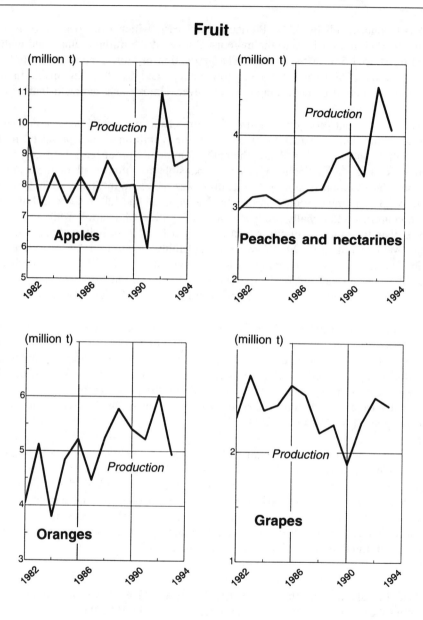

Figure 10

similar to the scheme in force between 1990 and 1993. The overrun of certain intervention thresholds during 1993/94 has resulted in the following prices being changed in relation to the Council decisions: apples (-7%), peaches (-17%), lemons and nectarines (-20%).

218. Under the existing rules, minimum price arrangements have been introduced for imports of bitter cherries intended for processing. A mechanism for controlling imports of garlic from China has also been set up, while garlic imports from Taiwan and Vietnam were suspended. The Commission also introduced (for blackcurrants only) the border controls provided for in certain circumstances under the minimum price arrangements for imports at reduced tariff rates of soft fruit originating in certain East European countries.

Products processed from fruit and vegetables

World market and European market

219. As in the case of fresh products, the world market for products processed from fruit and vegetables is greatly influenced by exports to the Union but, thanks to preservation and processing possibilities, the imbalance between supply and demand is less evident than for fresh fruit and vegetables. However, specific problems do arise occasionally.

220. Although there are no full and up-to-date figures on world prices, it is still possible to identify certain trends during the 1993/94 marketing year. The chronic shortage of frozen raspberries has kept prices very high, often more than twice the minimum price applied to reduced-duty imports from Eastern Europe. The market for frozen strawberries has also continued to firm up, although there do not appear to be problems in the supply of this product, which is so important to the food industry. The smaller market for frozen blackcurrants has been in surplus for two years. Except for tomatoes and orange juice, dried grapes are the subject of the greatest volume of world trade in this sector. The Union's import prices for both dried grapes and prunes suggest a stable world market. This is also the case for dried figs, for peaches and for pears in syrup, which remain important products.

The main developments in legislation and policy

221. Although the common organization of the market in this sector in principle covers all products processed from fruit and vegetables, institutional activities are focused on certain products: mushrooms (duty-free quotas), soft fruit (minimum prices for imports from countries benefiting from reduced duties), dried grapes (minimum price and storage and production aid), dried figs (storage and production aid), prunes, pineapples, peaches and pears in syrup (production aid). Raspberries for processing and dried grapes are also the subject of specific measures aimed at improving product quality and marketing. In the case of dried grapes, this measure was adopted by the Council in May 1994.

222.   Trends in processing aid during the 1993/94 marketing year in relation to the levels in force during 1992/93 were as follows: prunes: − 4.2 %; pineapples: + 13 %; dried figs: + 2.19 %; peaches: + 5.6 %; pears: + 4.4 %. Minimum producer prices remained the same except in the case of peaches, for which they were increased by 3.8 % in view of the trend in the basic price for fresh peaches. For dried grapes, the basic amount of aid per hectare was increased from ECU 2 045 to ECU 2 306. This aid is paid only for areas under specialized cultivation which meet certain criteria as to yields.

Bananas

223.   The common organization of the market in bananas was set up with a view to completing the single market by 1 January 1993. The internal market could not function correctly while the national agreements governing banana imports remained in place. These national agreements therefore had to be replaced by a single Community system for bananas, laying down common rules on the organization of the market. The system had to reconcile the conflicting needs and requirements of the main groups concerned with the Community banana market.

224.   Furthermore, any proposal to create a common organization of the market in bananas clearly had to meet certain objectives, i.e. the need not only to comply with obligations to producers and consumers in the European Union, in accordance with Article 39 of the Treaty of Rome, and to the ACP States in accordance with the Lomé Convention, but also to comply with the European Union's obligations under GATT, and avoid damaging the interests of suppliers in third countries.

225.   The common organization of the market in bananas entered into force on 1 July 1993.[1] It lays down provisions for bananas produced in the Union and bananas originating in third countries.

226.   A tariff quota of 2 million tonnes (net weight) was opened for imports of bananas from third countries and non-traditional ACP bananas. Under this tariff quota, imports of bananas from third countries are subject to duty of ECU 100/tonne, while non-traditional ACP bananas are exempt from duties. For imports in excess of this tariff quota, the tariff is ECU 850/tonne for imports from third countries and ECU 750/tonne for imports of non-traditional ACP bananas. Traditional ACP banana imports are not included in the quota and are set out in the annex to Regulation (EEC) No 404/93. They are exempt from duties.

---

[1]   Regulation (EEC) No 404/93, 13.2.1993, OJ L 47, 25.2.1993, p. 1.

227.   In order to maintain Community production, Regulation (EEC) No 404/93 provides for compensatory aid to offset loss of income as a result of the new arrangements. The compensation for Community bananas marketed during the second half of 1993 amounted to ECU 24.5/100 kg. The aid amounts to ECU 27.3/100 kg for bananas produced in Madeira, because of the very difficult production conditions. Total Community production in the second half of 1993 was 295 254 tonnes. The aid therefore totalled ECU 72.8 million.

228.   All the detailed rules for the application of the common organization of the market have been adopted. They concern the arrangements for banana imports,[1] safeguard measures,[2] cessation of production,[3] compensatory aid,[4] producers' organizations[5] and quality standards.[6]

229.   Lastly, an agreement on the Community import arrangements for bananas was signed on 15 April 1994 at Marrakesh in Morocco with four of the five Latin-American countries who are GATT signatories. In addition to certain specific arrangements for part of the exports from those countries to the European Union, the agreement provides for the tariff quota to be increased by 100 000 tonnes in 1994 and 100 000 tonnes in 1995 and for the tariff to be reduced from ECU 100 to ECU 75 per tonne within the quota.

Wine

230.   The European Union is the leading wine economy in the world, accounting on average for 60 % of production and 55 % of consumption. The percentages in 1992 were 69 % and 58 % respectively. In decreasing order of production, the other main producers are Argentina, the United States, the former Soviet Union, the East European countries as a whole (Bulgaria, Hungary, Romania, the former Yugoslavia, the Czech Republic and Slovakia) and South Africa.

231.   Taking imports and exports together, international trade in wine involves approximately 30 % of world production, which in recent years has stood at around 285 million hl. With average annual consumption of around 235 million hl, the world surplus can thus be put at 50 million hl, the Community being responsible for two thirds. The latter's structural surplus is mainly sent for distillation.

---

[1]   Regulation (EEC) No 1442/93, 10.6.1993, OJ L 142, 12.6.1993, p. 6.
[2]   Regulation (EEC) No 1662/93, 29.6.1993, OJ L 158, 30.6.1993, p. 16.
[3]   Regulation (EEC) No 1639/93, 28.6.1993, OJ L 157, 29.6.1993, p. 5.
[4]   Regulation (EEC) No 1858/93, 9.7.1993, OJ L 170, 13.7.1993, p. 5.
[5]   Regulation (EC) No 919/94, 26.4.1994, OJ L 106, 27.4.1994, p. 6.
[6]   Regulation (EC) No 2257/94, 16.9.1994, OJ L 245, 20.9.1994, p. 6.

232. In 1993, there was a sharp rise in trade from the Community to third countries; 12 552 000 hl was exported compared with 9 954 000 hl in 1992. Spain (4 221 000 hl), Italy (3 479 000 hl) and France (2 985 000 hl) are the main exporters, the main trading partners being the EFTA countries (28 %), the United States (22 %) and Canada (8 %).

233. Wine imports into the Community remained stable in 1993 at 2 658 000 hl. The main importers are Germany and the United Kingdom, with 45 % and 33 % respectively of Community imports. The East European countries are the main suppliers (accounting for 55 % of wine imports into the Community). Australian wine is fast improving its share of the market as its exports to the Community continue to rise (5.7 % in 1990, 7.8 % in 1991 and 11.1 % in 1992). Around 8 % of wine imported from third countries comes from the United States.

234. Community production in the 1993/94 wine year is estimated provisionally at 159 million hl for all wines (table wine, quality wine psr and other wine), which is down on 1992/93 (191 million hl). The guide prices were set at ECU 3.17/°/hl for table wine of types AI, RI and RII. Table wine prices varied considerably depending on the region of production; they rose by 15 to 20 % during the wine year, ranging from 55 % to 100 % of the guide prices.

235. Direct human consumption of wine appears to be stable as compared with preceding wine years (127 million hl), as does industrial utilization for the production of vermouth and vinegar (3.7 million hl).

236. Quantities to be distilled under Community intervention measures (preventive, compulsory and support distillation) remain significant at 22 million hl compared with 33 million hl in 1992/93.

237. To increase the effectiveness of preventive distillation, it was decided to concentrate such distillation in the first few months of the wine year and to introduce a security to ensure that wine is delivered to the distillery. Compulsory distillation was opened for a total of 18.2 million hl, which resulted in a very low price level equal to 26 % of the guide price. Support distillation was opened for 3 million hl at 82 % of the guide price.

238. Stocks with producers and traders at the beginning of the wine year amounted to 122 million hl. At the end of the year they should be down to 108 million hl.

239. From 22 December 1993 to 15 February 1994, long-term storage contracts (nine months) covered 9 823 330 hl (table wine: 6 977 904 hl; must: 2 505 560 hl; concentrated and rectified concentrated must: 339 866 hl). Producers were allowed to apply for early withdrawal of up to 90 % of the quantities under contract, the aid then being paid for the actual period of storage.

Wine

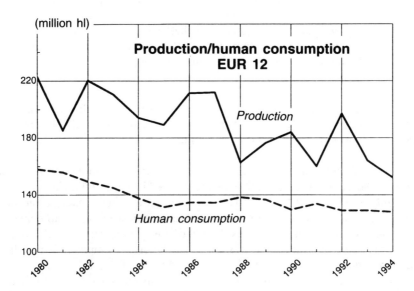

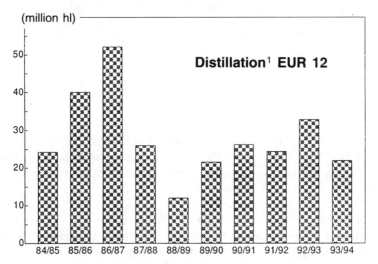

[1] Compulsory and optimal distillation, including Portugal from 1979 to 1991.

NB: The figures for 1992 include Portugal.

Figure 11

240.   Grubbing-up in exchange for premiums (the permanent abandonment of wine-growing areas) has continued to increase (320 000 ha having been abandoned between 1988 and 1993); however, this drop in the Community area under vines has not as yet brought about balance between demand and supply, but has simply compensated for the drop in consumption and the rise in yields. Furthermore, the complex distillation system, in particular compulsory distillation of low-priced table wine, has not proved sufficient to discourage wine producers and has not led to the restoration of sound market conditions.

241.   The medium-term outlook for the wine industry is for a high surplus as a result of the foreseeable stagnation of exports in the medium term, the fall in consumption and insufficiently controlled production in this context.

242.   Faced with this situation, the Commission has presented to the Council a proposal for a reform of the common organization of the market in wine, following on the discussion paper of July 1993 (COM(93) 380). The objective is to achieve balance on the market by channelling Community expenditure towards the adjustment of supply to demand instead of financing the elimination of the surplus. In addition to the objective of achieving quality and increasing the value of wine, this new system will involve a simplification of the distillation system based on national references, stricter standards for the enrichment of wine and regional viticultural adaptation programmes leaving much scope for subsidiarity by letting the regions and the Member States work out the content in partnership with the Commission. This should involve the identification of surplus areas under vines to be grubbed, incorporation of this procedure into a more general approach to rural planning (regional priorities as regards the environment, land use and rural development), etc.

Cotton

243.   The world area sown to cotton in 1994/95 is around 32 million hectares producing an estimated 19.2 million tonnes (85.7 million bales),[1] i.e. the same as in 1993/94. An increase in both production and consumption is expected in the new marketing year, with stable but considerable stocks at the end of the year.

244.   Prices for cotton fibre on the world market, which stood at the relatively low level of 54.90 cents/pound at the beginning of 1993/94, rose from December 1993 to reach 76.35 cents/pound at the beginning of the 1994/95 marketing year. Unginned cotton is not traded internationally but the Community, with a spinning capacity much greater than its fibre production, imports considerable quantities: from 1986 to 1990 over 1 million tonnes and 943 000 tonnes in 1993. The United States, the former Soviet Union, Syria, Pakistan, India,

---

[1]   One bale = approximately 217.7 kg.

Paraguay, Israel, Argentina and China, in certain periods, are the main suppliers. Intra-Community trade remains limited but is increasing.

245. In the European Union, cotton is a minor crop in terms of area and number of producers but plays a very important socio-economic role in the areas of Greece and Spain where cultivation is concentrated. The Community area under cotton increased from 311 500 hectares in 1991 to 397 000 hectares in 1992, 383 300 hectares in 1993 and 422 600 hectares in 1994 (382 600 hectares in Greece and 40 000 in Spain) giving an estimated Community production volume of 1 170 000 tonnes of unginned cotton (1 050 000 tonnes in Greece and 120 000 tonnes in Spain), as against 1 085 000 tonnes in 1993 and 985 000 tonnes in 1992. Self-sufficiency in cotton fibre is 25 to 30 %, with consumption at around 1.25 million tonnes.

246. The Community aid scheme involves an annual guide price (ECU 101.46/100 kg for 1994/95) and aid equal to the difference between that price and the world price granted to ginneries which pay a minimum price to the producer. If the volume of unginned cotton produced exceeds a maximum guaranteed quantity (MGQ) (701 000 tonnes), the guide price and the aid are reduced proportionately within a certain limit (18.5 % in 1994/95). Production has exceeded the MGQ every year since 1986/87, entailing reductions in the aid, and guide price reductions of 25 % in 1990/91, 7 % in 1991/92 and 15 % in 1992/93 and 1993/94. As a result of the stabilizer mechanism, the area sown has stabilized at around 75 000 to 80 000 ha in Spain.

247. However, in 1993 and 1994 the areas under cotton in Spain dropped substantially (only 32 000 ha and 40 000 ha respectively) because of a serious shortage of irrigation water resulting from the persistant drought. In Greece, by contrast, the areas increased considerably (223 000 ha in 1991/92, 321 200 ha in 1992/93, 352 000 ha in 1993/94 and 382 000 ha in 1994/95). The contradictory trends in these two Member States can be partially explained by the fact that the devaluation of the drachma had cancelled out the negative impact of the price reductions resulting from the stabilizers. Aware of the problems caused by the stabilizers for the Spanish producers, the Council has undertaken to introduce a fairer system from 1995/96.

Seed flax and hemp

248. The European Union produces both fibre flax, grown primarily for fibre but also giving a high seed yield, and seed flax, grown exclusively for seed. The seed is used without further processing or crushed to obtain oil (for industrial applications) and cake used for animal feed.

249. The Union imports large quantities of seed (some 325 000 tonnes per year), mainly from Canada. In 1990 only 42 500 hectares were sown but areas increased considerably in 1991 (121 000 ha, of which 103 500 ha were in the United Kingdom) and 1992 (266 500 ha, with 155 000 ha in the United Kingdom, 90 000 ha in Germany and 12 200 ha in France).

250. In order to control production, a better balance between the aid granted for flax seed and that for other current crops has been sought. From the 1993/94 marketing year, therefore, non-fibre flax has been added to the arable crops (cereals, oilseeds, protein crops) receiving aid per hectare under the reform decided in May 1992. A compensatory payment, multiplied by the regional yield for cereals, was granted. This payment was fixed at ECU 85 for 1993/94 and at ECU 87 for subsequent years.

251. Since there was uncertainty regarding the aid scheme at the time of sowing, the area under seed flax fell in 1993 to 204 600 hectares, with 155 800 ha in the United Kingdom, 29 200 ha in Germany and 10 700 ha in France. Uncertainty as to the amount of the compensatory payment for 1994 has resulted in a further drop in the area under this crop in the Community to 89 000 ha. Since the new scheme applies only to seed flax, the loss of aid for seed from fibre flax was offset by a corresponding increase in the flat-rate aid. In addition, aid for hemp seed was abolished and the flat-rate aid per hectare for hemp was increased in the same way as for flax to offset the loss of income for producers.

Silkworms

252. Silkworms are reared in Greece, Italy and, to a lesser extent, in France. Rearing accounts for only a tiny part of the Community's agricultural activity, and of world silk production, but is of some importance in Thrace, Veneto and Marche.

253. World production of raw silk has risen in the last few years. It reached 104 000 tonnes in 1994 (as against an average of 68 000 tonnes from 1979 to 1981). The main producers (accounting for 98 000 tonnes) are all in Asia: China (70 000 tonnes), India (16 000 tonnes), Korea (5 000 tonnes) and Japan (4 000 tonnes) together account for 90 % of world production.

254. The Community industry is finding it difficult to sustain its level of activity: increased rearing costs are not always balanced by the trend in market prices. In 1993, 3 300 boxes were started, as against 3 400 boxes in 1992. These produced 52 000 kg of cocoons. The aid for 1994/95 was fixed at ECU 110.36 per box.

Fibre flax and hemp

255. The world fibre flax area is around 1.2 million hectares, giving 500 000 to 550 000 tonnes of fibre. There is no trade in straw flax between the European Union and third countries but fibre imports sometimes reach considerable levels capable of disturbing the Community market. The Union has a deficit in medium and low-quality fibre, which it imports from Eastern Europe, Egypt and China, but is the only producer of good and superior qualities of fibre, which it exports worldwide.

256. After falling sharply in the last few years (78 900 ha in 1990, 55 000 ha in 1991 and 44 200 in 1992), Community areas sown to fibre flax increased to 51 800 ha in 1993 and around 88 200 ha in 1994. France, Belgium and the Netherlands are the main producers. Germany and the United Kingdom started to produce fibre flax again a few years ago and the area under this crop grew considerably in the United Kingdom in 1994 (17 400 ha). The flax straw, except for 2 000 ha that goes to paper mills, is converted to fibre by some 150 retting and scutching firms in north-west France (30), Belgium (110), the Netherlands, Germany and Denmark. Yields for 1994/95 were in general somewhat below the average of 8 tonnes of straw flax per hectare, giving 1 500 kg of fibre, 1 000 kg of seed and 3 500 kg of shives (used to make fibreboard).

257. The world hemp area is about 400 000 ha. In the Community it is a very marginal crop, confined to France, except for a very small amount grown in Spain. Certain other Member States (the United Kingdom and the Netherlands) are trying to start producing this crop. In 1994 about 7 800 ha were cropped. Trade with third countries is very limited.

258. The market in flax fibre, depressed for some years, picked up towards the end of 1992/93 due partly to a marked decline in the Community crop over the last three years and partly to an upturn in consumption. The price of scutched flax began to pick up at the end of 1992/93. There has been a return to balance, which lasted throughout 1993/94.

259. Fibre flax aid for 1994/95 was fixed at ECU 774.86/ha. The amount varies depending on the production zone and the harvesting method and according to the traditional yield in seed per hectare. ECU 44.42 was withheld from the aid to finance measures to promote the use of flax. Aid for hemp was fixed at ECU 641.60/ha.

Tobacco

260. In 1993, world production of tobacco remained stable in relation to the 1992 harvest (+ 0.3 %), totalling 8.32 million tonnes. With 41.4 % of world production, China remains by far the world's biggest producer, followed by the United States, Brazil and India. The

European Union, with only 4 % of world production, is in fifth place: in 1993 its production of leaf tobacco totalled 337 968 tonnes, which was less than in 1992, when the harvest accounted for 5 % of world production.

261. Despite this stabilization in production, prices continued to fall this year on the markets of Malawi and Zimbabwe, which are considered indicative of world price trends. This downward trend in prices (particularly for flue-cured and light air-cured varieties), which had already begun in 1992, is the result of several years of growing world production and stocks.

262. World tobacco consumption continues to grow (+ 2.7 % between 1992 and 1993). This increase is even more marked in many developing countries, while consumption in the developed countries is tending to drop.

263. Community tobacco exports amounted to 209 738 tonnes in 1993, the main exports being of the oriental varieties, sought after for their aromatic qualities, and the dark air-cured varieties which have a market in certain low-income countries. The European industry used about 620 821 tonnes of baled tobacco, of which 417 430 tonnes (67.2 %) were imported.

264. The 17 % drop in Community production of raw tobacco in relation to 1992 is the direct result of the application, for the first time in 1993, of the reform of the common organization of the market in raw tobacco, adopted in mid-1992. This reform establishes a quota scheme by Member State and group of varieties (there are now 8 groups of varieties, instead of the 34 groups under the old system), and puts an end to intervention and export refunds.

265. The total quota was set at 370 000 tonnes for 1993 and Community production fell short by 8 %. This reduction is significant for certain varieties such as flue-cured in Greece, which, having escalated to 71 526 tonnes in 1992 as a result of speculation, fell to 37 921 tonnes in 1993, i.e. a reduction of 47 %. A comparable trend can be seen in the production of hybrid Guedertheimer and Forchheimer Havanna in Italy, which is down 30 % in relation to 1992.

266. The Commission has continued to implement the reform, preparing accompanying measures and strengthening control measures.

Hops

267. The world area planted to hops is around 92 000 ha, of which almost 72 000 ha are situated in member countries of the International Hop Growers Convention (IHGC) and producer Member States of the European Union. There is also considerable production in

China and the former Soviet Union, for which there are no exact figures, only very rough estimates.

268.   In general the areas under hops in the IHGC and the European Union have remained relatively stable, with variations not exceeding 5 % in either direction, with the exception of New Zealand, where there was a 22.14 % increase, and Bulgaria, where there was a reduction of the same magnitude. The dramatic increase in Portugal can be explained by the fact that the entire area under hops there was the subject of a varietal conversion plan, part of which was replanted in 1993.

269.   The 1993 harvest, about 2 700 000 Ztr, [1] was 300 676 Ztr (i.e. 12.53 %) higher than in 1992. Quality was also very good, with an alpha acid content of 6.9 %, producing a total of 9 300 tonnes. This bumper harvest far exceeds the average of the last 20 years, mainly due to record-breaking harvests in Germany and the United States, which outstripped all forecasts.

270.   World beer production, which drops from year to year, was estimated at 1 100 million hectolitres in 1994. Since for this volume it is usually necessary to use 7 590 tonnes of alpha acid with hopping of 6.5 g per hl, the quantity of hops produced was in principle more than adequate, and even left a surplus, to cover needs. It should be noted that hopping of 6.5 g of alpha acid per hectolitre of beer requires a supply of around 6.9 g/hl at the breweries, the difference being lost during the storage and processing of the hops.

271.   In addition, the breweries seem still to have sufficient stocks to cover another 8 to 11 months' production. It should be added that less alpha acid is needed nowadays than before, as a result of a move towards the production of less bitter beers and permanent technical progress.

272.   Within the Union, hops are cultivated in seven Member States (Belgium, Germany, Spain, France, Ireland, Portugal and the United Kingdom), with Germany accounting for 80 % of the 28 675 ha of Community land planted to hops. The total area increased by 121 ha, or 0.42 %, in relation to the 1992 harvest.

273.   In quantitative terms, the 1993 harvest was far superior to that of 1992. The total of 1 033 893 Ztr represented average production per hectare of 1.80 t/ha, or 37.31 Ztr. Quality was good, particularly in Germany, and alpha acid content had a Community average of 6.47 % for the three categories of varieties, the equivalent of 3 344 tonnes of acid 116 kg per hectare — for 1994 beer production.

---

[1] 1 Ztr (Zentner) = 50 kg.

274. As is to be expected when there is surplus production, the market was very quiet during 1993/94. Prices were very low on the free market from the beginning of the marketing season in September 1993, with a few differences between production regions and varieties. In comparison with previous years, fewer contracts were concluded in advance with contract prices falling ever lower, except in the United States, where the situation seems less dramatic than in Europe.

275. Under the common organization of the market in hops, the Union gives aid of two types. Firstly, aid is granted to growers to enable them to attain a reasonable income level. For the 1993 harvest, the Council set this at ECU 395 per hectare for aromatic varieties, ECU 435 per hectare for bitter varieties and ECU 307 per hectare for other varieties, including experimental varieties. Secondly, special aid is granted to hop producers for converting part of their bitter varieties to others more suited to market requirements. By March 1994, the Commission had approved conversion plans for 3 263 hectares. The conversion programme will continue until 31 December 1994.

276. The estimates indicate a poorer Community harvest in 1994 than in 1993.

Seeds

277. The common market organization for seeds provides for production aids for basic seed and certified seed belonging to some 40 different species of agricultural plant seeds, including fodder seeds, rice and flax. With respect to hybrid maize and hybrid sorghum for sowing, a reference price is fixed for imports from third countries. If import prices fall short of the level fixed, a countervailing charge is applied.

278. In 1994 the total area in the Community sown to seeds qualifying for Community aid was 309 153 ha, approximately 8.7 % down on 1993. At 130 199 ha and 125 459 ha respectively, the area cultivated for fodder grass seeds and fodder legume seeds fell by around 13.2 % and 8.4 % as against 1993, continuing the downward trend of recent years.

279. The area sown to rice seeds totalled 16 161 ha, increasing by about 0.8 % in relation to 1993.

280. Because of the increase in areas sown to fibre flax in Belgium and France, the area sown to oilseeds increased to 36 004 ha, up about 18.1 % on 1993.

281. The area cultivated for hybrid maize seed reached 56 448 ha, increasing by about 4.9 % from the 1993 total. This was primarily because more land was sown with simple hybrids in France.

282.   Seed imports into the European Union continued to exceed exports to third countries, confirming the trend over the past decade. About 47 000 tonnes of fodder seed were imported in 1993/94, while some 21 000 tonnes were exported. During the same period, imports of hybrid maize seed totalled 55 180 tonnes, with simple hybrids reaching 40 193 tonnes (around 72.8 % of the total).

Flowers and live plants

283.   This market organization covers a wide range of products: bulbs, live plants (ornamental and seedlings), cut flowers and foliage. The market organization includes quality standards and simple customs duties, with no specific import protection measures other than any safeguard measures which might prove necessary.

284.   Over the last few years production and trade have increased in both the European Union and other countries.

285.   Around 115 000 ha are used for ornamental horticulture, of which about 22 000 ha for bulb production, principally in the Netherlands.

286.   Community imports from third countries totalled 213 000 tonnes in 1992, with a value of ECU 673 million, which represents an increase of about 60 % on the figures for 1988. About half of this quantity was fresh cut flowers, the Union being the biggest market in the world for this product. The majority of the flowers (approximately 80 %) benefit from an exemption on customs duties under agreements concluded with third countries, such as the generalized preference system for Colombia and other Central and South American countries and the agreements concluded with the ACP States under the Lomé Convention.

287.   Four Mediterranean countries (Israel, Morocco, Jordan and Cyprus) are granted tariff reductions within set quotas, provided that the import prices for certain cut flowers (roses and carnations) are not below a certain percentage of the Community price. Under the new Mediterranean policy, the quotas are increased by 3 % a year for the first three countries and by 5 % a year for Cyprus.

288.   Colombia is losing its place as the Union's second most important supplier of fresh cut flowers (21 387 tonnes) to Kenya (21 811 tonnes), while Israel remains in first place (25 029 tonnes).

289.   Community exports to third countries were around 360 000 tonnes in 1993, with a value of ECU 1.03 billion, the principal exports in order of importance being live plants and seedlings, bulbs, fresh cut flowers and foliage.

290.   The external trade balance for the entire sector is positive, with a surplus of ECU 362 million in 1993, although this figure is down 12 % (ECU 50 million) on 1992. However, more fresh cut flowers and foliage were imported than exported. The trade deficit for flowers in 1993 amounted to 44 000 tonnes, with a value of ECU 26 million.

Animal feedingstuffs

291.   Large quantities of agricultural produce are used for animal feed, including much of the European Union's output of cereals and oilseeds and virtually all of its permanent grassland and fodder production from arable land. Three quarters of all the Union's UAA (utilized agricultural area) is used for this purpose. Furthermore, feeding costs can account for up to 70 % of the production costs of pigmeat and poultrymeat.

292.   Overall demand[1] has been stagnant or declining since 1985: the fall in the cattle sector (milk and meat) has not been offset by increases in the pig and poultry sectors. Aggregate supply[2] is composed half of feedingstuffs which are not generally marketed (grass, hay, silage) (around 180 million tonnes — marketable units), mainly used for ruminants. The other half, which can be used for all animals, consists of feedingstuffs which are marketed (cereals, substitutes, oilcakes, etc.) where competition (prices, nutritional value) is extremely intense. In recent years, the proportion of cereals in animal feed has declined and has been replaced by substitutes and oilcakes.

293.   1993/94 was the first year of implementation of the reform of the CAP: the sharp drop in the average price of cereals on the internal market increased the proportion of cereals used in feedingstuffs and reduced that of substitutes and oilcake, most of which are imported.

294.   Total animal consumption of the key marketable products[3] in the Community is estimated at 177 million tonnes in 1993/94 (+ 2.5 million tonnes over 1992/93). This consumption is made up of, on the one hand, indigenous products, estimated at 125 million tonnes (up 6 million tonnes compared with the previous year, caused mainly by greater use of indigenous cereals and oilseeds) and, on the other hand, products imported from non-member countries estimated at 55 million tonnes, down 3.5 million tonnes on the previous year as a result of several trends in imports:

(i)   + 0.5 million tonnes of cereals for animal feedingstuffs;

---

[1]   This includes all marketable and non-marketable animal feed.
[2]   Estimate based on the EUR 10 feed balance sheet expressed as FU (feed units) equivalent to the energy provided by 1 kg of average barley. *Source:* Eurostat.
[3]   Covering most of the marketable feedingstuffs used in the Community by the compound feed industry and by farmers (farm consumption and purchases of raw materials) and set out in the table 'Animal consumption of key marketable products: EUR 12'. *Source:* DG VI.

(ii)   − 2.0 million tonnes of substitutes and other energy sources (mainly − 0.4 million tonnes of manioc, − 0.8 million tonnes of corn gluten feed and other maize byproducts and − 0.8 million tonnes of molasses);

(iii) − 1.9 million tonnes of oilseed cake (including − 1.2 million tonnes of soya); and

(iv) − 0.7 million tonnes of other oilcake.

295.   Exports of key products over the same period remained stable overall, at 2.6 million tonnes.

296.   As regards substitutes subject to import quotas:

(i) the quota utilization rates in 1993 were around 97 % for manioc from Thailand and 100 % for Indonesia; the rate for other GATT member States increased from 59 % in 1991 to 96 % in 1993. The quota for countries not members of GATT, including China, fell from 100 % in 1991 and 1992 to 65 % in 1993. In addition, utilization of China's sweet potato quota fell from 100 % in 1991 and 1992 to 70 % in 1993;

(ii) the manioc voluntary restraint agreement with Thailand, which was concluded for 1990-94, should be renewed at the end of 1994.

297.   Total industrial production of compound feedingstuffs in the Community [1] in 1993 increased to 115 million tonnes, up 2.3 million tonnes or 2.1 % in relation to 1992. This rise is mainly due to the significant increase in the production of feedingstuffs for pigs, which increased to almost 40.9 million tonnes (up 3.9 %) and, to a lesser extent, the growth in production of feedingstuffs for cattle (up 1.2 %) and poultry (up 0.8 %). In terms of total production of compound feedingstuffs by Member State, the largest increases were recorded in Ireland, Denmark, Belgium and France, while the decreases occurred mainly in Spain. Production in Italy remained at 1992 levels.

298.   The factor determining feed composition is the prices of raw materials and their movement relative to one another.

299.   The weighted average price of the key marketable products in the Community fell by about 8 % in 1993/94 in relation to the previous marketing year, with cereal prices falling by about 16 % on the internal market and substitutes by 6 % and protein prices increasing by 10 %, mainly as a result of the poor soya harvests in the United States in 1993. The following table sets out the trend in average prices in the Community since 1984/85.

---

[1] Provisional figures not including Luxembourg, see Table 4.13.7.3, in the statistical annex. *Source:* European Feed Manufacturers' Association (FEFAC).

Animal consumption of key marketable products: EUR 12 (estimate)

(million t)

Products	Rate of import duties	1990/91 Animal consumption				1991/92 Animal consumption				1992/93 (p) Animal consumption				1993/94 (e) Animal consumption			
		Orig. EC	Im-ports	Ex-ports	Total	Orig. EC	Im-ports	Ex-ports	Total	Orig. EC	Im-ports	Ex-ports	Total	Orig. EC	Im-ports	Ex-ports	Total
Cereals																	
Common wheat	L	24.6	—	—	24.6	24.1	—	—	24.1	22.8	—	—	22.8	26.1	—	—	26.1
Barley	L	32.6	—	—	32.6	29.8	—	—	29.8	29.2	—	—	29.2	28.0	—	—	28.0
Maize	L	18.5	1.2	—	19.7	18.9	1.6	—	20.5	20.3	1.1	—	21.4	22.1	1.5	—	23.6
Other	L	7.9	0.5	—	8.4	7.5	0.4	—	7.9	7.2	0.4	—	7.6	8.1	0.5	—	8.6
Total cereals		83.6	1.7	—	85.3	80.3	2.0	—	82.3	79.5	1.5	—	81.0	84.3	2.0	—	86.3
Cereal substitutes listed in Annex D	6% B/L	19.2	18.0	0	37.2	18.8	18.1	0	36.9	18.8	19.3	0	38.1	19.3	18.1	0	37.4
of which: manioc	E/L	0	6.4	0	6.4	0	6.8	0	6.8	0	6.9	0	6.9	0	6.5	0	6.5
sweet potatoes	E	0	0.6	0	0.6	0	0.6	0	0.6	0	0.6	0	0.6	0	0.5	0	0.5
corn gluten feed	E	1.2	5.7	0	6.9	1.3	5.0	0	6.3	1.3	5.9	0	7.2	1.4	5.5	0	6.9
bran	L	10.5	0.1	0	10.6	10.5	0.1	0	10.6	10.5	0	0	10.5	10.5	0	0	10.5
maize germ cake	E	0.2	1.3	0	1.5	0.2	1.2	0	1.4	0.2	1.3	0	1.5	0.2	1.0	0	1.2
citrus pellets	E	—	1.6	0	1.6	—	1.6	0	1.6	—	1.6	0	1.6	—	1.8	0	1.8
dried sugar beet pulp	E	5.3	1.0	0	6.3	4.8	1.0	0	5.8	4.7	0.8	0	5.5	5.0	0.7	0	5.7
other (brewing residues and other fruit waste)	E	2.0	1.3	0	3.3	2.0	1.8	0	3.8	2.1	2.2	0	4.3	2.2	2.1	0	4.3
Molasses	L	1.3	3.3	0.2	4.4	1.4	3.2	0.1	4.5	1.1	3.8	0.1	4.8	1.2	3.0	0.1	4.1
Animal and vegetable fats (added to feedingstuffs)	4-17% B	0.8	0.6	—	1.4	0.8	0.6	—	1.4	0.8	0.6	—	1.4	0.8	0.6	—	1.4
Total energy-rich feeds		21.3	21.9	0.2	43.0	21.0	21.9	0.1	42.8	20.7	23.7	0.1	44.3	21.3	21.7	0.1	42.9
Seed cakes and seed (oilcake equivalent)		7.0	27.0	0.9	33.1	7.1	28.3	0.9	34.5	6.1	30.6	1.6	35.1	5.4	28.7	1.5	32.6
of which: soya	E	1.8	20.3	0.9	21.2	1.3	20.8	0.9	21.2	1.0	22.5	1.3	22.2	0.6	21.3	1.2	20.7
rape	E	3.1	0.6	—	3.7	3.7	0.9	—	4.6	3.1	1.2	0.2	4.1	3.1	1.0	0.3	3.8
sunflower	E	2.1	1.3	—	3.4	2.1	1.9	—	4.0	2.0	2.0	—	4.0	1.7	1.8	—	3.5
other	E	0	4.8	—	4.8	0	4.7	—	4.7	0	4.9	0.1	4.8	0	4.6	—	4.6
Protein crops	2-5% B	4.9	0.6	0	5.5	4.3	0.7	0	5.0	4.2	1.0	0.1	5.1	5.2	1.0	0.1	6.1
Dried fodder, etc.	0-9% B	3.8	0.2	0	4.0	4.1	0.4	0.1	4.4	4.6	0.6	0.2	5.0	5.0	0.5	0.5	5.3
Fish meal and bone meal	0-2% B	2.8	0.9	0.4	3.3	2.8	0.8	0.4	3.2	2.8	0.8	0.6	3.0	2.8	0.8	0.7	2.9
Skimmed-milk powder	L	0.8	—	1.3	0.8	0.8	—	1.4	0.8	0.8	—	2.5	0.8	0.7	—	2.5	0.7
Total protein-rich feeds		19.3	28.7	1.3	46.7	19.1	30.2	1.4	47.9	18.5	33.0	2.5	49.0	19.1	31.0	2.5	47.6
Total animal consumption of key products		124.2	52.3	1.5	175.0	120.4	54.1	1.5	173.0	118.7	58.2	2.6	174.3	124.7	54.7	2.6	176.8
Key product index (base 100 = 1990/91)																	
• Consumption index					100.0				98.9				99.6				101.0
• Demand index					100.0				99.0				99.9				101.2

(P): provisional; (e) = estimate. L = levy; B = binding under GATT; E = exempt.

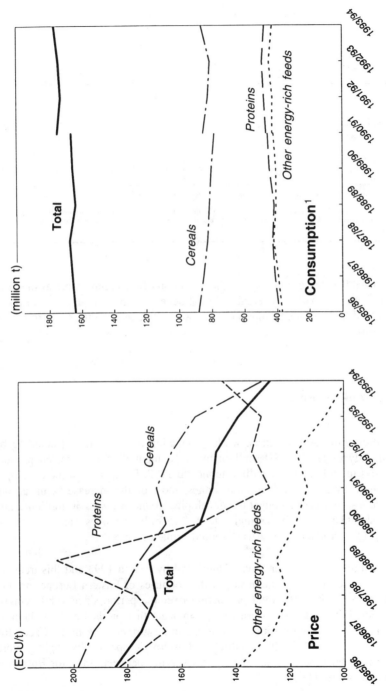

Weighted average price and animal consumption of key marketable products

1985/86-1993/94

Figure 12

1 Including the former GDR from 1990/91.

Weighted average price of key marketable products[1]
EC average (estimate)

(ECU/t)

	1984/85	1990/91	1991/92	1992/93	1993/94	% change 1992/93 1993/94
I — All cereals	202.6	169.0	163.4	154.6	130.5	− 15.6
II — Substitutes and other energy-rich feeds	146.1	113.4	117.4	106.3	99.9	− 6.0
III — Oilcake and other proteins	215.4	127.5	133.9	130.0	142.6	+ 9.6
IV — Total weighted average price	194.9	148.5	147.1	138.6	127.2	− 8.2

[1] Indicator of the trend in average prices of the key marketable products and their principal constituents, weighted by the share of each in total consumption.
For each cereal an average EC price is calculated by weighting the arithmetical average price in each Member State by animal consumption in each.
For oilcakes and substitutes, which are for the most part imported, the average cif Rotterdam price has been used.
Source: DG VI.

300. For 1994/95, the second stage in the reduction of the common prices under the reform of the CAP, the internal market in cereals will not benefit from the high prices of proteins as it did in 1993/94, and livestock consumption of cereals will depend on the trend in import prices and currency parities.

Milk and milk products

301. According to FAO forecasts, world production of cows' milk (including buffalo milk) will fall slightly in 1994 to 518 million tonnes compared with estimated production of 520 million tonnes in 1993 and 524 million tonnes in 1992. Production in the developing countries should increase by 2 % to 179 million tonnes, most of the increase being accounted for by India, the largest producer amongst the developing countries, with 30 million tonnes. In China production should increase by around 3 to 4 % (5 million tonnes), and in Central and South America by 1 %. Production in Africa should remain stable.

302. Production in the former Soviet Union fell by 3 % in 1993 and this trend will continue in 1994 but should be less pronounced. In the countries of Eastern Europe, production fell by an average of 5.1 % in 1993, total production over the period since 1989 being down by 41 million tonnes. Forecasts for Eastern Europe indicate that production in Poland and Hungary will remain stable although it should fall by around a further 17 % in the Czech Republic. The production of cows' milk in the countries of Eastern Europe fell by slightly more than 22 % over the period under observation, the figures varying between 48 % for Bulgaria and 10.6 % for the former Soviet Union.

303. In order to meet their hard currency needs, these countries are exporting at unmatchable prices which are sometimes well below the minimum GATT prices. They are likely to confirm their position on world markets given their considerable agricultural potential, the significant devaluation of their currencies, the depression of demand on their domestic markets which show no sign of recovery in the short term and their pressing need for foreign currency.

304. Whilst West European countries are moving towards a liberalization of the dairy industry, some countries of Eastern Europe are moving in the opposite direction with the introduction of support within the country and protection against imports.

305. Deliveries of cows' milk in the OECD countries held at around 215 million tonnes until 1989. In 1990, they increased to 225 million tonnes, explained essentially by the incorporation of the five new German *Länder* and the increase of deliveries in the United States (2.8 %) and the European Union (1 %). Between 1991 and 1993, they stabilized at around 220.6 million tonnes. There was a further slight increase in 1994 to 222.5 million tonnes. The United States, which had seen an annual increase of 1.5 % between 1984 and 1992, stabilized deliveries in 1993 and 1994. Deliveries in the EFTA countries between 1991 and 1994 were stable at around 12.5 million tonnes. Three other OECD member countries, Japan, Australia and New Zealand, increased deliveries of cows' milk, production in New Zealand and Australia responding rapidly to the increase in world prices and to very favourable climatic conditions. Milk production in Japan continues to increase in line with domestic demand.

306. Community milk production increased slightly in 1994 (+ 0.4 %) and deliveries remained stable for the third consecutive year at around 103 million tonnes.

307. Butter production, which in 1992 had fallen by 150 000 tonnes (– 8 %), only fell by 0.7 %, or 11 000 tonnes, in 1993. It is forecast to remain at the same level of 1.6 million tonnes in 1994.

308. The total production of milk powder continued to fall in 1994 (– 0.6 %) after a slight recovery during the previous year but reached barely 2.2 million tonnes. Production of skimmed-milk powder in 1992 was the lowest since the introduction of the quota system, but in 1994 production reached a new low of 1.1 million tonnes. This was partly offset, however, by an increase of 1.5 % in the production of other milk powders.

309. Production of concentrated and evaporated milk fell by 2 % in both 1993 and 1994 after having increased by 2.2 % in 1992 against the downward trend of 2.1 % per year since 1984.

310. Casein production fell by 19 % in 1993 and by 5 % in 1994 to 110 000 tonnes.

311.   Cheese production continues to grow, although by only 0.9 % in 1993 compared with an annual rate of growth of 2.3 % per annum since the introduction of the quota system. Production should, however, rise by around 2 % in 1994.

312.   Consumption of dairy products within the Union remains stable overall, but the trend is towards products with a reduced butterfat content. Total liquid milk consumption in 1993 increased by 1.5 %, due essentially to a growth in the consumption of semi-skimmed and skimmed milk. Consumption of these two types of milk is currently almost equal to that of whole liquid milk, although in 1986 consumption of whole milk was double that of reduced-fat milk. This confirms the trend, observed for several years, away from whole milk to reduced-fat milk. Similarly, 'light' yoghurts and *fromages frais* continue to proliferate. Cheese consumption continues to grow by 1.8 % per year and, again, there is more and more demand for reduced-fat cheeses.

313.   In contrast, the butter market continues to contract as consumers switch to competing yellowfat products containing less (or no) butterfat. Butter consumption in 1994, however, fell by only 0.8 %, contrasting with falls of 11.43 % in 1989, 6.3 % in 1990 and 3.6 % in 1991. This change in the trend of the demand for butter is the result of the 2 % reduction in the intervention price for butter in 1993 and should be reinforced by the 3 % reduction on 1 July 1994.

314.   Community stocks, particularly of butter and skimmed-milk powder, are at very low levels, the latter reaching an all-time low at the end of July 1993 at slightly more than 29 000 tonnes.

315.   The Council decided to maintain quotas unchanged for the 1994/95 and 1995/96 reference periods. Similarly, the system of quotas in the new German *Länder* was extended. The increase in the Spanish quota was definitively confirmed. In Italy, the total guaranteed quantity for 1994/95 is 9 212 190 tonnes, including a special reserve of 347 701 tonnes for the allocation of reference quantities to producers successfully appealing to the authorities. The increase in Greek quotas is still provisional pending the next Commission report to the Council in 1995.

316.   The intervention price for butter was reduced by 3 %[1] while that for skimmed-milk powder remained unchanged. As a result, the target price for milk fell by 1.5 %. Refunds for dairy products were reduced twice in 1994, the total reduction for cheese being more than 10 %.

317.   The European Union's overall share of the world market fell by 3 percentage points in 1993 to 44.2 % or 13.1 million tonnes milk equivalent. The Union remains the world's largest

---

[1]   See point 17.

exporter in the sector, followed by New Zealand (16.7 %), Australia (9 %) and the United States (8 %).

318.   Butter exports fell by 16 % in 1993 to 201 000 tonnes.

319.   After a slight fall in 1992, cheese exports grew again in 1993 by 12.5 % to reach 524 000 tonnes. The Union has a 53 % share of the world cheese market.

320.   The Union exported 283 000 tonnes of skimmed-milk powder (– 27 %) and 588 000 tonnes of other milk powders (+ 1.3 %), accounting for 27 % and 55 % respectively of world trade. Exports of skimmed-milk powder fell after a 55 % increase in the previous year which had led to a considerable reduction in intervention stocks.

321.   The minimum GATT prices for butter and butteroil were suspended for one year from 1 May 1994. Prices had been below the minimum GATT prices since the second half of 1993 for both these products, but prices rose after their suspension.

International prices and GATT minimum prices[1]

(USD/t)

Year	Butter	GATT	Butteroil	GATT	Cheese	GATT	SMP[2]	GATT
1985	950-1 050 1 000-1 050	1 000	1 200-1 400 1 200-1 400	1 200	1 100-1 250 1 150-1 275	1 000	600-680 600-650	600
1986	1 050-1 150 800-1 100	1 000	1 250-1 350 800-1 300	1 200	1 100-1 200 1 000-1 100	1 000	680-720 680-720	600
1987	750-1 100 900-1 150	1 000	950-1 250 1 100-1 300	1 200	900-1 200 1 000-1 300	1 030	760-840 890-1 150	680
1988	1 150-1 350 1 350-1 500	1 100	1 200-1 400 1 300-1 500	1 325	1 250-1 500 1 800-2 050	1 200 1 350	1 150-1 550 1 750-2 050	900 1 050
1989	1 800-2 000 1 650-1 900	1 250	2 000-2 300 1 800-2 150	1 500	1 900-2 100 1 900-2 200	1 350 1 500	1 800-2 100 1 350-1 640	1 050 1 200
1990	1 350-1 550 1 350-1 500	1 350	1 600-1 900 1 600-1 800	1 625	1 700-2 000 1 550-2 000	1 500	1 200-1 700 1 300-1 500	1 200
1991	1 350-1 400 1 450-1 850	1 350	1 600-1 800 1 675-2 250	1 625	1 600-1 900 1 600-2 100	1 500	1 200-1 400 1 450-1 800	1 200
1992	1 350-1 600 1 350-1 800	1 350	1 625-1 950 1 625-2 200	1 625	1 750-2 100 1 800-2 100	1 500	1 550-1 900 1 775-2 170	1 200
1993	1 350-1 500 1 150-1 550	1 350	1 625-1 800 1 475-1 800	1 625	1 750-2 100 1 675-2 000	1 500	1 650-2 000 1 200-1 800	1 200
1994	1 000-1 450 1 000-1 600	3	1 475-1 700 1 475-1 875	3	1 650-1 900 1 650-1 900	1 500	1 250-1 660 1 250-1 800	1 200

[1] Where two sets of prices are indicated for each year these refer to the periods January to June and July to December respectively.
[2] SMP = skimmed-milk powder.
[3] Minimum GATT prices for butter and butteroil suspended from 1 May 1994 for one year.

322. The restriction of subsidized exports provided for in the GATT Agreement, signed in Marrakesh in April 1994, only affects cheese and 'other products', since the quantities of butter and milk powder exported are currently below the GATT requirements.

Milk deliveries[1]

(million t)

	1987	1988	1989	1990	1991	1992	1993	1994
OECD (24 countries)	214.1	213.4	213.8	217.7	215.6	216.2	216.4	218.5
of which: EUR 12[2]	101.7	99.1	99.2	100.7	100.1	98.9	98.5	98.8
Former GDR[2]	—	—	—	7.2	5.0	4.4	4.4	4.4
USA	63.7	64.9	64.5	66.3	66.0	67.5	67.3	67.8

[1] Production minus farm use and direct sales.
[2] For purposes of comparison, the former GDR has been excluded from OECD and EC totals for 1990 onwards. However, the GDR is included in statistics on Eastern Europe for 1987-89.

Milk production[1]

(million t)

	1987	1988	1989	1990	1991	1992	1993	1994
Eastern Europe[1]	148.0	151.1	154.1	144.5	136.3	119.6	113.5	112.8
of which: former Soviet Union	103.4	106.6	108.4	108.2	104.0	90.5	87.8	87.5

[1] For the purposes of comparison, the former GDR has been excluded from OECD and EC totals from 1990 onwards. However, the GDR is included in statistics on Eastern Europe for 1987-89.

World market exports in milk equivalent[1]

	1987		1988		1989		1990		1991		1992		1993*	
	1 000 t	%	1 000 t	%	1 000 t	%	1 000 t	%	1 000 t	%	1 000 t	%	1 000 t	%
EUR 12[2]	13 834.3	47.6	16 528.0	52.5	14 188.1	50.8	12 451.6	47.2	12 837.5	45.3	13 856.5	47.1	13 115.6	44.2
New Zealand	3 899.1	13.4	4 649.8	14.8	3 806.4	13.6	4 532.5	17.2	5 244.4	18.5	5 146.4	17.5	4 948.0	16.7
Australia	1 617.5	5.6	1 646.9	5.2	1 746.6	6.3	1 911.8	7.2	2 377.7	8.4	2 345.1	8.0	2 745.9	9.3
USA	3 514.0	12.1	2 706.8	8.6	1 805.0	6.5	328.6	1.2	884.2	3.1	1 577.2	5.4	2 488.7	8.4
Canada	840.5	2.9	879.3	2.8	602.6	2.2	625.0	2.4	727.8	2.6	562.4	1.9	347.2	1.2
EFTA	2 525.8	8.7	1 703.7	5.4	1 768.7	6.3	2 144.3	8.1	1 803.2	6.4	1 487.7	5.1	1 571.9	5.3
Eastern Europe + CIS	2 005.4	6.9	2 201.7	7.0	2 311.3	8.3	2 569.4	9.7	2 852.8	10.1	3 124.2	10.6	3 023.8	10.2
Other countries	826.8	2.8	1 152.3	3.7	1 676.5	6.0	1 821.1	6.9	1 607.4	5.7	1 307.7	4.4	1 418.9	4.8
Total	29 063.4	100	31 468.6	100	27 905.1	100	26 384.3	100	28 335.1	100	29 407.2	100	29 660.0	100

[1] Except for casein and fresh products on the basis of Community coefficients.
[2] Including the five new *Länder* from 1991.
[3] Figures for third countries are provisional.

Milk

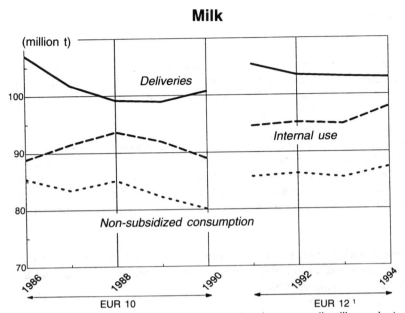

NB: Consumption has been calculated on the basis of an overall milk products balance in terms of milk equivalent (referring to fat content).
Consumption for 1988 incuding 3.245 million tonnes for animal feed.

[1] Excluding the five new *Länder*.

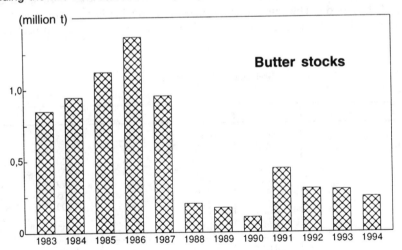

NB: Public and private stocks at 1 January; for 1994, at 31 December.

Figure 13

Beef and veal

323. The improvement in the world market situation for beef and veal in most countries which produce and/or import outside the European Union and Eastern Europe was confirmed in 1993 and should continue in 1994. After two consecutive years of decline, world production of beef and veal should pick up again in 1994 and beyond.

324. The stabilization in sustained economic growth has encouraged a gradual upturn in the consumption of beef and veal in major consumer countries like the United States, Japan and certain countries of South-East Asia, where there is a growing need for meat imports, but also in the main producer countries of Latin America. However, the sharp drop in consumption in the countries of Oceania has freed up ever increasing exportable surpluses.

325. The world market in beef and veal continues to be split into two areas, those free of foot-and-mouth disease and those which are not. World trade in beef and veal fell in 1993 and is likely to do so again in 1994.

326. In 1993 and 1994, export prices for beef and veal remained relatively high on foot-and-mouth free markets and international prices also firmed up on the other markets, particularly in the Atlantic area. Buoyant prices for both exports and imports resulted in relatively favourable prices on the domestic markets of the main countries participating in the world beef and veal trade. The current investment in cattle herds in many countries can therefore be expected to continue, eventually resulting in an increase in world beef and veal production.

Beef and veal production

(1 000 t carcass weight)

	1992	1993	% change	1994	% change
USA	10 611	10 584	− 0.3	10 986	+ 3.8
CIS	6 856	6 270	− 8.5	6 125	− 2.3
Argentina	2 487	2 508	+ 0.8	2 520	+ 0.5
Brazil	4 590	4 719	+ 2.8	4 931	+ 4.5
Australia	1 838	1 796	+ 2.3	1 779	− 0.9
Japan	592	593	+ 0.2	605	+ 2.0
China	2 525	2 610	+ 3.4	2 800	+ 7.3
India	2 398	2 458	+ 2.5	2 475	+ 0.7
EU	8 396	7 743	− 7.8	7 920	+ 2.3
World	54 028	53 428	− 1.1	54 280	+ 1.6

327. There was a rapid change in cattle-rearing structures in the European Union during the 1980s which resulted in the disappearance of a million farm holdings.

328. The European Union produces some 15% of total world production of beef and veal (second only to the USA) and through its exports in particular accounts for about 25% of world trade in beef and veal. Imports are covered by special multilateral and bilateral agreements for the most part, including the agreement relating to the ACP/EC Convention which was expanded on 1 March 1993 and the agreement which took effect on 1 July 1993 with the countries of Central and Eastern Europe providing for improved access, amounted to more than 500 000 tonnes in 1993.

329. In 1993, exports of beef and veal amounted to 1.23 million tonnes, resulting in a net surplus of almost 730 000 tonnes.

330. As part of the cyclical trend in prices and production, prices for beef and veal and for female animals in particular remained relatively firm.

331. Production dropped by almost a million tonnes or 11.2% between 1991 and 1993. The current cyclical drop (around 13%) in the numbers of cattle slaughtered thus resulted in a cyclical production ebb of 7 743 000 tonnes in 1993.

332. Annual consumption, at around 22 kg/per person, remained stagnant because of competition from cheaper meats, the depressed economic climate and specifically because of the impact of health or veterinary problems affecting the image of beef and veal at consumer level.

333. Because of the firmer market prices for beef and veal and the planned reductions in the intervention price, buying-in was suspended from the end of July 1993 for young bull carcasses and from mid December 1993 for bullock carcasses. For intervention purposes, the maximum weight currently fixed for such carcasses is 340 kg. The stocks of meat in intervention dropped from 1 090 000 tonnes at the end of 1992 to less than 210 000 tonnes deadweight at the end of September 1994.

334. The intervention price for beef dropped by 5.3% at the beginning of July 1994 to stand at ECU 304.31/100kg carcass weight for R3 category adult male animals and the main premiums already in force were increased to compensate for the policy change which had been decided in May 1992.

Sheepmeat and goatmeat

335. World production of sheepmeat and goatmeat is currently estimated at approximately 10 million tonnes annually. Of the world sheep and goat population of 1 700 million head, over 60% are located in Asia and Africa while China, with 210 million head and production of 1.3 million tonnes, is the world's leading producer.

336. World trade, however, is dominated by New Zealand which exports up to 400 000 t meat each year and Australia whose exports, often in the form of live animals, reach over 300 000 t annually. The European Union is the world's second largest producer and its largest importer and consumer.

337. Wool rather than meat production is the main purpose of sheep and goat production in most of the world but, although there has been an improvement in wool prices, world wool stocks remain high and there is little prospect of any growth in numbers in coming years.

Production of sheepmeat and goatmeat

(1 000 t)

	1992	1993	Variation %	1994 (estimate)	Variation
EU	1 177	1 156	− 1.8	1 140	− 1.4
New Zealand[1]	585	486	− 16.9	405	− 16.7

[1] Year ending 30 September.

338. On the Community market production is expected to fall by 1.4 % to 1 140 000 t in 1994 due mainly to decreases in UK, French and Irish production. Consumption may fall to 1 360 000 t with the decline in annual UK consumption estimated at over 80 000 t since the demise of the variable slaughter premium in 1991.

339. Intra-Community trade may contract this year to considerably less than 300 000 t due to production decreases in the main exporting Member States (and the effects of the refusal of ferry companies to transport live sheep from the UK and Ireland to continental Europe). France will remain the most important recipient of internal trade.

340. Market prices in 1994 have been influenced by the decline in overall supply and also by severe weather conditions in spring which reduced the availability of lamb for the important Easter market and led to near record prices in the north of the Union at that time. The average Community market price for lamb reached ECU 277 per 100 kg, up 7 % on 1993.

341. Imports into the Union are expected to reach 250 000 t in 1994. New Zealand and Australia will probably fulfil their voluntary restraint agreement quantities of 205 000 t and 17 500 t respectively. However, following the reduction in supplies from Eastern Europe in 1993 due to foot-and-mouth disease, it is probable that the more general destocking there in recent years may reduce imports from these suppliers to well below traditional levels.

342. Adaptations agreed by the Union with its trading partners under the voluntary restraint agreements for the period up to the end of 1994 have had the effect of reducing possible imports under these arrangements by approximately 45 000 t and of reducing the levy on imports from 10 % to zero.

343.   The level of the Community ewe premium set for 1993 was ECU 20.90 per ewe and the rural world premium applicable in less favored areas at ECU 5.5 per ewe. Private storage stocks of lamb amounted to just 1 900 t at the end of 1993.

344.   In 1994 to date, two advances on the ewe premium have been fixed totalling ECU 11.83 per ewe. Although the conditions for the opening private storage tendering have been fulfilled throughout the year no product, as yet, has been aided.

345.   The outlook for 1995 is for stability or a slight fall in production as producers continue to adapt to the premium quota regime. Consumption should change in accordance with production. Sheep numbers will remain stable at approximately 98 million head and intra-Community trade may also remain stable. A similar trend may be forecast in the medium term.

Pigmeat

346.   China remains the foremost pigmeat producer followed by the European Union. According to forecasts, production in 1994 should drop in all the main producer countries except China. In the United States, the world's third largest producer, production should be slightly down in 1994, after a considerable increase in 1993.

347.   Production in the European Union in 1994 will probably be at the same level as in 1993 because of a fall in the number of animals slaughtered during the second half of the year. In 1993 pigmeat production reached 15.2 million tonnes, 6% up on the previous year. Despite the drop in pigmeat prices, which started in the autumn of 1992 and continued throughout 1993, production continued to rise in 1993 and the first six months of 1994. From April 1994, production slowed down and the various market support measures helped improve prices.

348.   In 1993 and 1994 animal health conditions were a cause for concern because of the persistent outbreaks of classical swine fever in Germany and Belgium. The Commission introduced special market support measures for these two countries in the form of buying-in of lard pigs and piglets withdrawn from the market and mainly sent for rendering. Under the measures in Germany, 907 000 lard pigs and 175 000 piglets were bought in between 29 October 1993 and mid-August 1994. In Belgium, 293 000 lard pigs and 142 000 piglets were bought in. The lard pigs bought in represent 100 000 tonnes of pigmeat withdrawn from the Community market. The 317 000 piglets represent 25 000 tonnes of meat not placed on the market.

349.   The depressed pigmeat prices, the need to free up markets in the European Union and increased demand in certain Central and Eastern European countries made it possible to export 730 000 tonnes of pigmeat in 1993. Exports are expected to drop in 1994.

350.   Imports, mainly from the countries of Eastern Europe, fell in 1993 to around 20 000 tonnes, despite the existence of reduced-levy import agreements between the European Union and several Central European and developing countries. The forecasts suggest that imports should increase in 1994 because of the favourable concessions granted to the countries of Eastern Europe under the association agreements and the introduction of a 7 000 tonne quota under GATT.

Poultrymeat

351.   During the last six years world production of poultrymeat has increased steadily by an average of 3.6 % a year. In the United States the rate of increase has been even higher (5 %). In the other main producer regions, production has increased in Brazil and China (annual rate of increase: 10.9 %) but has fallen in Eastern Europe, Russia included, and in Japan.

Poultrymeat production

(1 000 t)

Year	United States	Brazil	China	Japan	USSR or Russia	Hungary	EUR 12	Others	World production
1986	8 262	1 680	1 879	1 421	2 988	445	5 443	7 165	29 283
%	28.2	5.7	6.4	4.9	10.2	1.5	18.6	24.5	100.0
1987	9 105	1 865	2 040	1 465	3 126	470	5 784	7 416	31 271
1988	9 272	1 997	2 744	1 471	3 107	465	5 997	7 640	32 693
1989	9 931	2 139	2 820	1 423	3 233	420	6 108	6 279	32 353
1990	10 645	2 416	3 229	1 391	3 169	426	6 336	6 604	34 216
1991	11 204	2 691	3 952	1 357	1 751	320	6 756	8 033	36 064
1992	11 885	2 932	4 540	1 367	1 577	320	6 922	8 710	38 253
1993	12 417	3 195	5 100	1 365	1 420	320	6 924	8 934	39 675
%	31.3	8.1	12.9	3.4	3.6	0.8	17.5	22.5	100.0
1994	12 984	3 485	5 600	1 335	1 350	330	7 023	9 108	41 215

1993: Estimates.   1994: Forecasts.
Since 1991 the Community figures include the five new German *Länder*.
Until 1990, USSR; since 1991, Russia.
Sources: Eurostat, USDA.

352.   The world market continued to expand slightly in 1994 thanks in particular to growing demand in the Far East. The United States retained first place in the league of exporting countries due in particular to its exports of low-value cuts and to its various promotional programmes. In the first half of 1994, exports from the United States increased by 215 000 tonnes (+ 52 %) over the equivalent period in 1993, while Community exports increased by

Meat

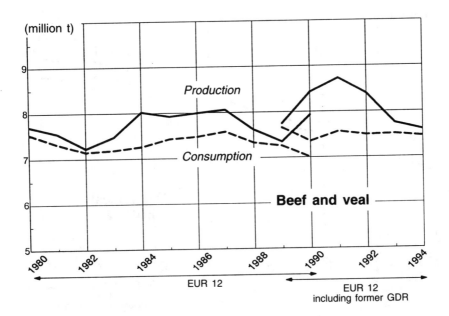

Production

Consumption

Beef and veal

EUR 12

EUR 12
including former GDR

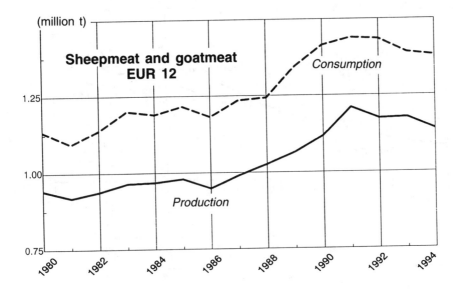

**Sheepmeat and goatmeat
EUR 12**

Consumption

Production

Figure 14

Meat

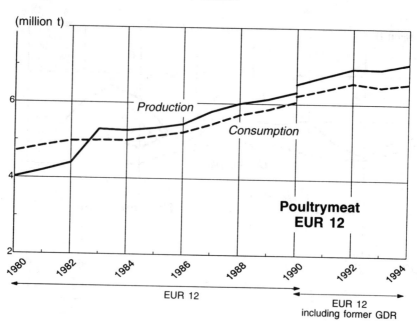

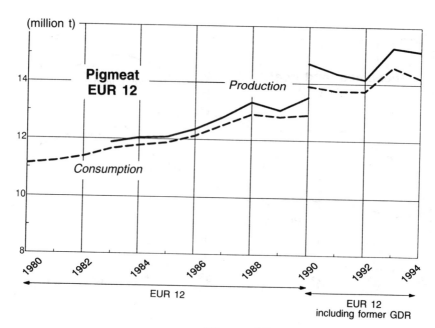

Figure 15

33 000 tonnes (+ 14 %) during the first five months of 1994, after increasing by 28 % in 1993.

353. Total production of poultrymeat on the Community market should increase by 1.4 % in 1994. This increase is slightly higher than in 1993 but lower than in previous years. The cost indices for chickens/feed were higher in the first eight months of 1994 than they had been in 1993.

354. The rules governing trade with non-member countries have been adapted to the world market situation by differentiating export refunds according to destination so as to maintain trade flows. Since 1 August 1994 there has also been a differentiation between hens and broilers. The reform of the common agricultural policy and the situation on the internal and world markets have led to reduced export refunds since mid-1993.

355. In addition to the import quotas at reduced levies provided for under the generalized system of preferences and the association agreements (Poland, Hungary, the Czech Republic, Slovakia and, since January 1994, Romania and Bulgaria), 15 000 tonnes of boned chicken and 2 500 tonnes of turkeymeat can be imported at zero duty from 1 July 1994.

Eggs

356. World production again increased in 1994. The annual rate of increase over the last six years has been 1 %. In some of the major producer countries and regions, especially in the European Union, production has tended to slow down or even decline over the medium term. Only China has a high rate of expansion: 7.5 % a year since 1988.

Egg production

(billion units)

Year	United States	Mexico	Eastern Europe	Japan	USSR or Russia	China	EUR 12	Others	World production
1987	70		34	37	82	118	81	90	512
%	13.7		6.6	7.2	16.0	23.0	15.8	17.6	100.0
1988	69		34	40	85	139	82	89	538
1989	67	18	20	40	85	141	79	46	496
1990	68	18	19	40	82	159	81	43	510
1991	69	20	18	42	47	185	83	75	539
1992	71	20	17	43	43	204	82	74	553
1993	71	20	16	43	40	215	81	72	550
%	12.8	3.6	2.9	7.7	7.2	38.5	14.5	12.8	100.0
1994	72	20	16	43	39	225	81	73	569

1993: Estimates.   1994: Forecasts.
Since 1991 the Community figures include the five new German *Länder*.
Until 1990, USSR; after 1991, Russia.
Sources: Eurostat, USDA.

357. World trade has been falling constantly for several years now, following the setting up of production units in North Africa and the Middle East. The main importer countries are currently Japan (egg products) and Hong Kong (eggs in shells). Community exports increased by 26 % in 1993 but fell by 13 % during the first three months of 1994.

358. On the Community market, 1993 was a good year which resulted in a further increase in the number of laying hens. An increase in the production of eggs for consumption can therefore be expected in 1994 (+ 1.4 %). Since April 1994, the financial situation of producers has become less favourable than it was the year before.

359. The common organization of the market is similar to that for poultrymeat.

360. As regards trade, the reform of the common agricultural policy and the internal and world market situation resulted in a fall in export refunds from mid-1993.

361. The association agreements concluded with Poland, Hungary, the Czech Republic, Slovakia and Bulgaria involve a 60 % reduction in the levies on certain types of egg.

Potatoes

362. Potatoes are one of the products for which no market organization has been established. With a view to the completion of the single market the Commission presented a proposal two years ago for a common market organization for potatoes.

363. With a total cultivated area of 1 366 000 hectares, potatoes constitute a substantial crop in the European Union. They are grown in all the Member States, although, because of climatic and soil conditions, they are more widely grown in northern regions.

364. The Union is self-sufficient in potatoes, with the exception of early varieties. These are imported in winter and early spring from Mediterranean countries when no, or only limited, Community production is available. The main suppliers are Egypt, Morocco and Cyprus. During the past few years an annual average of some 400 000 tonnes of early potatoes have been imported from third countries.

365. 1993 was a good year for potato growers, who have seen the area under potatoes fall after four years of continual increase. This trend has resulted in a drop in production and a rise in prices.

366. In 1994, following the drought which hit many growers this summer, a poorer quality crop is expected, with price rises, as well as technical difficulties in harvesting.

Honey

367. At 126 000 tonnes, the European Union is the third largest honey producer after the former Soviet Union and China. It is also the world's largest importer, with 48 % of the market. Consumption, estimated at 250 000 tonnes in 1992, shows that the Union has a deficit in honey. Self-sufficiency is around 50 %. However, between 1986 and 1992 Community production increased by 49 % and imports fell slightly (7 %).

368. The sector is highly heterogeneous because of the 435 000 known beekeepers, only 13 000 can be considered professional. According to estimates, of the five Member States with more than one million hives, Spain has the largest number of hives and of professional beekeepers in relation to the total number of beekeepers.

369. In the report which it presented to the Council and Parliament in June 1994, the Commission stated that it did not consider it appropriate to create a new market organization specifically for honey or to introduce a generalized income support scheme in view of the state of knowledge and the current situation of beekeeping. However, the Commission proposes to improve production conditions through national programmes to control varroasis, rationalize transhumance and introduce a network of regional beekeeping centres. In addition, measures will be introduced to implement a quality policy, in particular defining specifications regarding the botanical and geographical origin of honey and corresponding harmonized analysis methods.

'Non-food' set-aside

370. Following the Council's decisions on CAP reform, the Commission adopted two regulations in 1993 which set out the detailed rules of application for cultivating, firstly, annual raw materials on set-aside land, and secondly pluriannual raw materials for use in the manufacture within the Union of products not primarily intended for human or animal consumption.

371. A wide range of raw materials, including cereals, certain oilseeds and protein crops, forest trees with a maximum harvest cycle of 10 years, and certain shrubs and bushes are allowed under the scheme. Eligible 'non-food' uses include biofuels for motor vehicles, biomass, plastics, paper, chemicals and pharmaceuticals.

372. The regulation relating to annual raw materials was amended in March 1994 to include sugarbeet in the list of raw materials which may be grown on set-aside land for 'non-food' purposes. However, the land upon which sugarbeet is grown within the context of the scheme loses the right to compensation in respect of the obligation to set aside.

373. Data suggests that sowings on set-aside land for 'non-food' purposes in the marketing year 1993/94 for harvest in 1994/95 amounted to some 686 000 hectares in the Union, as compared to around 264 000 hectares in the previous year.

374. Around 37 % of sowings in the marketing year 1993/94 were undertaken in France, 24 % in Germany, 16 % in UK, and 10 % in Italy.

375. Of the 686 000 hectares sown in the marketing year 1993/94, around 603 000 hectares were put down to rapeseed and sunflower seed, a significant proportion of the former being used for biofuel.

376. Clause 7 of the Memorandum of Understanding on certain oilseeds between the European Union and the United States within the framework of the GATT stipulates that 'should the byproducts made available as a result of the cultivation of oilseeds on land set aside for the manufacture within the Union of products not primarily intended for human or animal consumption exceed one million metric tonnes annually expressed in soya bean meal equivalents, the Union shall take appropriate corrective action within the framework of the CAP reform'.

377. The 603 000 hectares of rapeseed and sunflower seeds (both of which are covered by the Memorandum of Understanding) which were grown on set-aside land in the marketing year 1993/94 represent around 600 000 tonnes of soya bean meal equivalents.

378. The Commission is currently examining a system to monitor production of soya bean meal equivalents as byproducts, and if the need arises, to take appropriate corrective action.

Scheme for starch production

379. About 6.6 million tonnes of starch are produced in the European Union, consisting of 3.5 million tonnes of maize starch, 1.5 million tonnes of wheat starch and 1.6 million tonnes of potato starch.

Production refunds

380. Production refunds are granted on about 40 % of all the starch produced in the European Union. The purpose of the measure is to supply starch at world prices to manufacturers of certain non-food products to enable them to compete on an equal footing with imported products, which enjoy free access to the Community market.

381. In 1992/93, refunds were paid for about 2.4 million tonnes (6% more than the previous marketing year), of which 1.2 million tonnes were produced from maize, 0.75 million tonnes from potatoes and 0.45 million tonnes from wheat.

Special scheme for potato starch

382. From the 1995/96 marketing year, potato-starch production will be subject to production quotas. This measure was adopted by the Council in July 1994 following an undertaking to restrict production if a production threshold of 1.5 million tonnes is exceeded. Production overran this limit, so that the five producer Member States (Denmark, Germany, Spain, France and the Netherlands) will keep production within bounds mainly on the basis of historical production: the allocation of quotas among the starch manufacturers is based on the average quantity of potato starch produced during the 1990/91, 1991/92 and 1992/93 marketing years or, if the Member State prefers, on the basis of the quantity of potato starch produced in 1992/93 only, and on investments made by the starch manufacturers before 31 January 1994.

383. Any production in excess of the quota must be exported from the Union without qualifying for any export refund or the payment of a minimum price or compensation to the farmer.

384. The quota has currently been fixed for three years. Before 1 November 1997, the Commission will present a report to the Council on the allocation of the quotas, together with appropriate proposals, if necessary, for a new three-year quota, taking account of any restructuring of the market in the sector and catering for new producers.

IV — Rural development

Horizontal measures

385. This heading covers all measures applicable throughout Community territory relating to the improvement of the conditions under which agricultural and forestry products are produced, processed and marketed.

386. On 21 November 1994, the Council adopted Regulation (EC) No 2843/94 [1] amending Regulations (EEC) No 2328/91 relating to production structures and (EEC) No 866/90 relating to processing and marketing structures. The amendment will give the Member States greater freedom in the choice of the specific means of implementing that objective and greater flexibility regarding aid for environmental protection, animal welfare and young farmers.

Modernization and improvement

387. Approximately 36 000 farms receive investment aid. This aid has been gradually restricted to combat agricultural surpluses. The emphasis is now on individual investment plans to enhance competitiveness, improve production conditions and diversify.

EAGGF Guidance Section commitments for investment aid [1]

(ECU million)

Member State	1993	1992	1991	Average 1987-91
Belgique/België	6.832	4.937	3.886	6.508
Danmark	6.018	4.540	3.332	5.249
Deutschland	13.507	12.023	12.211	29.966
Elláda	43.107	38.416	34.069	16.683
España	64.196	56.329	43.191	24.150
France	42.489	33.492	23.375	46.877

[1] OJ L 302, 25.11.1994, p. 1.

(ECU million)

Member State	1993	1992	1991	Average 1987-91
Ireland	14.657	16.401	15.032	9.848
Italia	23.770	20.865	15.242	11.740
Luxembourg	1.816	2.295	1.607	1.000
Nederland	1.588	1.180	0.493	5.656
Portugal	46.897	40.819	32.679	20.988
United Kingdom	5.185	6.468	9.607	17.400
Total	270.062	237.765	194.724	196.065

[1] Regulation (EEC) No 2328/91.
Source: EAGGF Guidance Section.

Young farmers

388. Aid to young farmers, which includes the setting-up premium and investment aid granted under Articles 10 and 11 of Regulation (EEC) No 2328/91, seems to be levelling out in terms of overall commitments. It is still indispensable in order to encourage a younger farming population and provide a new generation of farmers to take over from those retiring.

EAGGF Guidance Section commitments for aid to young farmers[1]

(ECU million)

Member State	1993	1992	1991	Average 1987-91
Belgique/België	11.012	4.668	5.997	2.898
Danmark	4.164	4.202	3.514	1.966
Deutschland	19.836	13.678	11.480	7.101
Elláda	3.072	2.082	1.618	0.595
España	21.728	10.375	7.518	17.015
France	91.068	86.419	100.875	41.970
Ireland	2.372	2.533	2.954	1.508
Italia	9.506	8.246	10.876	4.473
Luxembourg	1.112	1.322	1.214	0.549
Nederland	1.797	4.426	0.597	0.178
Portugal	17.190	13.863	10.936	6.177
United Kingdom	0.112	0.127	0.159	0.97
Total	182.969	151.941	157.738	84.527

[1] Regulation (EEC) No 2328/91.
Source: EAGGF Guidance Section.

Less-favoured rural areas

389. Specific aid measures for farmers in less-favoured farming areas head the list of EAGGF Guidance Section commitments under Regulation (EEC) No 2328/91.

390. In 1994 mountain and hill areas and other less-favoured farming areas continued to receive compensatory allowances to facilitate a continued agricultural presence and maintain the population. This aid, intended to compensate for the generally higher production costs, is widely applied.

391. More than 1.1 million agricultural holdings receive such allowances.

392. The less-favoured farming areas total 55 % of all the Community utilized agricultural area.

EAGGF Guidance Section commitments for compensatory allowances[1]

(ECU million)

Member State	1993	1992	1991	Average 1987-91
Belgique/België	2.470	2.413	2.207	2.352
Deutschland	146.795	88.071	88.037	79.759
Elláda	29.289	35.119	62.886	38.924
España	18.134	65.600	62.539	37.676
France	131.311	69.268	73.273	59.808
Ireland	103.520	86.634	63.709	45.233
Italia	44.487	31.910	29.838	23.946
Luxembourg	4.618	2.464	2.319	2.201
Nederland	0.290	0.042	0.218	0.136
Portugal	16.376	51.942	29.915	21.822
United Kingdom	44.533	39.567	44.158	40.352
Total	541.823	473.230	459.099	352.208

[1] Regulation (EEC) No 2328/91.
Source: EAGGF Guidance Section.

Measures concerning the processing and marketing of agricultural and forestry products

393. The application of measures to improve the conditions under which agricultural and forestry products are processed and marketed, provided for in Regulation (EEC) No 866/90, [1] as amended by Regulation (EC) No 3669/93 [2] to adapt it to the new phase of work of the Structural Funds (1994-99), called for the presentation by the Member States of plans to serve as a basis for drawing up the Community support frameworks (CSFs). For areas covered by Objective 1, such plans, and indeed the remainder of the Objective 5a measures, were incorporated in the relevant regional development plans.

[1] OJ L 91, 6.4.1990, p. 1.
[2] OJ L 338, 31.12.1993, p. 26.

394.   The Member States could also present single programming documents (SPDs) giving the information required in the plans and aid applications. In such cases, the Commission is to adopt a single decision covering the necessary items. All the Member States except Italy have chosen this type of presentation.

395.   In 1994, by Commission Decision 94/173/EC, [1] the criteria for selecting projects were updated to bring them into line with the new guidelines for the common agricultural policy and the other Community policies.

Regional measures

396.   Rural development policy continued to be implemented through measures financed by the three Structural Funds in the Objective 1 and 5b regions.

Objective 1

397.   Objective 1 covers regions whose development is lagging behind as defined in Regulation (EEC) No 2052/88 [2] and whose gross domestic product (GDP) is less than 75 % of the Community average.

398.   During the first phase (1989-93), EAGGF Guidance Section aid in these regions went principally to agricultural holdings, the diversification of income sources and the maintenance of economic activity in rural areas. The July 1993 review extended the scope of EAGGF action to new areas, such as the improvement of rural living conditions, the renovation of villages, a policy of quality products and product promotion, support for applied research and financial engineering measures.

399.   In 1994, almost all the programmes from the first phase of the reform of the Structural Funds were finalized.

400.   Following the amendments to the regulations on the Structural Funds for the second phase (1994-99), the areas eligible under Objective 1 underwent significant changes. These changes are due not only to the incorporation of five new *Länder* into the Union but also the reclassification of certain regions such as Hainaut in Belgium, Flevoland in the Netherlands, certain *arrondissements* in Nord in France, Merseyside and Highlands and Islands in the United Kingdom and Cantabria in Spain.

---

[1]   OJ L 79, 23.3.1994, p. 29.
[2]   Amended by Council Regulation (EEC) No 2081/93, OJ L 193, 31.7.1993.

401. The areas eligible under Objective 1 thus amount to 45.5 % of Community territory (38 % during the first phase) while the population covered is 26.6 % of the Community population (21.7 % during the first phase).

402. The financial contribution from the three Structural Funds to rural development during the second phase amounts to ECU 15 billion, i.e. around 16 % of the total Community contribution to the regions concerned (ECU 93 810 million).

403. Community assistance for rural development comes almost entirely from the EAGGF Guidance Section (ECU 13 600 million or 90 % of the total), in most cases for Objective 5a or almost exclusively agricultural measures (compensatory allowances, food industry, etc.). The part of the Community contribution allocated to integrated local development measures in rural areas thus generally remains very limited.

404. All the Community support frameworks (CSFs), all the single programming documents (SPDs) and most of the operational programmes (OPs) have already been approved.

405. The SPDs, a new device in the second phase to simplify and speed up the procedure, have been used by several Member States, in particular for small regions (Flevoland, Belgian Hainaut, Merseyside, etc.). These documents combine the data in the development plan and the operational programme (aid application).

406. The following table and synopsis summarize the main programmes approved in 1994.

407. In Belgium, the province of Hainaut was selected for structural aid under Objective 1 in the second phase. The strategic choices made for contributions from the EAGGF Guidance Section to the modernization of Hainaut comprise horizontal measures falling under Objective 5a, aid for forestry, the search for new agricultural products, increasing the value of quality products (quality mark or designation of origin), the equestrian industry and rural tourism.

408. In Germany, the new *Länder* were classified as Objective 1 regions. With the exception of Berlin, all rural development action is concentrated on a multifund programme by *Land*. EAGGF regional measures will focus on infrastructures, land consolidation, diversification of farming including related activities, direct marketing, rural tourism and environmental measures. Such measures will be grouped within an integrated measure for the development and renovation of villages.

409. As regards Greece, the agricultural section of the CSF focuses on improving competitiveness, modernizing agricultural structures and infrastructures and rural develop-

Objective 1 (1994-99)
Indicative allocations shown in the CSFs or SPDs at 1994 prices, not including Community initiatives

Member State	Structural Funds ECU million	ERDF ECU million	ERDF %	ESF ECU million	ESF %	EAGGF ECU million	EAGGF %	FIFG ECU million	FIFG %	Assistance per inhabitant in ecus	Inhabitants %	Area %
Belgique/België	730.0	515.9	70.7	166.7	22.8	47.0	6.4	0.4	0.1	608	12.8	12.4
Deutschland	13 640.0	6 820.0	50.0	4 092.0	30.0	2 644.5	19.4	83.5	0.6	853	20.0	30.0
Elláda	13 980.0	9 489.5	67.9	2 560.5	18.3	1 800.0	12.9	130.0	0.9	1 398	100.0	100.0
España	26 300.0	15 944.2	60.6	6 047.0	23.0	3 313.8	12.6	995.0	3.8	1 140	59.4	77.7
France	2 190.0	1 194.9	54.6	525.5	24.0	431.4	19.7	38.2	1.7	890	4.3	16.8
Ireland	5 620.0	2 562.0	45.6	1 953.0	34.8	1 058.0	18.8	47.0	0.8	1 606	100.0	100.0
Italia	14 860.0	9 660.0	65.0	2 739.0	18.4	2 228.0	15.0	233.0	1.6	704	36.3	40.8
Nederland	150.0	80.0	53.3	40.0	26.7	21.5	14.3	8.5	5.7	600	1.8	3.4
Portugal	13 980.0	8 723.9	62.4	3 148.7	22.5	1 894.2	13.5	213.2	1.5	1 357	100.0	100.0
United Kingdom	2 359.8	1 331.8	56.4	747.2	31.7	245.9	10.4	34.9	1.5	690	5.9	18.7
Total	93 809.8	56 322.3	60.0	22 019.6	23.5	13 684.3	14.6	1 783.7	1.9	1 028	26.6	45.5

ment. The main measures laid down in the 13 regional multifund operational programmes relate to infrastructures, land improvement, irrigation, land consolidation, forestry, stockfarming, rural development and renewable sources of energy. Around 70 % of the EAGGF funds for the national operational programme relate to measures under Objective 5a, the remainder being intended in particular for completing irrigation projects and water retention basins in the preceding CSF and for forestry measures.

410. In Spain, the strategic choices laid down for contributions from the EAGGF Guidance Section comprise in particular measures under Objective 5a, support for agriculture and rural development through the improvement of rural infrastructures, safeguarding the environment and forestry measures, and diversifying and increasing the value of agricultural products.

411. In France, the strategic choices laid down for contributions from the EAGGF Guidance Section comprise in particular measures under Objective 5a, diversifying and increasing the value of agricultural products and support for agricultural and rural development.

412. The Irish CSF will be implemented in the form of nine operational programmes, three of which will be part-financed by the EAGGF Guidance Section, namely the operational programme for agriculture, rural development and forestry, the sub-programme for the food industry in the operational programme for industrial development, and the operational programme for urban local and rural development.

413. In Italy, EAGGF allocation, which is substantially above that for the first phase, will provide a higher rate of part-financing than in the previous period. The strategic choices made comprise in particular measures for the application of Objective 5a, diversifying and increasing the value of agricultural products, support for agricultural and rural development, assistance for the development of services to agriculture and agricultural advisory services.

414. In the Netherlands, in addition to measures under Objective 5a (16.7 % of total EAGGF Guidance Section funds), other EAGGF measures relate to the encouragement of new agricultural activities, the improvement of water management for agricultural purposes, the concentration and expansion of technological research and development in agriculture, the establishment of a training centre for sustainable agriculture and various measures relating to the renovation of villages, rural tourism and infrastructures for the food sector.

415. The Portuguese CSF provides for the continuation of the major measures from the preceding phase, namely agricultural infrastructures, support for agricultural holdings, forestry, training and organization, and the processing and marketing of agricultural and forestry products. The agricultural section focuses on improving the competitiveness of agricultural holdings and diversification of activity in rural areas. The Portuguese CSF comprises 16 operational programmes, which have already been approved. The EAGGF

Guidance Section assists in four operational programmes on modernization of the economic base, enhancing development potential, Madeira and the Azores.

416. In the United Kingdom, three regions are concerned, namely Northern Ireland, Highlands and Islands and Merseyside. As regards Northern Ireland, a major share of the EAGGF Guidance Section funds is allocated to measures under Objective 5a (around 58 % of the total). Other measures relate to improving competitiveness, quality, marketing and management in the food sector, and various measures covering research projects, development and training, the protection and improvement of the rural environment, and diversification outside farming. In Highlands and Islands, the EAGGF Guidance Section assists measures under Objective 5a (53.7 % of the total), measures for improving efficiency on holdings and local marketing of production, the renovation of villages, and various measures concerning in particular the enhancement and protection of the environment, better forest management and the development of the forestry and forest products industry. The measures financed by the EAGGF Guidance Section in Merseyside relate solely to measures under Objective 5a.

Objective 5b

417. Objective 5b concerns the development of rural areas in difficulty which do not fall within the scope of Objective 1. It applies in nine Member States, namely Belgium, Denmark, Germany, Spain, France, Italy, Luxembourg, the Netherlands and the United Kingdom.

418. In the period 1994-99, the rural areas eligible for assistance from the Union under Objective 5b feature a low level of socio-economic development, assessed in terms of the gross domestic product per inhabitant, and must also meet various criteria, in particular as regards the share of total jobs accounted for by agriculture, the level of farm income and population density.

419. After discussion with the Member States within the framework of partnership, the list of eligible areas was approved formally by the Commission Decision of 26 January 1994. [1] The areas selected have a population of around 28 523 000, i.e. 8.24 % of the Community total. As compared with coverage in the period 1989-93, the population affected is 72 % higher.

---

[1] Decision 94/197/EC, OJ L 96, 14.4.1994.

Population and area covered by Objective 5b

Member State	Population (1 000 inhabitants)		Area (km^2)	
	1989-93	1994-99	1989-93	1994-99
Belgique/België	256	448	3 446	6 831
Danmark	107	360	1 762	11 707
Deutschland	4 441	7 725	50 541	95 450
España	992	1 731	63 359	84 972
France	5 830	9 759	185 853	280 201
Italia	2 904	4 828	34 660	75 269
Luxembourg	3	30	164	830
Nederland	443	800	2 413	5 350
United Kingdom	1 627	2 841	60 618	68 814
Total	16 604	28 523	402 816	629 425

420. Total funds for Objective 5b amount to ECU 6 667 million. Of that amount, 9 % has been set aside for Community initiative programmes, i.e. undertaken on the Commission's own initiative. A sum amounting to ECU 6 134 million at 1994 prices has been earmarked for the single programming documents. On 28 February 1994, [1] the Commission approved the indicative breakdown between the Member States of appropriations allocated for single programming documents. That breakdown is given in the following table.

Appropriations allocated for single programming documents

Member State	ECU million at 1994 prices
Belgique/België	77
Danmark	54
Deutschland	1 227
España	664
France	2 238
Italia	901
Luxembourg	6
Nederland	150
United Kingdom	817
Total	6 134

421. The operational programmes with a link with Objective 5b involve 72 areas. The integrated, multi-sector approach used in these programmes will provide new impetus for the rural economy and help to maintain the present population. This integrated approach is based on coordinated cooperation between the three Structural Funds (EAGGF, ERDF and ESF), the active participation by all (local, national and international) partners and the concentration of financial resources on the development of local potential. The priority objectives are as follows:

[1] Decision 94/203/EC, OJ L 97, 15.4.1994.

(i) diversification of the primary sector;

(ii) development of non-agricultural sectors;

(iii) development of tourism;

(iv) conservation and development of the national environment;

(v) development of human resources.

Community initiatives and innovative measures

Leader Community initiative

422. The very numerous and positive reactions to the Green Paper on the Community initiatives emphasize the need to continue, step up and improve coordination of the approach to the Leader Community initiative, setting sights higher and laying down stricter requirements for projects to be financed.

423. After the European Parliament delivered a favourable opinion, the Commission, in its notice to the Member States of 1 July 1994,[1] laid down guidelines for global grants and integrated operational programmes for which the Member States are invited to submit applications for assistance in the framework of a Community initiative for rural development.

424. Pursuant to the principle of subsidiarity, the Commission will negotiate, in partnership with the Member State concerned, the content of the programmes at regional and national level and will no longer play a direct part in the selection of projects, as was the case for the first phase.

425. The Community contribution is raised to ECU 1 400 million, of which ECU 900 million is for Objective 1. Following the Commission Decision of mid-July, the Member States have been notified of the indicative contribution from the Funds. The recipients under the initiative are local action groups and public and private bodies in rural areas.

The main objective of Leader II will be to encourage innovative projects put forward by local players from both the public and private sectors. Eligible measures may be grouped under three headings, i.e. acquisition of skills, rural innovation programmes and transnational cooperation.

Innovative measures

426. The measures planned under Article 8 of Regulation (EEC) No 4256/88 fall into several different categories:

---

[1] OJ C 180, 1.7.1994, p. 48.

(i)   pilot and/or demonstration projects;

(ii) technical assistance;

(iii) assessment studies;

(iv) measures for disseminating results.

427. With a view to the efficient and transparent management of this instrument for promoting rural development and the adaptation of agricultural structures, new rules on implementation have been adopted, involving:

(i)   the publication of a call for proposals for pilot and demonstration projects;

(ii) the publication of an invitation to tender for priority studies requested by the Commission;

(iii) in certain specific cases, the appraisal of projects by independent experts.

Accompanying measures

Environment

428. Most of the 158 programmes submitted by the Member States under Regulation (EEC) No 2078/92 have been approved by the Commission. This Regulation continues and extends the measures provided for in Article 19 of Regulation (EEC) No 2328/91, which came into force before the reform of the CAP.

429. The programmes submitted by the Member States set out a variety of approaches to solving environmental problems in agriculture. In accordance with the principle of subsidiarity, the programmes are approved by the Commission as long as the conditions and the principal objectives laid down in Regulation (EEC) No 2078/92 are observed.

430. Owing to the high level of spending forecast in the original programmes, the Commission had to devise a system for closely monitoring the expenditure that the Member States undertake in order to find speedy solutions to any overspending that might occur. In addition, the budget for Regulation (EEC) No 2078/92 was increased substantially.

The Community aid scheme for early retirement

431. The Community aid scheme for early retirement instituted by Regulation (EEC) No 2079/92 aims to encourage younger farmers, who can improve the viability of the remaining holdings, to take over from elderly farmers.

432.   To date, nine Member States have presented draft schemes for early retirement from farming, which have been approved by the Commission. The United Kingdom, the Netherlands and Luxembourg do not currently apply the measure.

433.   Taken together, the programmes' objective is the retirement of around 184 200 farmers and 7 500 farm workers. The area released should amount to almost 3 million hectares. It is estimated that around 5 % of the land will be used for non-agricultural purposes, such as forestry and the creation of ecological reserves. The remainder of the area released will be taken over by farmers, who will use it to extend their holdings or set up as farmers practising farming as their main occupation.

Forestry measures in agriculture

434.   Pursuant to Regulation (EEC) No 2080/92 instituting a Community aid scheme for forestry measures in agriculture, 17 programmes (11 national and 6 regional) and 19 sub-programmes (for Italy) were approved by the Commission in 1994.

435.   Still under examination are five regional sub-programmes for Basilicata (Italy), Bavaria (Germany), the Loire Region, Île-de-France and Midi-Pyrénées (France).

Other forestry measures

436.   Regulation (EEC) No 1615/89 establishing a European forestry information and communication system (Efics) was extended by Council Regulation (EC) No 400/94 of 21 February 1994.[1] The latter covers the period 1993-97 and has been allocated funds amounting to ECU 3.9 million.

437.   Under Council Regulation (EEC) No 2158/92 of 23 July 1992 on protection of the Community's forests against fire, the Commission approved the lists of high- and medium-risk areas and delivered a favourable opinion on forest protection plans. On 11 April 1994 Commission Regulation (EC) No 804/94[2] laying down certain detailed rules for the application of Council Regulation (EEC) No 2158/92 as regards forest-fire information systems was adopted.

438.   Under Council Regulation (EEC) No 3528/86 on the protection of the Community's forests against atmospheric pollution, two implementing Regulations were approved, namely Commission Regulation (EC) No 836/94 of 13 April 1994,[3] which provides for the analysis

[1] OJ L 54, 25.2.1994, p. 5.
[2] OJ L 93, 12.4.1994, p. 11.
[3] OJ L 97, 15.4.1994, p. 4.

and recording of the chemical content of needles and leaves within the Community network of observation plots, and Commission Regulation (EC) No 1091/94 of 29 April 1994, [1] which establishes a new network of permanent observation plots for intensive surveillance of forest ecosystems.

Agricultural research

439. The specific Community research and technological development programme in the field of the competitiveness of agriculture and management of agricultural resources (1989-93) [2] (CAMAR) reached completion in 1994. The 80 research and technological development projects have largely achieved their objectives, in particular as regards adaptation to changing market conditions, strengthening of agricultural structures, maintenance of incomes, conservation of natural resources and the countryside and development of agricultural information services.

440. The implementation of the AIR specific programme [3] continued in 1994. This programme covers research, technological development and demonstration projects in agriculture, horticulture, forestry, fisheries and aquaculture.

441. By the end of 1994, 139 projects were under way (45 projects commencing in 1992, 60 in 1993 and 34 in 1994). A further 48 projects were identified for funding in 1994. In addition, 25 seminars and workshops were provided support, and 25 researchers received grants for training in another Member State.

Genetic resources in agriculture

442. On 20 June 1994 the Council approved Regulation (EC) No 1467/94 on the conservation, characterization, collection and utilization of genetic resources in agriculture. A first Community programme was adopted for a period of five years, with a budget of ECU 20 million. The objectives are to help ensure and improve the conservation, characterization, documentation, evaluation and utilization of potentially valuable plant and animal genetic resources in the European Union. This will be carried out by means of the setting-up of a permanent inventory of existing genetic resources, coordination between Member States, support for new actions and various accompanying measures.

---

[1] OJ L 125, 18.5.1994, p. 1.
[2] OJ L 58, 7.3.1990.
[3] OJ L 265, 21.9.1991.

V — Financing of the CAP in 1994

The financial consequences of the reform of the CAP

443. The 1994 financial year is the first in which the main budgetary effects of the reform of the CAP were felt. The Commission established that the fair compensation offered to producers to offset the lower prices would initially entail higher budget expenditure, which is warranted if the situation of the CAP can be improved. In 1997 when the new provisions will be fully operational, EAGGF Guarantee Section expenditure for that year will be much less than would have been the case if the system applying prior to the reform of the CAP had been maintained.

444. Any forecast of expenditure on agriculture must be undertaken with caution. Many unforeseeable factors, both internal and external, including the ecu/dollar conversion rate, may affect expenditure during the period concerned. Difficulties in establishing forecasts are aggravated to a large extent by the major shift in the CAP, which calls for a fundamental adaptation of the mechanisms hitherto in force.

Budgetary discipline

The guideline

445. Like previous years, 1994 will continue to be subject to the requirements of budgetary discipline and in particular to compliance with the guideline resulting from the agreement reached at the February 1988 European Council meeting in Brussels and extended to 1999 on the same basis at the Edinburgh European Council of December 1992. In order to curb CAP expenditure, the guideline limits its annual rate of increase.

446. The general trend in EAGGF Guarantee Section expenditure since 1989 can be summarized as follows:

EAGGF Guarantee Section

(ECU million)

	1989	1990	1991	1992	1993	1994
Guideline	28 624	30 630	32 511	35 039	36 657	36 465
Expenditure financed within the guideline	24 406	25 069	30 961	31 119	34 590	32 960 [1]
Margin	4 218	5 561	1 550	3 920	2 067	3 505 [1]
Total expenditure[2]	25 873	26 454	31 784	31 950	34 590	32 960

[1] Provisional.
[2] All types of expenditure have been financed within the guideline as from 1993.

447.   The guideline for 1994 was set at ECU 36 465 million; the initial budget for this financial year provided for appropriations amounting to that sum, not including the appropriations entered in respect of the monetary reserve (ECU 1 billion). That amount was reduced by ECU 1 678 million following the adoption of supplementary and amending

Trend in EAGGF Guarantee Section expenditure

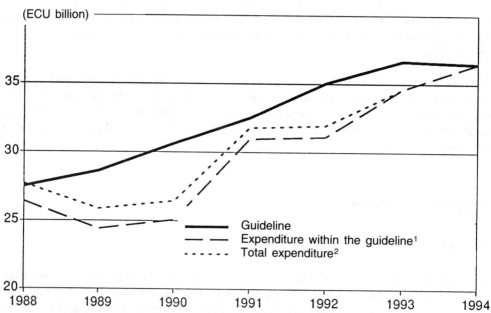

Sources: EAGGF, Eurostat and DG VI.
[1] Budget appropriations for 1994.
[2] All types of expenditure have been financed within the guideline as from 1993.

Figure 16

budget No 2/1994. Thus the appropriations allocated to the EAGGF Guarantee Section for 1994 amount to ECU 34 787 million, leaving a margin of ECU 1 678 million with respect to the guideline.

The monetary reserve

448. The ECU 1 billion appropriation entered as a provisional monetary reserve is intended to offset the impact of significant and unexpected fluctuations in the US dollar/ecu exchange rate recorded on the market as compared with the parity used when the budget was drawn up. That reserve is not included in the guideline.

449. If the trend in the US dollar/ecu parity is favourable, the savings made in the Guarantee Section are transferred to the monetary reserve up to a limit of ECU 1 billion. In the opposite case, funds are transferred from the reserve to the Guarantee Section headings adversely affected by the trend. A threshold of ECU 400 million is laid down, below which transfers to or from the monetary reserve do not take place.

450. The monetary reserve is only used exceptionally, as illustrated by the fact that own resources amounting to one twelfth of the general budget excluding the monetary reserve are called on each month. Thus own resources corresponding to that reserve are only called on when they are needed.

451. In accordance with the decisions of the Edinburgh European Council, the monetary reserve may also be used to cover any excessive costs of agrimonetary origin, which increases the risk of its being used up; if this were to occur, the Council would need to take special measures to allocate further resources to the EAGGF Guarantee Section.

452. In 1994, since agricultural expenditure including agrimonetary costs and those due to fluctuations in the ecu/dollar parity did not exceed the guideline, it was not necessary to call on either the monetary reserve or the 'Edinburgh mechanism' (specific Council measures).

The EAGGF in the general budget

453. In the general budget of the European Union for the 1994 financial year, totalling ECU 73 444 million (in payment appropriations), ECU 36 465 million (not including the monetary reserve), i.e. 49.7 % against 54.4 % in 1993, were allocated to the EAGGF Guarantee Section. This fall in the EAGGF Guarantee Section's share of total Union expenditure which comes in the wake of the development of the other common policies, is illustrated by the following figure.

Trend in EAGGF Guarantee Section expenditure as compared with total expenditure of the European Union

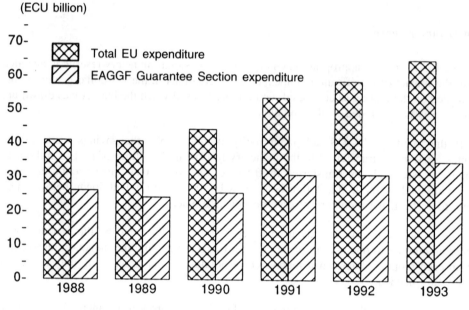

Sources: EAGGF, Eurostat and DG VI.

Figure 17

The EAGGF and its financial resources

454.   The EAGGF forms an integral part of the European Union's budget; its appropriations are therefore decided in accordance with budgetary procedures as with other Community expenditure.

455.   The agricultural policy also generates revenue in the form of sums collected under the common organizations of the markets. This revenue, which forms part of the Union's own resources, consists in:

(i) levies, which are variable charges on imports of agricultural products covered by the common organizations of the markets and coming from third countries; such charges are intended to compensate for the difference between prices on the world market and the prices agreed within the Community;

(ii) levies collected under the common organization of the market in sugar; these are divided into production levies on sugar and isoglucose production, sugar storage levies and additional elimination levies; they ensure that farmers and sugar manufacturers finance the cost of disposing of surplus Community sugar over and above the Community's domestic consumption.

Trends in revenue

Receipts from the Community's own resources under the common agricultural policy

(ECU million)

Type of receipt	1990	1991	1992	1993	1994[1]	1995[2]
Levies	1 173.4	1 621.2	1 206.8	1 029.1	1 023.4	861.3
Sugar levies	910.6	1 141.8	1 002.4	1 115.3	1 242.2	1 251.4
of which:						
Production[3]	504.6	770.0	606.8	698.4	745.3	744.1
Storage costs	406.0	371.8	388.6	416.7	400.0	410.4
Other	0.0	0.0	7.0	0.2	96.9	96.9
Total	2 084.0	2 763.0	2 209.2	2 144.4	2 265.6	2 112.7

[1] 1994 budget.
[2] Preliminary draft budget for 1995.
[3] Including the elimination levies, amounting to ECU 94.1 million in 1988, ECU 89.3 million in 1989, ECU 84.9 million in 1990, and ECU 16.5 million in 1991 (last year of application) and the additional elimination levy, amounting to ECU 110.8 million in 1988, ECU 175.5 million in 1989, ECU − 19.9 million in 1990, ECU 30.7 million in 1991 and ECU 0.9 million in 1992. This levy was not applied in 1993.

456.   It should be noted that other sources of agricultural income are considered to be the result of measures to stabilize the agricultural markets and are therefore directly deducted from agricultural expenditure for the financial year in question in the sectors concerned. Under the market organization for milk and milk products, producers pay an additional levy if milk production quotas are exceeded. This revenue, which does not form part of the Community's own resources, is considered to be a result of the measures to stabilize the agricultural markets and is paid into the budget chapter where the expenditure is incurred.

The EAGGF Guarantee Section

Expenditure

457.   Essentially the EAGGF Guarantee Section finances expenditure on the common organization of the agricultural markets, comprising:

(i) refunds on exports to third countries;

(ii) intervention to stabilize the agricultural markets.

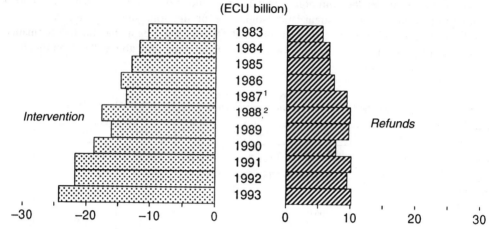

Trend in EAGGF Guarantee Section expenditure by type

(ECU billion)

Intervention

Refunds

1983
1984
1985
1986
1987[1]
1988[2]
1989
1990
1991
1992
1993

−30   −20   −10   0   0   10   20   30

Sources: EAGGF, Eurostat and DG VI.

[1] Expenditure in respect of the 1987 budget (10 months).
[2] Expenditure in respect of the 1987 budget (11.5 months).

Figure 18

458.   Depending on the product, the latter may take the form of production aid or premiums, price compensation, compensation for the withdrawal of products from the market or storage aid.

459.   However, over the past few years and in particular as a result of the reform of the CAP, the EAGGF Guarantee Section has been used to finance, in whole or in part, measures which go beyond the management of the agricultural markets in the strict sense, such as the distribution of agricultural products to deprived persons in the Community, measures to combat fraud and to promote quality, the set-aside of arable land and appropriations intended to cover agricultural expenditure incurred under the new measures whose purpose is to compensate for the geographical isolation of the French overseas departments (Poseidom), Madeira and the Azores (Poseima), the Canary Islands (Poseican) and the Aegean islands. Mention should also be made, in connection with the reform of the CAP, of the measures to aid producers of certain arable crops, measures relating to environmental protection and the upkeep of the countryside, aid for early retirement from farming and aid for forestry measures on agricultural holdings.

Breakdown of appropriations by product group

460. A detailed breakdown of EAGGF Guarantee Section appropriations for the 1994 budget by product group and type appears in Tables 3.4.3 and 3.4.4 (statistical annex to the report).

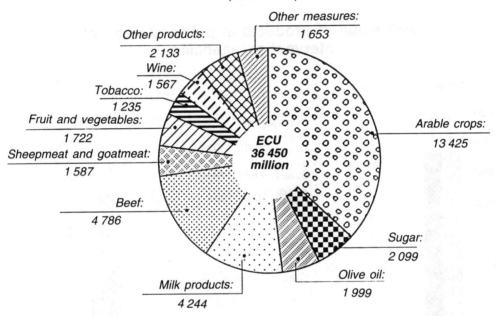

Budget appropriations for 1994

(ECU million)

Other measures: 1 653

Other products: 2 133

Wine: 1 567

Tobacco: 1 235

Fruit and vegetables: 1 722

Sheepmeat and goatmeat: 1 587

Beef: 4 786

Milk products: 4 244

Olive oil: 1 999

Sugar: 2 099

Arable crops: 13 425

ECU 36 450 million

Sources: EAGGF, Eurostat and DG VI.

Figure 19

Public storage

461. As indicated in Table 3.4.5 (statistical annex to the report), the book value of public stocks fell slightly during 1993 from ECU 2 813.6 million at 31 December 1992 to ECU 2 344.4 million at 31 December 1993, a drop of 17 % after a 5 % rise between 1991 and 1992.

462.   Changes in quantities in storage varied greatly, and included:

(i)   a major increase in cereal stocks (+ 11 %), principally common wheat, barley, maize and grain sorghum; stocks of rye and in particular durum wheat fell;

(ii)   a significant fall in stocks of milk products (– 10 %), in particular stocks of milk powder and, to a lesser extent, of butter;

(iii)   a sharp fall in beef stocks (– 38 %).

463.   During the first few months of 1994, that fall was confirmed and in June 1994 the value of the stocks was dropping, with the exception of olive oil, milk powder, alcohol and tobacco.

Book value of products in public storage at intervention agencies

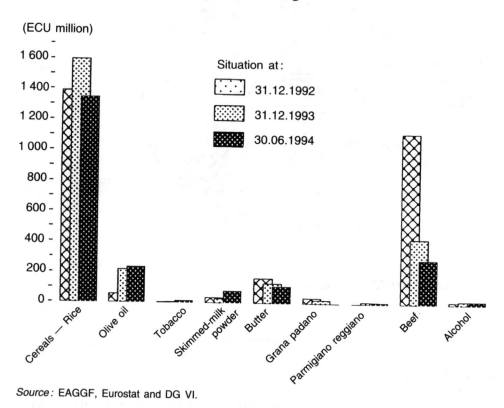

Source: EAGGF, Eurostat and DG VI.

Figure 20

Clearance of accounts

464.   The clearance of the EAGGF Guarantee Section accounts represents the definitive recognition of expenditure incurred by Member States under the common agricultural policy, after verification of the annual claims and after on-the-spot checks in respect of various files and the administrative and physical checks which must be carried out in sufficient number by the competent departments of the Member States.

465.   In 1994 the Commission scrutinized 1992 expenditure amounting to ECU 32.3 billion. At the same time checks began at the disbursing agencies on expenditure claimed for the 1993 financial year and at the bodies responsible for checking the compliance of claims with the Community regulations.

466.   On 1 July 1994 the Commission presented proposals to the Council for implementing the reform of the clearance of accounts procedure. The aim of the reform is to shift a large share of clearance work to preventive investigations, which will result in constructive recommendations to improve the verification procedures introduced by the Member States.

467.   Details of the clearance procedure are published every year by the Commission in its financial report on the EAGGF Guarantee Section. [1]

Fraud prevention and control

468.   In 1994 the Commission continued its activities in the fight against fraud and irregularities to the detriment of the EAGGF. This work was organized under the three following main headings:

(i)   prevention, which includes applying legislative measures on control and improving the regulatory framework and methodology;

(ii)   investigations and special checks;

(iii)   the introduction and implementation of specific control measures and the use of suitable effective techniques and methods.

Application of legislative measures regarding control

469.   The Commission has been closely monitoring the application of Community provisions on stricter control. In this connection Commission officials have made several on-site visits to the Member States to study, with the services concerned, the introduction and proper

---

[1]   Twenty-third Financial Report 1994 (COM(94) 464 final).

implementation of the integrated administration and control system in the context of the reform of the CAP, established by Council Regulation (EEC) No 3508/92. [1]

470. There is provision for the Community to co-finance 50% of the cost for three years. The Commission has proposed to extend this co-financing for a further year.

471. The *a posteriori* administrative checks provided for in Regulation (EEC) No 4045/89 [2] are carried out more comprehensively and systematically, and are based in particular on risk analysis by product group and by company and on better utilization of human and technical resources. The Commission has proposed to amend the abovementioned regulation to improve the conditions governing checks.

Improvement of the regulatory framework and methodology for control

472. In 1994 the Commission continued its efforts to simplify the agricultural legislation to make control more effective. Risk analysis is applied more systematically in the various sectors of the CAP. Accordingly, Council Regulation (EC) No 163/94 [3] amending Council Regulation (EEC) No 386/90 on the monitoring carried out at the time of export of agricultural products receiving refunds or other amounts [4] was adopted.

473. In order to protect the financial interests of the Union better, the Commission proposed to the Council that measures be adopted to rule out from eligibility for Community aid granted by the EAGGF Guarantee Section operators who do not provide the necessary guarantees, on account in particular of fraud committed. There is provision for a system of notifications between the Member States and the Commission. The Council adopted the Commission proposals on 12 December 1994. [5]

Investigations and special checks

474. The Commission conducted several investigations in the Member States, in close cooperation with the national authorities, pursuant to Regulations (EEC) Nos 729/70 and 595/91 concerning irregularities and the recovery of sums wrongly paid in connection with the financing of the common agricultural policy and the organization of an information system in this field. The national services of the Member States also conducted several major investigations.

475. At the same time, the Member States detected irregularities and reported them to the Commission under Articles 3 and 5 of Regulation (EEC) No 595/91.

---

[1] OJ L 355, 5.11.1992.
[2] OJ L 338, 30.12.1989.
[3] OJ L 24, 29.1.1994.
[4] OJ L 42, 26.2.1990.
[5] Regulation (EC) No 3094/94, 12.12.1994, OJ L 328, 20.12.1994.

Specific control measures

476.   In 1994 the agencies responsible for monitoring the olive oil industry carried out a large number of checks on olive oil production and marketing, and detected several irregularities.

477.   Community bodies responsible for monitoring wine and fruit and vegetables also conducted several inspection visits resulting in the discovery of irregularities and helped to reduce irregularities in these markets. At the same time, remote sensing helped significantly to improve the effectiveness of checks in the production aid field in particular.

Expenditure on agricultural markets in 1994

478.   The provisional uptake of appropriations for 1994 [1] is ECU 32 960 million, i.e. an under-utilization in relation to the SAB of ECU 1 827 million. This difference is explained as follows:

(i)   ECU 233 million is accounted for by savings linked to changes in the dollar/ecu rate;

(ii)   ECU 634 million is accounted for by expenditure not implemented for which appropriations should be carried forward in 1995;

(iii)   almost ECU 960 million is accounted for by new economic conditions or by overestimates of forecasts by the Member States.

479.   Thus the main chapters showing under-utilization are arable crops (income generated by the sale of intervention stocks), sugar (fewer exports), olive oil (delays in payment of production aid 1993/94), fruit and vegetables (forecast expenditure overestimated by some Member States), beef/veal (very little buying-in and suckler cow premiums less than allowed for), sheepmeat and goatmeat (delays in payment) and, finally, accompanying measures (start-up of programmes slower than planned).

The EAGGF Guidance Section

480.   Structural Fund reform, operational from 1 January 1989, has gradually affected EAGGF Guidance financing. A growing proportion of Community assistance is granted as part-financing of operational programmes (52 % of the 1993 total, 45 % in 1992, 40 % in

---

[1] The EAGGF financial year ends on 15 October 1994.

1991). This is the preferred form of Structural Fund financing, involving multiannual programming and devolving a great deal of responsibility to the Member States.

481. On 20 July 1993 the Council renewed the Structural Fund legislation in preparation for a second programming period running from 1994 to 1999. The changes introduced were designed to consolidate the principles governing action in the first period following reform and also to simplify procedures to some extent. For instance, the provisions governing Objective 5a measures applicable throughout the Community were amended to make programming procedures less cumbersome (e.g. in the case of processing and marketing of agricultural products, covered by Regulation (EEC) No 866/90) and to harmonize the financial mechanisms.

482. In 1993 Community aid schemes (indirect measures), where the Member State is also responsible for implementation, still absorbed a significant proportion of Guidance Section funding. This category includes Regulation (EEC) No 2328/91 (farm improvement plans, compensatory allowances, etc.), by itself accounting for 35 % of total assistance granted in 1993.

483. Direct funding of individual projects, on the other hand, now plays a very limited role: demonstration and pilot projects and part of the integrated Mediterranean programmes.

Financing

484. EAGGF Guidance Section expenditure by Member State for the first Fund reform period (1989 to 1993) was as follows:

EAGGF Guidance Section expenditure (commitment appropriations)

(ECU million)

Member State	1989	1990	1991	1992	1993
Belgique/België	31.58	23.05	30.49	28.18	41.70
Danmark	17.29	16.92	18.04	23.52	19.99
Deutschland	127.15	183.28	200.19	253.76	348.72
Elláda	235.30	270.16	274.20	392.20	402.85
España	203.89	301.83	514.15	633.60	412.91
France	179.77	382.93	425.26	554.36	633.50
Ireland	121.74	124.77	168.50	194.51	165.71
Italia	263.61	269.26	326.51	375.87	624.98
Luxembourg	3.58	4.60	6.67	6.36	9.01
Nederland	20.66	10.71	20.50	21.90	19.53
Portugal	179.39	241.61	313.40	289.77	313.95
United Kingdom	78.03	96.55	110.24	100.78	99.49
Total[1]	1 461.99	1 925.68	2 408.16	2 874.81	3 093.40[2]

[1] Not including commitments for Regulation (EEC) No 1852/78 (fisheries).
[2] Including ECU 975 000 of multi-country funding.

485.   It is also interesting to note the breakdown by Structural Fund Objective. The EAGGF Guidance Section provides aid under:

(i)   Objective 1: regions lagging behind in development;

(ii)   Objective 5a: agricultural structures in all regions

(iii) Objective 5b: rural development in certain limited areas.

Expenditure trend by Objective

(ECU million)

	1989	1990	1991	1992	1993
Objective 1 (regions lagging behind)	862.13	1 081.16	1 440.83	1 634.68	1 599.22
Objective 5a (agricultural structures)	516.20	743.81	631.25	701.33	923.88
Objective 5b (rural areas)	26.86	44.00	260.15	475.80	508.64
Transitional measures[1]	56.80	56.70	75.93	63.00	61.65
Total	1 461.99	1 925.68	2 408.16	2 874.81	3 093.40[2]

[1] Expenditure under old measures that cannot be assigned to one of the present Objectives.
[2] Including ECU 975 000 of multi-country funding.

New German *Länder*
Expenditure charged to the EAGGF Guidance Section incurred pursuant
to Regulation (EEC) No 3575/90

(ECU million)

	1989	1990	1991	1992	1993
New *Länder*	—	—	120.29	236.37	263.95

486.   Expenditure on Objective 1 has increased since 1989 to stabilize in 1993, at the end of the first period.

487.   Expenditure on Objective 5a has also grown but to a lesser extent, particularly if 1993 is considered exceptional owing to commitment and payment for the first time of advances on compensatory allowances. That on Objective 5b increased markedly from 1989 to 1993, reflecting the accent put on rural development.

488.   Expenditure on the transitional measures, which are being wound up, was lower than in 1991 and 1992.

489.   Expenditure under Objectives 1 and 5b takes in both action covered by Community support frameworks and that taken under the Community initiatives. Expenditure on the latter in 1993 was very much lower than in 1992 since commitments for the most important rural development initiative, Leader, were almost all made in 1991 and 1992. For Objective 1,

Community initiatives accounted for ECU 63.6 million (4 %) and for Objective 5b ECU 1.5 million (0.3 %).

Budget execution

490. For the EAGGF Guidance Section as a whole execution was complete both for commitment appropriations (99.8 %) and payment appropriations (99.9 %).

491. Appropriations available in 1993 amounted to ECU 3 099 million in commitments (including ECU 27.8 million of re-entered appropriations carried over and ECU 3.2 million in additions by transfer) and ECU 2 953.6 million for payments (taking account of ECU 811.1 million re-entered and a reduction by transfer of ECU 423.5 million).

492. For 1994 appropriations amount to ECU 3 410 million in commitments (17.4 % of total Structural Fund appropriations) and ECU 2 864 million for payments. Adjustments by transfer raise these amounts to ECU 3 359 million for commitments and ECU 2 829 million for payments. In contrast to previous years these amounts include the allocations assigned pursuant to Regulation (EEC) No 3575/90 to the new German *Länder*. Allocations to programmes for the outermost regions are again included under the 'Structural Funds' heading. The commitment appropriations total does not include fisheries, which for Guidance Section financial management purposes has been separate from agriculture since 1 January 1990.

493. The budget for 1994, the first year of the new programming period (1994-99), has been amended in line with the budget projection established for that period and the outcome of negotiations with the Member States in the partnership framework.

VI — External relations

Activities within GATT

494. Following the success of the negotiations in December 1993, the Uruguay Round itself was formally concluded on 15 April 1994 with the adoption of the Marrakesh Declaration and the signature of the Final Act and the agreement establishing the World Trade Organization (WTO). Along with other signatories to the agreement, the European Union is now committed to implementing the results of the Uruguay Round.

495. As regards the agreement on agriculture, the Union has indicated in its schedule the dates on which its commitments in the different areas will become applicable.

496. Since the conclusion of the negotiations, the Union has been intensively preparing the implementation of the outcome of the Round and has forwarded to the Council a single package of measures to that end.

497. Two other issues were settled during the final negotiations of the Uruguay Round — those concerning bananas and imports of Chilean apples — both of which were the subject of GATT panels established in 1993.

498. In March 1994 a framework agreement on bananas was reached between the Community and four Latin American countries. This agreement includes increases in the global tariff quota and has been incorporated into the Community's GATT commitments.

499. Regarding apples, discussions between the Commission and Chile led, in March 1994, to an agreement set out in an exchange of letters and to Chile withdrawing its request for a GATT panel.

500. The Community participated as an intervening party in the panel against the USA on its tobacco legislation on domestic content requirements and importer assessments. Negotiations with the USA have also been requested following its notification to unbind tariff commitments on tobacco.

501. In January 1994 the Union was successful in a panel against Brazil concerning countervailing duties on milk powder. The Union had sought the panel after Brazil failed to demonstrate injury before imposing countervailing duties on Community exports of that product.

502.   The Commission has been actively involved in the agricultural aspects of negotiations with a number of countries such as China, Taiwan and Russia for their accession to GATT.

Relations with the United States

503.   The Blair House Agreement on corn gluten feed was implemented by the European Union through a modification of the Combined Nomenclature and the establishment of certification requirements and a monitoring system for imports from the USA.

504.   An agreement was signed on the mutual recognition of certain spirit drinks: for the Union, Scotch whisky, Irish whiskey, Cognac, Armagnac, Calvados and Brandy de Jerez; for the US, Tennessee whisky and Bourbon whisky. Negotiations will begin in due course on extending mutual recognition to other spirit drinks.

505.   Although considerable progress has been made, full implementation of the EC/US agreement on slaughterhouses has been postponed to 1 July 1995. In the meantime US slaughterhouses will continue to be listed under the interim measures of the agreement.

506.   Discussions in the wine sector are continuing, with the Union seeking improvements in the protection of a number of its appellations of origin and the USA seeking EC authorization for exports to the Union of wine made according to US oenological practices.

507.   After a long period of discussion, the USA established a tolerance level for procymidone which allows for import of Community wine made from grapes which have been treated with this pesticide.

508.   Discussions are continuing on possible mutual recognition of standards for organically farmed produce.

Relations with Canada

509.   Following the GATT agreement of April 1994, relations between the European Union and Canada became noticeably less dynamic as each worked on measures for implementing its multilateral commitments.

510.   However, the European Union did obtain from Canada an agreement not to restrict its imports of European pasta, as it had decided to do at the beginning of the year.

Relations with industrialized countries

511.   Voluntary restraint agreements in the sheepmeat sector with New Zealand and several other countries, which were due to expire at the end of 1993, were extended for a further year pending the completion and implementation of the Uruguay Round. Similarly, arrangements for access to the UK market for New Zealand butter on special terms were extended for a further year.

512.   The Commission signed an agreement with Australia, on the basis of guidelines laid down by the Council, on trade in wine based on the principles of non-discrimination and reciprocity.

513.   Bilateral contacts with Japan and the Republic of Korea concentrated on the lowering of non-tariff barriers and of market access in general.

514.   Throughout 1994 the Commission took part in negotiations to draft a new wheat trade convention. The previous convention dates from 1986 and remained in force until 30 June 1991 but it was extended until 30 June 1995. The future International Grains Convention will cover all grains trade and will be in force from July 1995.

The Commission continued to be an active member of several specialized international agricultural organizations including the International Olive Oil Council and the International Wheat Council. In the case of the former, a new agreement was agreed and ratified by a majority of the members.

Relations with the countries of Central and Eastern Europe

515.   In 1994, the European Union started negotiations for the conclusion of a Europe Agreement with Slovenia in addition to the agreements already in existence since 1992 and 1993 with Poland, Hungary, the Czech Republic, Slovakia, Romania and Bulgaria.

516.   In the agricultural sections, the agreements contain mutual tariff concessions with a gradual increase over a five-year period as well as cooperation and technical assistance in the field of agriculture and food production.

517.   The agreements concluded in 1992 and 1993 have been amended following the decision of the Copenhagen Summit in June 1993. The timing of the gradual increase in EU trade concessions was brought forward by six months for all the countries concerned. Further adaptation to the agreements will be necessary due to the implementation of the Uruguay Round and the enlargement of the Union. It must be noted, however, that only a small amount of the preferential tariff quotas made available to these countries has been taken up. The Commission has decided to carry out a study examine the reasons for this.

518. As to the Baltic States (Estonia, Latvia, Lithuania), the Copenhagen Summit recommended the conversion of the existing cooperation agreements into trade agreements. These agreements were concluded in July 1994 and are expected to come into force on 1st January 1995. In the meantime, the Commission, following a specific recommendation of the Council, started preliminary talks on the conclusion of Europe Agreements.

Relations with the former USSR

519. In October 1992, the Council adopted directives for the negotiation of partnership and cooperation agreements with the New Independent States. These agreements contain provisions for cooperation in the agricultural and agro-industrial sector. In 1993, the Community opened negotiations with the Russian Federation, Ukraine, Belarus and Kazakhstan and in 1994 with Kyrgyzstan. On the occasion of the Corfu European Summit in June 1994, the first agreements with the Russian Federation and Ukraine were signed.

520. Pursuant to a 1991 Council decision, an EU credit line of ECU 1 250 million for the purchase of food and medical supplies was opened for the new independent States of the former Soviet Union. This programme was implemented throughout 1993 and saw the inclusion of the Russian Federation and Kazakhstan in January 1993. In 1994, Azerbaijan and Uzbekistan, the last two Republics, signed loan agreements with the European Union.

Relations with the Republic of South Africa

521. Following the resumption of official relations with South Africa, the European Union decided in April 1994 to give firm support to the process of racial integration and social reform planned for that country.

522. Specifically, the Union granted the equivalent of ECU 110 million within the generalized preferences system, which did not, however, include the agricultural sector in 1994. Possibilities for the years to come are still being studied.

523. There is already an agreement in principle between the European Union and South Africa to initiate at the end of 1994 talks intended to lead to an agreement in the wine-growing sector, the attraction of which will be that it will provide for reciprocal protection of designations of origin.

Mediterranean countries

524. 1994 is the third year of application of the new Mediterranean policy the objective of which is to foster economic development in Algeria, Cyprus, Egypt, Israel, Jordan, the Lebanon, Malta, Morocco, Tunisia and Syria. The import tariff quotas agreed to by the EU have increased by 5 % (based on the original quotas) and 3 % for some sensitive products. These quotas will increase again in 1995 at the same rate, therefore completing the provisions of the NMP.

525. New cooperation or association agreements are being discussed with Tunisia, Morocco and Israel. The Council mandate foresees limited expansion of the agricultural concessions. These new agreements should apply from 1996 onwards.

526. Following the conclusion of the Uruguay Round, certain adjustments to the current arrangements are being discussed with some Mediterranean countries.

527. The Customs Union between the EU and Turkey may be postponed by one year to 1996. However, it will not include the agricultural sector where only limited adjustments to the current arrangements are foreseen.

528. A new agreement with Egypt, comparable to those signed with the Maghreb countries, is being envisaged. It will include the current agricultural concessions.

Food aid

529. Apart from emergency food aid to which ECU 36 million were allocated, the Community continued to apply its own traditional food aid programme restricted to developing countries.

530. In May, the Commission adopted a decision fixing the total quantities of food aid for 1994 and drew up a list of products.

531. This list comprised 1 387 500 tonnes of cereals (this amount includes the minimum 927 000 tonnes Community contribution to the 1986 Food Aid Convention), 50 000 tonnes of milk powder and equivalent products, 4 000 tonnes of butteroil, 68 000 tonnes of vegetable oil and 17 000 tonnes of sugar and other products (legumes, beans, dried fish, etc.) worth a total of ECU 51 million.

532. In addition, the Commission has continued its programme of distributing food from intervention stocks to the poorest citizens of the European Union.

International organizations

533. Since its membership of the FAO in November 1991, the Union has fully participated in numerous meetings of the various FAO bodies dealing with agricultural policies and commodities and food issues.

534. The Union has continued to collaborate actively in the work of the Organization for Economic Cooperation and Development (OECD), participating in a number of committees including the Committee for Agriculture and its working parties, the Environmental Committee and the Group of the Council on Rural Development. The OECD Council, which met on 7 and 8 June 1994, also addressed the subject of agriculture. In their final communiqué, the ministers stressed the need to pursue the reform of the agricultural policies on the basis of the principles that the Council established in 1987 and 1992.

Enlargement negotiations

General

535. The negotiations on the enlargement of the European Union, which had commenced in early 1993, were concluded with Austria, Sweden and Finland on 1 March 1994 and with Norway on 16 March 1994. The final chapter on institutions was however only settled by the Union at Ioannina on 27 March 1994 enabling the Accession Conferences to formally agree on all the negotiating chapters on 30 March. The political agreements reached were then consolidated into a single Treaty and supporting Act of Accession. This was signed by all the plenipotentiaries at the Corfu Summit on 24 June 1994 following assent by the European Parliament on 4 May. Positive referenda in Austria in June, Finland in October and Sweden in November (the referendum was negative in Norway in November), together with ratification by the 15 national Parliaments, will enable these three countries to join the Union on 1 January 1995 as originally foreseen. Consequently, they will play their full part in the forthcoming Intergovernmental Conference, scheduled for 1996.

536. The agricultural chapter was one of the key issues for the negotiations. Agriculture and agricultural policy in the acceding countries had, apart from specific and limited bilateral agreements, never figured as such in the European Free Trade Area (EFTA) or in the European Economic Area (EEA). Their agricultural support and external protection had traditionally been high, although in recent years Sweden had made reforms which brought its support arrangements more into line with those of the Union. The negotiations, therefore, centred on requests for both permanent national support arrangements and substantial transitional arrangements but were distinguished by the agreement to apply the Community 'acquis' immediately upon accession and to preserve the principle of the single market.

Key results of the negotiations

(a) The transition arrangements

537. Given the degree of difference in price and support levels, the transition arrangements posed some of the main problems for the whole negotiation process with Finland and Austria. Conversely, Sweden, which already had favourable structures, particularly in the south, and which had already undertaken a major policy reform in recent years, was practically in line with the common agricultural policy. Sweden's position therefore enabled it to forego any transitional arrangements, at least with regard to price alignments. On the other hand, Finland in particular had very high agricultural prices aiming at self-sufficiency via very high price guarantees and heavy trade restrictions.

538. Its instruments, mostly complex, included, in particular, regionalized price supplements (aid directly linked to the product), aid for transport, State monopolies, etc., in other words a battery of measures substantially out of line with Community rules, especially since the CAP reform. The Union had to reject the traditional transitional system of accession compensatory amounts because this would require the maintenance of border controls which would be incompatible with the single market. Thus an immediate alignment of prices combined with degressive national aids over a transitional period of five years were agreed for both Finland and Austria. These will result in a radical change in the system of support in the three countries concerned.

539. These transitional degressive aids make it possible to:

(i) bridge the price differences within an enlarged Community without internal borders;

(ii) preserve, on a transitional and degressive basis, total support which will not jeopardize the maintenance of farming throughout the territory of the new member countries while enabling them to embrace the objectives and instruments of the CAP;

(iii) facilitate the structural adjustment of farming and related sectors in the two countries concerned, during the transitional period.

540. In the case of Finland, these arrangements have been supplemented by a provision for additional national aids, in the event of serious difficulties, which should facilitate the full integration of its agricultural sectors into the CAP.

541. Furthermore, to the extent that the immediate adoption of the single market might lead to serious market disturbances in Austria and Finland by virtue of trade with Member States, the Act of Accession provides safeguard arrangements requiring the Commission to take rapid action whilst not reintroducing border controls.

(b) The common organizations of the markets

542. Agreement on quotas for milk, sugar and tobacco, base areas for arable crops and reference herds for suckler cows, young male cattle and ewes were required which used criteria consistent with those which had been imposed on the existing Member States. Further details of the quotas and reference quantities agreed are annexed.

543. By contrast, the European Union was able to allow certain transitional derogations, including:

(i) a transitional period prior to implementation of the quality standards for fruit and vegetables (Austria, Finland and Sweden);

(ii) maintenance during a transitional period of national definitions of drinking milk (Sweden and Finland);

(iii) certain derogations to the quality standards for eggs in the case of direct sales (Finland);

(iv) national investment aids for pigmeat, eggs and poultry (Finland and Austria).

544. In any event, all products which do not comply with Community legislation are to be restricted to the national markets and the transitional derogations do not prevent free movement of Community goods.

(c) Structural policy and accompanying measures

545. The candidate countries have fully accepted the 'acquis communautaire' with regard to both regional policies and those concerning the improvement of production structures (Objective 5a).

546. The definition of agricultural less-favoured areas, including mountain and hill areas, will therefore be determined on the basis of the existing Community criteria. However, without departing from the 'acquis', the Union has agreed to take account also of the production conditions in the northern regions. Consequently latitude (indicating a short growing season), has been accepted as an equivalent to the altitude criterion for defining mountain areas, thereby ensuring equal treatment with the upland areas of the present Member States.

547. Taken as a whole a large part of the utilized agricultural area (UAA) of the new Member States is foreseen to be established as less-favoured areas (Austria, 70-75 % and Sweden 48-52 % of UAA).

548. Finland will additionally benefit from an agreement that all of its less-favoured areas (85 % of its UAA) will be treated as mountain areas. Austria, on the other hand, which had developed a similar support mechanism 'Grundbetrag' to the Union's LFA will, to the extent that it offers a higher rate of support, be able to maintain the measure for a period of 10 years.

549. A certain number of temporary derogations have also been accepted in the area of agricultural structures:

(i) for the Nordic countries, recognition of 'agri-forestry' family farms, provided they have a UAA of at least 15 ha;

(ii) for Austria and Finland, a derogation to allow investment aid in the pigmeat, egg and poultry sectors and for the installation of greenhouses;

(iii) for the food-processing industries of Finland and Austria, flexibility in the restriction for providing nationally-financed investment aids.

550.   The accompanying measures under the CAP reform will also play an important role in the new agricultural policy as applied to the future Member States. In particular, financial envelopes have been allocated for these measures especially for agri-environment support in order to ensure that it is implemented relatively widely.

(d) Nordic agriculture

551.   The two Nordic countries have traditionally modulated their support systems in order to compensate for the climate, geographical and demographical conditions which exist in their northern regions which result in increased costs, shorter growing seasons and reduced productivity. Such regionally-differentiated supports have been consistent with these countries' social policies aimed at maintaining an even distribution of employment opportunities and income. This has largely continued in Sweden despite its agricultural reforms of the early 1990s which led to a general decrease in support levels. In recognition of the severe handicaps of farming in these northern regions, it was agreed that those areas located north of 62° latitude and in certain adjacent areas where farming suffers from particularly harsh climatic conditions, long-term national aids could be granted to ensure that farming activities are maintained in these regions. The aids will be subject to Commission control and will have to meet certain conditions. In particular, they may not result in an increase in, or intensification of, agricultural production.

(e) Plant health and veterinary matters

552.   In the veterinary field, the entire 'acquis' has been accepted, subject to the temporary provision by the Union of additional guarantees with regard to certain animal diseases or zoo-noses and short transitional periods needed for the new Member States to bring their production establishments into line with EU public health requirements.

553.   Transitional derogations were also conceded for feedingstuffs, particularly with regard to the use of antibiotics.

554.   With regard to plant health and the quality of seeds and propagating material, the Union has agreed, in particular:

(i)   to temporarily recognize each of the four countries as a protected area for certain organisms harmful to plants;

(ii)   to grant a transitional period of not more than two years to Finland and Sweden to enable them to adjust their inspection arrangements for certain plant diseases;

(iii) to maintain for not less than five years, in Finland and Sweden, national legislation on the marketing of forest seeds, given that during that time the current Community arrangements will be reformed;

(iv) to include the varieties accepted in the candidate countries in the common catalogues as from 1 January 1996, provided those varieties comply with Community rules;

(v) to adopt certain technical amendments of the current legislation on certain standards.

General impact of the enlargement

555. In overall economic terms, and in terms of agriculture, the enlargement to include the three EFTA countries will be much less important than the first enlargement which brought in the United Kingdom, Denmark and Ireland (1973) and the 'Southern' enlargement which brought in Greece (1981), Portugal and Spain (1986). The increase in overall population will be only 6.3 % (an addition of 22 million) while the increase in economic wealth will be 7 % (based on 1993 GDP figures). Although the surface area of the Union will increase by 37 %, extending it just beyond the Arctic Circle and to the Russian border, the increase in utilized agricultural land will be only 6.9 %. Agriculture's share of GDP and employment is quite similar in the three acceding countries to the Union average. Their most important sector is milk (which corresponds to 8 % of EUR 12 production) with the accompanying production of beef. Their self-sufficiency in grain, beef, pigmeat and milk is also relatively high.

VII — Agricultural development

Statistical information

NB: For practical reasons the following pages employ the continental representation of numbers, i.e. one thousand two hundred and thirty-four point five is represented as 1 234,5 rather than the more conventional 1,234.5.

Foreword

Codification of the tables

The choices made for the revision of the tables are reflected in a new codification, established on the basis of the same principle for all the tables. Each of them has been given a code with four digits, the first of which designates the subject to which the table refers (see table of contents following this foreword):

1. Conversion rates,
2. Basic data,
3. Economic tables,
4. Tables on agricultural markets.

The second and third digits refer to specific aspects of the field concerned and their significance varies from one field to another.

For the tables concerning the agricultural markets (Tables 4) a standard codification for all the products has been used for these two digits:

(i) the second digit of the code designates the agricultural product concerned,

(ii) the third digit refers to the nature of the statistic presented:

   –.–.0.– livestock numbers,
   –.–.1.– area, yields and production (crop products) or slaughterings and production (livestock products),
   –.–.2.– world production,
   –.–.3.– external trade,
   –.–.4.– supply balance,
   –.–.5.– prices (producer prices, market prices, consumer prices),
   –.–.6.– market management,
   –.–.9.– various.

For certain sectors, all the possibilities are used (e.g. cereals). For other products only some are used (e.g. potatoes), either because the data needed are not available or because the features of these sectors in the Community do not justify such an exhaustive presentation in a general document such as this, which, for considerations of space, can provide only the most important information.

Remarks

1. Up to December 1987 this report used the SITC Rev. 2, which was worked out using the 6-digit Nimexe, while from January 1988 it uses the SITC Rev. 3, which has been drawn up using the 8-digit subheadings of the Combined Nomenclature.

   In particular, it should be noted that considerable divergences have arisen at subheading level between the Combined Nomenclature and the formerly used Nimexe, leading to a break in the goods-related time series between 1987 and 1988.

2. From 1991 data for the former German Democratic Republic is included in the figures for the Federal Republic of Germany and accordingly in the figures for the Community as a whole.

   The tables are indicated under 'Key to symbols and abbreviations' following this foreword by the symbol ∞.

3. As a result of gradual introduction of data for the ex-German Democratic Republic the % TAV rates calculated from one year to another may sometimes be inconsistent.

4. The new Intrastat system for collecting statistics on intra-Community trade was introduced in 1993. As a result, the data on intra-Community trade from 1993 onwards will no longer be comparable with the data for previous years.

Contents
Statistical data and tables

□ = New table

Remark: The following tables of *The Agricultural Situation in the Community — 1993 Report* have not been repeated: 3.5.2.1, 3.5.6.15, 4.5.4.1, 4.5.4.2, 4.5.4.3, 4.5.4.4, 4.5.4.5

Tables replaced: 3.5.6.13, 3.5.6.16

Key to symbols and abbreviations

Statistical symbols

—	Nil
0	Less than half a unit
×	Not applicable
:	Not available
.	Not fixed
..	No prices quoted
#	Uncertain
p	Provisional
*	Eurostat estimate
**	Commission estimate, Directorate-General for Agriculture
r	Revised
s	Secret
Ø	Average
» 1980 «	Ø (1979, 1980, 1981)
» 1985 «	Ø (1984, 1985, 1986)
1980/81	Marketing year, starting in 1980 and ending in 1981
%	Percentage
% TAV	Annual rate of change (%)
∞	Former German Democratic Republic included

Units

— *Currency*

ECU	European currency unit
EUA	European unit of account
u.a.	Gold parity unit of account
BFR	Belgian franc
DKR	Danish crown
DM	German mark
DR	Greek drachma
ESC	Portuguese escudo
FF	French franc
HFL	Dutch guilder
IRL	Irish pound
LFR	Luxembourg franc
LIT	Italian lira
PTA	Spanish peseta
UKL	Pound sterling
USD	US dollar
NC	National currency

— *Other units*

cif	Cost, insurance, freight
VAT	Value-added tax
Mrd	Thousand million
Mio	Million
t	Tonne
kg	Kilogram
hl	Hectolitre
l	Litre

ha	Hectare
UAA	Utilized agricultural area
LU	Livestock unit
ESU	European size unit
FU	Fodder unit
AWU	Annual work unit
TF	Type of farming
PPS	Purchasing power standard
NUTS	Nomenclature of territorial units for statistics

Geographical abbreviations

EC	European Communities
EUR 9	Total of the Member States of the EC (1980)
EUR 10	Total of the Member States of the EC (1981)
EUR 12	All EC Member States (1986)
BLEU/UEBL	Belgo-Luxembourg Economic Union
DOM	French overseas departments
ACP	African, Caribbean and Pacific countries party to the Lomé Convention
PTOM	Countries and overseas territories of Member States of the EC

Sources

Eurostat	Statistical Office of the European Communities
SITC	Standard international trade classification (Eurostat)
Nimexe	Nomenclature of produce for the Community's external trade statistics and trade between its Member States (Eurostat)
ESA	European system of integrated economic accounts (Eurostat)
FADN	Farm accountancy data network (Commission of the European Communities, Directorate-General for Agriculture)
OECD	Organization for Economic Cooperation and Development
FAO	Food and Agriculture Organization of the United Nations
UNRWA	United Nations Relief and Works Agency
IMF	International Monetary Fund
GATT	General Agreement on Tariffs and Trade
Fefac	European Federation of Manufacturers of Compound Feedingstuffs
Fediol	Federation of Seed Crushers and Oil Processors in the EEC
AIMA	Intervention Agency for the Agricultural Markets (Italy)
USDA	United States Department of Agriculture

Currency units used in this report

1. European Monetary System (EMS) — ecu

The EMS came into force on 13 March 1979 (Regulations (EEC) No 3180/78 and No 3181/78 of 18 December 1978). With this system, the ecu was introduced as the sole unit of account for the Community. The definition of the ecu is identical to that of the EUA (European unit of account, defined by Regulation (EEC) No 250/75 of 21 April 1975) except for a review clause allowing of changes in its composition. The ecu is a currency unit of the 'basket' type made up of specified amounts of the currencies of the Member States determined mainly on the basis of the size of the economy of each State. The central rates used in the system are rates fixed by the central banks around which the market rates of the EMS currencies may fluctuate within margins at any given time.

2. The ecu in the common agricultural policy

— Before 9 April 1979, the unit of account used in the agricultural sector was the unit of account (u.a.) as defined by Regulation (EEC) No 129/62 and the representative rates (green rates) were fixed by the Council.

— On 9 April, the ecu was also introduced into the CAP (Regulation (EEC) No 652/79) and its use was subsequently renewed by Regulations (EEC) No 1264/79, No 1011/80, No 1523/80 and No 876/81. The agricultural prices and the common amounts are expressed in ecus. The conversion rates (representative rates) of the common amounts are expressed in ecus. The agricultural conversion rates of the common prices into national currencies are fixed by the Commission.

— At the time of changeover from the u.a. to the ecu, on 9 April 1979, the common agricultural prices and amounts expressed in u.a. and converted into ecus were adjusted by a coefficient of 1,208953. Conversely, the green rates were adjusted by a reciprocal coefficient of 1/1,208953, leaving actual price levels unchanged. For example, 100 u.a. × 3,40 = DM 340 becomes ECU 121 × 2,81 = DM 340.

— For the recording of world market prices, offer prices are converted at the representative market rate.

According to context, different currency units have been used in this publication. The statistical series in terms of value are also calculated:

— at constant exchange rates, i.e. at the exchange rates obtaining during a specific period (e.g. 1980). These rates are used to eliminate the influences of exchange-rate changes on a time series;

— at current exchange rates (notably for external trade).

To assist the user of this publication wishing to convert units of account into national currencies and conversely, Tables 1.0.1, 1.0.2 and 1.0.3 give the rates to be used. Fuller information is given in specialized publications of the Commission of the European Communities.

Observations on statistical method

A — Statistics on external trade — explanatory note

Council Regulation (EEC) No 1736/75, of 24 June 1975, on the external trade statistics of the Community and statistics of trade between Member States, includes provisions to ensure that data are not recorded twice:

(i) when goods from a non-member country are first brought into a Member State, that Member State must record the import according to the origin of the goods;

(ii) if the goods are then subject to a legal operation (for example clearance for consumption) and subsequently imported into another Member State, the latter must record the goods according to the Member State from which they were received.

However, to satisfy national requirements, the Member States may, if they wish, operate in parallel with the above system the arrangements they applied previously; this means that a Member State's national data may be substantially different from the data supplied by Community sources.

For the calculation of the intra-Community trade of the Community as a whole in the supply balances, there were two possibilities: the sum of the Member States' intra-Community exports (calculation on the basis of goods leaving) or the sum of the Member States' intra-Community imports (calculation on the basis of entries). Eurostat has chosen the second alternative. Also, exports to non-member countries in the supply balances of the Community as a whole are calculated by deducting intra-Community trade from Member States' total exports.

As a result, there may be discrepancies between the external trade data given in the supply balances and those given in the specific external trade tables.

Users must also allow for a break in the series of Community external trade statistics in 1977, the date on which Regulation (EEC) No 1736/75 entered into force.

A last point is that, while the data relating to the external trade of the Community of Twelve from reference year 1985 use the same source for all the Member States (Community statistics), those which refer to a previous period may have been obtained from the Community statistics for the Community of Ten and from other sources for the new Member States.

B — Annual rate of change (% TAV)

1. The annual rate of change (symbol: % TAV) is used throughout this report for the calculation over periods of time of changes in a given aggregate. It measures the compound annual average increase or reduction, as a percentage, of the variable concerned from a base year (T in the following equations).

2. The annual rate of change is calculated as follows:

$$100 \times \text{Anti-log} \left[\log \left(\frac{\text{statistic for year } T+N}{\text{statistic for year } T} \right) \div N \right] - 100 = \% \text{ TAV}$$

Where the annual rate of change is calculated over only two successive years, N = 1 and the formula becomes:

$$100 \times \left[\frac{\text{statistic for year } T+1}{\text{statistic for year } T} \right] - 100 = \% \text{ TAV}$$

3. The following series illustrates the use of this formula:

	1970	1971	....	1975	1976
Series =	100 000	112 000		161 051	177 156
		$\dfrac{1971}{1970}$		$\dfrac{1975}{1970}$	$\dfrac{1976}{1975}$
% TAV		12,0 %		10,0 %	10,0 %

Most of the statistics in the tables have been provided by the Statistical Office of the European Communities (Eurostat). For longer and more detailed series, the user should refer to the following Eurostat publications:

Classification of Eurostat publications

Theme

1. General statistics (midnight blue)
2. Economy and finance (violet)
3. Population and social conditions (yellow)
4. Energy and industry (blue)
5. Agriculture, forestry and fisheries (green)
6. Foreign trade (red)
7. Services and transport (orange)
9. Miscellaneous (brown)

Series

A Yearbooks
B Short-term trends
C Accounts, surveys and statistics
D Studies and analyses
E Methods
F Rapid reports

1.0.1 Indicative currency parities (¹)

(ECU 1 = ... NC)

Since	Belgique/Belgïe Luxembourg BFR/LFR	Danmark DKR	BR Deutschland DM	Elláda DR	España PTA	France FF	Ireland IRL	Italia LIT	Nederland HFL	Portugal ESC	United Kingdom UKL
1	2	3	4	5	6	7	8	9	10	11	12
Central rates (¹)											
21.3.1983	44,3662	8,04412	2,21515	...	...	6,79271	0,717050	1386,78	2,49587	...	...
18.5.1983	44,9008	8,14104	2,24184	...	...	6,87456	0,725690	1403,49	2,52595	...	...
22.7.1985	44,8320	8,12857	2,23840	...	...	6,86402	0,724578	1520,60	2,52208	...	...
7.4.1986	43,6761	7,91896	2,13834	...	...	6,96280	0,712956	1496,21	2,40935	...	...
4.8.1986	43,1139	7,81701	2,11083	...	...	6,87316	0,764976	1476,95	2,37833	...	...
12.1.1987	42,4582	7,85212	2,05853	...	...	6,90403	0,768411	1483,58	2,31943	...	...
19.6.1989	–	–	–	...	133,804	–	–	–	–	...	...
8.1.1990	42,1679	7,79845	2,04446	...	132,889	6,85684	0,763159	1529,70	2,30358	...	...
8.10.1990	42,4032	7,84195	2,05586	...	133,631	6,89509	0,767417	1538,24	2,31643	...	0,696904
6.4.1992	42,0639	7,77921	2,03942	...	132,562	6,83992	0,761276	–	2,29789	178,735	0,691328
14.9.1992	41,9547	7,75901	2,03412	...	139,176	6,82216	0,759300	1636,61	2,29193	177,305	...
17.9.1992	40,6304	7,51410	1,96992	...	143,386	6,60683	0,735334	–	2,21958	176,844	...
23.11.1992	40,2802	7,44934	1,95294	...	142,150	6,54988	0,809996	–	2,20045	182,194	...
30.1.1993	40,2123	7,43679	1,94964	...	154,250	6,53883	0,808628	–	2,19672	180,624	...
14.5.1993	–	–	–	...	–	–	–	–	–	192,854	...
Green central rates (²)											
21.3.1983	44,3662	8,04412	2,21515	...	...	6,79271	0,717050		2,49587	...	...
18.5.1983	44,9008	8,14104	2,24184	...	...	6,87456	0,725690		2,52595	...	...
Marketing year 1984/85	46,4118	8,41499	2,31728	...	...	7,10590	0,750110		2,61095	...	...
12.7.1985		8,41501	–	...	...		0,750111		2,61096	...	...
7.4.1986	47,3310	8,58163	–	...	...	7,54546	0,772618		2,61097	...	...
4.8.1986	47,3307	8,58155	–	...	...	7,54539	0,839794		2,61094	...	...
12.1.1987	47,7950	8,83910	–	...	...	7,77184	0,864997		2,61097	...	...
Marketing year 1987/88	48,2869	8,93007	2,34113	...	...	7,85183	0,873900		2,63785	...	...
8.1.1990	48,2868	8,93008	...	...	...				–	...	...
8.10.1990	48,5563	8,97989	2,35418	...	...	7,89563	0,878776	1751,67	–	...	...
14.9.1992	–	–	–	...	...	–	–	1761,45	–	...	...
17.9.1992	–	–	–	...	...	–	–	1889,20	–	...	...
30.1.1993	–	–	–	...	...	–	0,976413	–	–	...	...

Source: EC Commission, Directorate-General for Agriculture.

(¹) Currencies within the exchange-rate mechanism of the European Monetary System.

(²) Representative market rate of 'fixed currencies'.

1.0.2   Conversion rates ([1])

1	2	1985	1990	1991	1992	1993	% TAV	
							$\dfrac{1992}{1985}$	$\dfrac{1993}{1992}$
		3	4	5	6	7	8	9
Belgique/België	BFR 1 000 = ECU ...	22,265	23,571	23,684	24,042	24,709	1,1	2,8
Danmark	DKR 1 000 = ECU ...	124,707	127,283	126,445	128,053	131,690	0,4	2,8
BR Deutschland	DM 1 000 = ECU ...	449,172	487,308	487,624	494,974	516,425	1,4	4,3
Elláda	DR 1 000 = ECU ...	9,457	4,965	4,440	4,048	3,723	− 11,4	− 8,0
España	PTA 1 000 = ECU ...	7,742	7,727	7,784	7,546	6,706	− 0,4	− 11,1
France	FF 1 000 = ECU ...	147,166	144,632	143,404	146,020	157,808	− 0,1	8,1
Ireland	IRL 1 000 = ECU ...	1 398,270	1 302,477	1 302,407	1 314,548	1 250,075	− 0,9	− 4,9
Italia	LIT 1 000 = ECU ...	0,691	0,657	0,652	0,627	0,543	− 1,4	− 13,4
Luxembourg	LFR 1 000 = ECU ...	22,265	23,571	23,684	24,042	24,709	1,1	2,8
Nederland	HFL 1 000 = ECU ...	398,246	432,504	432,717	439,595	459,726	1,4	4,6
Portugal	ESC 1 000 = ECU ...	7,677	5,522	5,599	5,724	5,309	− 4,1	− 7,3
United Kingdom	UKL 1 000 = ECU ...	1 697,860	1 400,853	1 426,509	1 355,656	1 282,071	− 3,2	− 5,4
USA	USD 1 000 = ECU ...	1 310,460	785,281	806,998	770,357	853,971	− 7,3	10,9

Source: EC Commission, Directorate-General for Agriculture.
([1]) Annual average of daily rates.

1.0.3   Agricultural conversion rates

(1 ECU = ... NC)

Since	Belgique/België Luxembourg BFR/LFR	Danmark DKR	BR Deutschland DM	Elláda DR	España PTA	France FF	Ireland IRL	Italia LIT	Nederland HFL	Portugal ESC	United Kingdom UKL
1	2	3	4	5	6	7	8	9	10	11	12
1.1.1993	48,5563	8,97989	2,35418	310,351	166,075	7,89563	0,878776	2 087,00	2,65256	209,523	0,939052
11.1.1993	..	..	..	..	..	..	..	2 133,00	..	..	0,951031
1.2.1993	..	..	..	..	..	..	..	..	..	..	..
3.2.1993	..	..	..	..	..	..	..	..	..	..	0,968391
1.3.1993	..	..	..	..	..	..	0,957268	2 156,72	..	212,128	0,980715
31.3.1993	..	..	..	..	..	..	..	2 207,67	..	..	..
1.4.1993	..	..	..	..	..	..	..	2 267,06	..	214,525	..
11.4.1993	..	..	..	314,4120	..	..	..	2 287,88	..	..	0,978559
21.4.1993	..	..	..	..	166,261	..	..	..	..	..	0,970726
27.4.1993	..	..	..	..	169,628	..	..	2 264,05	..	..	0,964017
1.5.1993	..	..	..	..	..	..	..	2 230,20	..	..	..
11.5.1993	..	..	..	..	..	..	..	2 226,76	..	..	..
18.5.1993	..	..	..	..	176,247	..	..	2 195,05	..	222,758	..
21.5.1993	..	..	..	..	176,451	..	..	2 194,16	..	..	..
28.5.1993	..	..	..	..	179,488	..	..	..	..	..	..
1.6.1993	..	..	..	..	182,744	..	..	..	..	..	0,959111
21.6.1993	..	..	..	315,843	..	..	..	2 191,70	..	..	..
1.7.1993	..	..	..	319,060	..	..	0,976426	2 166,58	..	..	..
21.7.1993	..	..	..	..	..	..	..	..	..	223,071	0,948645
24.7.1993	..	..	..	..	..	..	..	..	..	228,151	0,937041
30.7.1993	..	..	..	..	186,835	..	..	..	..	233,112	0,930787
1.8.1993	..	..	..	..	190,382	..	..	..	..	236,933	0,920969
11.8.1993	..	9,14292	..	..	..	7,95622	..	..	..	..	..
17.8.1993	..	9,34812	..	..	..	7,98191	..	..	..	..	..
21.8.1993	..	..	..	..	..	..	..	..	..	..	..
1.9.1993	..	..	..	322,728	..	..	..	..	..	..	..
14.10.1993	49,3070	..	..	328,567	..	..	..	2 222,98	..	..	..

30.12.1993	..	..	..	..	..	..	..	2 264,19	..	..	..
1.1.1994	..	..	..	331,890	..	..	..	..	..	..	..
11.1.1994	..	..	..	..	192,319	..	..	..	..	..	..
				334,226				2 274,93			
21.1.1994	..	..	..	337,814	..	..	..	..	..	..	..
1.4.1994	..	..	..	342,048	..	..	..	..	..	..	..
21.5.1994	..	..	..	..	..	..	..	..	..	239,331	..
1.6.1994	..	..	..	346,789	..	..	..	..	..	..	..
21.6.1994	..	..	..	..	..	..	..	..	..		0,932453
1.7.1994	..	..	..	..	..	..	..	..	..		0,946550
14.7.1994	..	..	..	..	..	..	..	2 294,57	..	..	..
1.8.1994	..	..	..	..	..	..	..	2 324,07	..	..	..
16.8.1994	..	..	..	..	..	..	..	2 339,97	..		0,953575
21.8.1994	..	..	..	349,469	..	..	..	..	..	..	..
21.9.1994	..	..	..								
11.10.1994	49,3070	9,34812	2,35418	349,469	192,319	7,98191	0,976426	2 339,97	2,65256	239,331	0,953575

Source: EC Commission, Directorate-General for Agriculture.

2.0.1   Basic data — key figures for agriculture in the EC

Features	Year	EUR 12	Belgique/Belgïe	Danmark
1	2	3	4	5
Total area (km²)	1993	2 368 188∞	30 518	43 093
Population (1 000 inhabitants)	1993	346 757∞	10 068	5 181
GDP/inhabitants (purchasing power standard — PPS)	1993	15 832∞	17 946	17 815
Inflation ([1])	1993	3,7*∞	4,4*	1,7
Unemployment rate (% of civilian working population)	1993	10,7∞	9,4	10,3
Trade balance (Mio ECU)	1993	54 211*∞	3 963 ([2])	6 868
Utilized agricultural area (1 000 ha)	1993	128 687*∞	1 412	2 976 (
Employment in the agriculture, forestry, hunting and fishing sector – number (1 000 persons) – share in employed civilian working population (%)	1993 1993	7 072* 5,5	95 ([6]) 2,5 ([6])	140 5,4
Number of holdings (1 000 holdings)	1989	8 171	85	81
UAA per holding (ha)	1989	14,0	15,8	34,2
Final production of agriculture (Mio ECU)	1992	197 990	6 559	6 459
Share of agriculture in the GDP (GVA/GDP) (%)	1992	2,5	1,8	3,0
Share of agriculture in total gross fixed capital formation (%)	1992	2,6*	1,3	3,9*
Share of imports of food and agricultural products in import of all products (%)	1993	11,2∞	13,0 ([2])	16,7
Share of exports of food and agricultural products in exports of all products (%)	1993	8,7∞	11,9 ([2])	29,4
External trade balance in food and agricultural products (Mio ECU)	1993	– 12 569∞	288 ([2])	4 728
Share of household consumption expenditure devoted to food, beverages and tobacco as proportion of total consumer expenditure of households (%)	1991	19,4*	18,7	21,0

Source: Eurostat and EC Commission, Directorate-General for Agriculture.

([1]) GDP price deflator.
([2]) BLEU/UEBL.
([3]) 1988.
([4]) 1989.
([5]) 1991.
([6]) 1992.

BR Deutschland	Elláda	España	France	Ireland	Italia	Luxembourg	Nederland	Portugal	United Kingdom
6	7	8	9	10	11	12	13	14	15
356 970∞	131 957	504 795	549 086	70 286	301 311	2 568	41 480	91 986	244 138
80 614∞	10 320	39 114	57 527	3 557	56 933	395	15 239	9 850	57 959
17 147∞	9 999	12 330	17 434	12 833	16 228	25 422	16 308	10 934	15 690
3,9∞	13,6*	4,4	2,3	3,6*	4,4	2,1*	1,6	7,4*	3,4
7,2∞	8,2*	21,8	10,8	18,4	11,1	2,6	8,8	5,1	10,4
38 070∞	− 9 017	− 13 702	6 027	6 031	27 767	3 963 (²)	11 226	− 5 854	− 17 168
17 162∞	5 741 (³)	26 389 (⁵)	30 217	4 444 (⁵)	17 215 (⁴)	127	1 997*	3 829	17 178
849	794	1 198	1 094	144	1 508	6,0 (⁶)	305 (⁶)	514	516
3,0	21,3	10,1	5,0	12,7	7,4	3,0 (⁶)	4,6 (⁶)	11,6	2,1
665	924	1 594	1 017	171 (⁵)	2 665	4	125	599	243
17,7	4,0	15,4	28,1	26,0	5,6	32,1	16,1	6,7	67,9
27 550	8 520	23 813	44 905	4 420	37 793	189	16 012	3 554	18 216
1,2	14,2	3,0	2,9	8,9	3,6	1,8	3,6*	3,2*	1,4
1,6	6,7	:	1,9	9,3	5,3	2,4	4,7	2,2	1,2
11,7∞	15,5	15,7	11,9	11,4	16,8	16,7 (²)	16,0	16,0	11,6
6,1∞	33,1	16,6	16,0	22,9	7,4	11,9 (²)	25,4	8,2	7,8
− 11 595∞	− 554	− 1 252	8 298	3 623	− 10 621	288 (²)	11 595	− 2 222	− 8 712
16,1	37,3*	21,0	18,3	34,8	20,3	18,6	15,1	33,1*	21,8

3.1.1   Shares of individual products in final agricultural production (1993)

1	EUR 12 ([3])	Belgique/België	Danmark	BR Deutschland ([2])
	2	3	4	5
Products subject to EEC market organizations				
Wheat	6,2	2,7	8,4	5,0
Rye	0,1	0,0	0,7	0,9
Oats	0,1	0,0	0,2	0,0
Barley	1,8	0,7	3,2	2,4
Maize	2,0	0,0	0,0	0,5
Rice	0,4	0,0	0,0	0,0
Sugarbeet	2,4	4,8	2,4	3,7
Tobacco	0,7	0,1	0,0	0,1
Olive oil	1,8	0,0	0,0	0,0
Oilseeds	0,9	0,1	1,4	0,8
Fresh fruit ([1])	4,7	3,9	0,5	7,4
Fresh vegtables	9,5	13,6	1,8	3,4
Other fruit and vegetables ([2])	1,9	0,1	0,9	0,1
Wine and must	5,7	0,0	0,0	3,8
of which : quality wine	0,0	0,0	0,0	0,0
Seeds	0,6	0,1	0,8	0,5
Textile fibres	0,5	0,1	0,0	0,0
Hops	0,1	0,1	0,0	0,4
Milk	16,6	14,4	25,0	24,0
Beef/veal	12,0	20,0	8,2	14,2
Pigmeat	11,7	19,3	30,9	17,8
Sheepmeat and goatmeat	2,1	0,2	0,1	0,4
Eggs	2,4	3,0	1,2	2,9
Poultry	4,8	3,6	2,5	2,3
Silkworms	0,0	0,0	0,0	0,0
Subtotal	89,0	86,8	88,2	90,6
Products not subject to EEC market organizations				
Potatoes	1,8	3,4	1,4	1,7
Other	9,2	9,8	10,0	7,7
Subtotal	11,0	13,2	11,4	9,4
Grand total	100,0	100,0	100,0	100,0
Value in Mio ECU	197 563	6 357	6 058	26 215

Source : Eurostat — Agricultural accounts and EC Commission, Directorate-General for Agriculture.

([1]) These are products listed in Annex II of Regulation (EEC) No 1035/72.
([2]) Dried pulses, citrus fruit.
([3]) 1992.

(%)

Elláda	España	France	Ireland	Italia	Luxembourg	Nederland	Portugal (³)	United Kingdom
6	7	8	9	10	11	12	13	14
4,3	3,8	7,6	1,4	4,8	2,0	0,8	1,6	10,8
0,0	0,1	0,0	0,0	0,0	0,1	0,0	0,3	0,0
0,0	0,1	0,1	0,2	0,1	0,3	0,0	0,1	0,3
0,3	3,3	2,0	2,1	0,4	2,2	0,2	0,1	3,6
3,0	0,6	4,7	0,0	3,4	0,0	0,1	2,1	0,0
0,6	0,5	0,1	0,0	1,4	0,0	0,0	1,0	0,0
1,5	2,1	2,9	1,7	1,6	0,0	1,9	0,0	2,1
5,3	0,5	0,2	0,0	1,0	0,0	0,0	0,5	0,0
7,9	6,1	0,0	0,0	4,3	0,0	0,0	4,5	0,0
0,1	1,2	1,9	0,0	0,5	0,4	0,0	0,2	1,4
7,5	6,7	3,3	0,2	4,8	1,2	1,5	5,0	2,1
12,8	15,8	7,1	2,7	14,3	1,0	10,8	9,9	7,6
2,5	4,4	1,2	0,0	3,1	0,0	0,1	2,5	0,7
2,2	2,7	11,4	0,0	7,8	9,4	0,0	5,8	0,0
0,0	0,0	8,3	0,0	0,0	0,0	0,0	0,0	0,0
0,1	0,1	1,7	0,0	0,0	0,0	2,1	0,0	0,3
12,2	0,3	0,1	0,0	0,0	0,0	0,0	0,0	0,0
0,0	0,0	0,0	0,0	0,0	0,0	0,0	0,0	0,1
11,2	7,6	17,9	33,2	11,2	45,5	23,0	13,7	23,9
3,3	7,4	15,0	39,1	9,8	27,1	10,7	9,7	14,6
3,0	11,3	6,4	5,9	6,8	7,4	13,8	13,9	7,4
6,5	4,8	1,3	5,0	0,7	0,0	0,5	3,4	5,4
2,3	3,1	1,7	0,7	2,5	1,0	3,3	3,2	3,5
2,9	4,5	6,7	3,0	5,8	0,1	3,8	8,6	7,3
0,0	0,0	0,0	0,0	0,0	0,0	0,0	0,0	0,0
89,7	87,0	93,3	95,2	84,3	97,7	72,6	86,1	91,1
2,6	1,9	1,3	1,6	1,2	1,2	1,9	4,7	2,9
7,7	11,1	5,4	3,2	14,5	1,1	25,5	9,2	6,0
10,3	13,0	6,7	4,8	15,7	2,3	27,4	13,9	8,9
100,0	100,0	100,0	100,0	100,0	100,0	100,0	100,0	100,0
8 486	17 003	43 997	4 419	31 272	196	15 222	3 934	18 002

3.1.2   Individual Member States' shares in final agricultural production (1992)

	Belgique/België	Danmark	BR Deutschland	Elláda
1	2	3	4	5
Products subject to EEC market organizations				
Wheat	1,8	4,0	11,2	3,5
Rye	0,5	14,7	83,0	1,0
Oats	1,4	3,8	0,9	1,6
Barley	1,5	5,8	18,4	0,7
Maize	0,0	0,0	3,4	6,8
Rice	0,0	0,0	0,0	4,7
Sugarbeet	6,2	2,7	21,9	2,9
Tobacco	0,2	0,0	1,9	46,7
Olive oil	0,0	0,0	0,0	21,8
Oilseeds	0,2	4,0	12,5	0,7
Fresh fruit ([1])	2,6	0,3	22,2	6,4
Fresh vegetables	4,6	0,6	5,0	5,8
Wine and must	0,0	0,0	9,4	1,6
Quality wine	–	–	–	–
Seeds	0,8	3,9	11,8	0,9
Textile fibres	0,5	0,0	0,0	77,6
Hops	2,7	0,0	73,7	0,0
Milk	2,8	4,7	20,2	2,5
Beef/veal	5,1	2,3	16,6	1,2
Pigmeat	7,1	9,8	21,2	1,3
Sheepmeat and goatmeat	0,2	0,1	2,5	13,5
Silkworms	0,0	0,0	0,0	0,0
Subtotal	3,5	3,5	14,9	4,4
Eggs	3,6	1,3	16,7	4,6
Poultry	2,3	1,6	6,7	2,8
Other fruit and vegetables ([2])	0,1	1,9	0,5	6,6
Subtotal	2,2	1,6	8,1	4,1
Products not subject to EEC market organizations				
Potatoes	4,1	3,5	13,2	4,2
Other	3,0	3,1	11,4	3,8
Subtotal	3,1	3,2	11,7	3,9
Grand total	3,3	3,3	13,9	4,3

Source: Eurostat — Agricultural accounts and EC Commission, Directorate-General for Agriculture.

([1]) These are products listed in Annex II of Regulation (EEC) No 1035/72.
([2]) Dried pulses, citrus fruit.

EUR 12 = 100

(%)

España	France	Ireland	Italia	Luxembourg	Nederland	Portugal	United Kingdom
6	7	8	9	10	11	12	13
6,5	37,9	0,8	14,8	0,0	1,3	0,5	17,7
4,8	4,8	0,0	0,6	0,1	1,4	4,1	1,4
10,0	23,5	5,7	19,0	0,3	1,0	2,4	27,9
11,1	35,3	3,8	4,2	0,1	0,8	0,1	22,3
6,9	67,0	0,0	28,4	0,0	0,1	1,9	0,0
23,9	5,4	0,0	63,3	0,0	0,0	4,8	0,0
8,6	25,4	1,6	16,1	0,0	6,9	0,0	9,5
8,8	6,6	0,0	28,7	0,0	0,0	1,3	0,0
35,9	0,0	0,0	36,7	0,0	0,0	4,6	0,0
8,5	35,9	0,0	12,5	0,0	0,2	1,6	14,8
18,8	13,8	0,1	31,1	0,0	2,1	1,9	4,3
20,2	15,1	0,6	31,2	0,0	0,1	0,2	7,2
8,0	49,6	0,0	31,8	0,2	0,0	1,8	0,0
–	100,0	–	–	–	–	–	–
1,9	57,6	0,0	0,0	0,0	28,0	0,0	5,4
19,6	1,9	0,0	0,0	0,0	0,3	0,0	0,0
5,7	3,7	0,0	0,0	0,0	0,0	0,0	15,7
5,4	22,4	4,3	12,8	0,2	10,7	1,5	12,4
6,8	27,3	7,0	14,0	0,2	7,2	1,5	10,8
12,6	13,8	1,2	11,6	0,1	12,3	2,1	6,4
29,2	13,0	4,9	6,3	0,0	2,0	2,9	21,3
0,0	0,0	0,0	0,0	0,0	0,0	0,0	0,0
11,3	26,9	2,6	18,4	0,0	5,8	1,4	9,2
15,4	14,9	0,6	18,0	0,0	10,4	2,4	12,1
11,5	31,1	1,6	21,0	0,0	6,2	3,2	13,4
28,5	21,6	0,0	35,3	0,0	0,0	1,2	4,8
16,1	24,8	1,0	23,2	0,0	6,0	2,6	11,3
13,7	11,0	2,0	14,9	0,1	11,9	4,6	17,3
14,6	12,0	0,8	29,6	0,0	28,8	3,5	5,5
14,5	11,8	1,0	27,2	0,1	26,0	3,7	7,5
12,1	22,7	2,2	19,1	0,1	8,1	1,8	9,2

3.1.3   Farm inputs: breakdown by Member State (1993)

(%)

1	Consumption of inputs Mrd ECU	Seeds and reproductive material	Animal feed	Fertilizers and soil improvers	Crop protection products	Pharmaceuticals	Energy and lubricants	Cattle	Farm implements, upkeep, repairs	Services	Other	VAT under-compensation	Share of inputs in production
	2	3	4	5	6	7	8	9	10	11	12	13	14
EUR 12 (^{1})	87 948	5,6	39,6	9,0	6,1	1,0	10,5	1,2	12,5	10,8	3,4	0,4	44,4
Belgique/België	3 924	5,9	43,8	5,5	4,4	1,8	7,4	3,8	9,2	7,7	10,3	0,0	58,8
Danmark	3 412	3,5	47,2	7,7	4,4	0,0	6,2	0,0	15,9	15,0	0,0	0,0	54,4
BR Deutschland (^{1})	14 576	5,9	30,3	7,8	4,6	0,0	16,0	1,2	17,4	15,3	1,5	0,0	52,4
Elláda	2 278	3,4	26,8	7,4	6,8	2,1	30,4	1,5	13,2	2,4	6,1	0,0	27,7
España	10 381	3,5	43,3	7,1	3,4	2,2	8,5	1,2	18,8	4,7	7,3	0,0	48,0
France	21 021	10,9	33,7	11,2	9,9	1,5	8,2	0,5	9,5	12,8	1,9	0,0	49,5
Ireland	1 801	2,9	40,7	16,8	2,8	4,4	11,7	0,5	8,0	6,9	5,3	0,0	41,7
Italia	9 479	4,7	50,6	9,7	6,4	0,2	14,2	0,0	0,0	8,8	5,5	0,0	29,0
Luxembourg	81	4,0	26,5	15,5	3,5	1,8	9,5	3,5	11,7	0,0	24,1	0,0	43,6
Nederland	8 059	6,6	46,4	3,4	2,2	0,0	11,0	2,2	13,4	13,8	1,1	0,0	50,4
Portugal (^{1})	1 864	0,0	48,3	0,0	16,9	0,0	10,6	9,6	3,6	5,5	5,5	0,0	52,4
United Kingdom	9 562	3,9	41,5	8,1	6,3	1,3	6,7	0,5	18,4	11,5	1,8	0,0	55,4

Source: Eurostat — Agricultural accounts and EC Commission, Directorate-General for Agriculture.
(^{1}) 1992.

3.1.4 Situation of the (1)
(a) final agricultural production
(b) consumption of inputs
(c) gross value-added of agriculture
(d) net value-added at factor cost

1993

1	2	NC (Mio)		% TAV on the basis of data in national currencies at 1990 prices		At current prices and rates of exchange			
		At current prices	At 1990 prices	1993/1985	1993/1992	PPS Mio	ECU Mio	As % of aggregate (EUR 12 = 100)	As % of final production by MS
		3	4	5	6	7	8	9	10
Final production	EUR 12 (4)	—	—	0,3	− 1,8	:	200 303	100,0	100,0
	Belgique/België	270 015	301 184	2,8	2,6	:	6 672	4,3	100,0
	Danmark	47 662	55 304	1,2	6,6	:	6 277	4,0	100,0
	BR Deutschland (4)	55 661	57 862	0,5	2,0	:	27 551	13 8	100,0
	Elláda	2 208 299	1 692 816	0,2	− 1,3	:	8 222	5,3	100,0
	España	3 228 100	3 337 351	0,4	− 1,2	:	21 647	13,9	100,0
	France	281 522	327 825	0,7	− 4,9	:	42 538	27,3	100,0
	Ireland	3 457	3 289	1,4	− 3,1	:	4 322	2,8	100,0
	Italia (2)	60 272	58 800	0,6	− 2,9	:	32 735	21,0	100,0
	Luxembourg	7 498	8 313	0,1	− 4,9	:	185	0,1	100,0
	Nederland	34 760	38 523	2,1	1,9	:	15 980	10,3	100,0
	Portugal (4)	620 855	661 940	1,5	− 4,8	:	3 554	1,8	100,0
	United Kingdom	13 450	13 079	3,1	− 3,0	:	17 244	11,1	100,0
Consumption of inputs	EUR 12 (4)	88 217	85 512	0,03	− 1,0	:	87 951	100,0	49,9
	Belgique/België	158 793	160 079	2,8	1,5	:	3 924	5,6	58,8
	Danmark	25 910	26 125	0,6	0,3	:	3 412	4,9	54,4
	BR Deutschland (4)	29 182	28 492	− 0,8	− 2,5	:	14 444	16,4	52,4
	Elláda	611 881	415 879	1,2	4,1	:	2 278	3,3	27,7
	España	3 228 100	1 506 193 (4)	1,9	2,6	:	21 647	14,8	48,0
	France	139 447	141 470	0,9	− 0,4	:	21 021	30,0	49,5
	Ireland	1 441	1 443	2,0	4,2	:	1 801	2,6	41,7
	Italia (2)	17 453	15 929	0,5	− 1,4	:	9 479	13,5	29,0
	Luxembourg	3 272	3 289	2,0	36,6	:	81	0,1	43,6
	Nederland	17 531	17 436	0,8	0,7	:	8 059	11,5	50,4
	Portugal (4)	325 623	312 124	1,1	− 6,7	:	1 864	2,1	52,4
	United Kingdom	7 458	6 753	− 0,4	0,0	:	9 562	13,7	55,4

Gross value-added at market prices

	EUR 12 (4)			−0,2	3,0	112 352	100,0	56,1
	—	—	:					
Belgique/België	111 222	141 105		2,5	4,0	2 748	2,5	41,2
Danmark	21 752	29 179		0,0	13,3	2 865	2,9	45,6
BR Deutschland (4)	26 479	29 370		2,1	6,8	13 106	11,7	47,6
Elláda	1 596 418	1 276 937		1,4	−3,0	5 944	5,6	72,3
España	142 075	1 870 634 (4)		−0,3	−2,2	11 266	10,7	52,0
France	2 016	186 355		−0,1	−7,8	21 417	22,9	50,5
Ireland	4 226	1 846		−0,1	−8,1	2 520	2,3	58,3
Italia	17 229	42 871		0,9	−3,3	23 256	24,9	71,0
Luxembourg	295 232	5 027		−0,1	−21,9	104	0,1	56,4
Nederland	5 992	21 087		1,0	2,9	7 921	7,4	49,6
Portugal (4)	13 436	349 816		0,9	−3,1	1 690	1,5	47,6
United Kingdom		6 326		0,1	−5,9	7 682	7,5	44,6

Net value-added at factor cost (3)

	EUR 12 (4)			2,2	−4,8	110 913	100,0	55,4
	—	—	:					
Belgique/België	98 376	—		0,9	−2,3	2 431	2,2	36,4
Danmark	17 381	—		−1,4	−2,3	2 289	2,1	36,5
BR Deutschland (4)	20 355	—		2,0	4,8	10 075	9,1	36,6
Elláda	1 804 147	—		12,6	8,9	6 718	6,0	81,7
España	2 127 800	—		4,3	16,1	14 269	12,3	65,9
France	171 622	—		1,2	−3,7	25 871	23,4	61,0
Ireland	2 396	—		4,9	2,2	2 995	2,8	69,3
Italia	48 892	—		3,4	−2,0	26 554	28,2	81,1
Luxembourg	3 615	—		−1,0	−5,0	89	0,1	48,2
Nederland	12 143	—		−1,3	−13,5	5 582	5,5	69,3
Portugal	313 173	—		6,6	−7,2	1 663	1,7	54,2
United Kingdom	6 334	—		6,0	16,2	8 121	6,6	47,1

Source: Eurostat — Agricultural accounts.

(1) The figures are calculated from series according to recording net of VAT.
(2) In thousand million lire.
(3) TAV at current prices.
(4) 1992.

3.1.5   Final agricultural production, crop production and livestock production (^{3})

1	2	1993		% TAV (^{1})	
		Mio NC	Mio ECU	$\dfrac{1993}{1985}$	$\dfrac{1993}{1992}$
1	2	3	4	5	6
Final production	EUR 12 (^{4})	–	200 303	0,3	– 5,9
	Belgique/België	270 015	6 672	2,8	2,6
	Danmark	47 662	6 277	1,2	6,6
	BR Deutschland (^{4})	55 661	27 551	0,5	2,0
	Elláda	2 208 299	8 222	0,2	– 1,3
	España	3 228 100	21 647	0,4	– 1,2
	France	281 522	42 438	0,7	– 4,9
	Ireland	3 457	4 322	1,4	– 3,1
	Italia	60 272 (^{2})	32 735	0,6	– 2,9
	Luxembourg	7 498	185	0,1	– 4,7
	Nederland	34 760	15 980	2,1	1,9
	Portugal (^{4})	620 855	3 554	1,5	– 4,8
	United Kingdom	13 450	17 244	3,1	– 3,0
Crop production	EUR 12 (^{4})	–	96 131	0,6	2,5
	Belgique/België	100 832	2 491	3,2	– 0,3
	Danmark	13 977	1 841	1,0	17,1
	BR Deutschland (^{4})	20 883	10 337	3,1	11,1
	Elláda	1 537 294	5 724	0,1	– 2,6
	España	185 900	1 247	– 0,5	– 3,0
	France	139 356	21 007	0,9	– 7,4
	Ireland	382	478	0,3	– 18,2
	Italia	35 645 (^{2})	19 359	0,9	– 4,2
	Luxembourg	1 373	34	2,7	– 23,9
	Nederland	15 286	7 027	5,4	2,8
	Portugal (^{4})	257 753	1 475	– 0,5	– 11,5
	United Kingdom	4 922	6 310	3,1	– 4,5
Livestock production	EUR 12 (^{4})	–	100 917	0,4	1,3
	Belgique/België	168 465	4 163	2,5	4,6
	Danmark	33 685	4 436	1,3	2,9
	BR Deutschland (^{4})	34 665	17 158	– 1,0	– 3,4
	Elláda	670 709	2 497	0,6	1,8
	España	1 318 600	8 842	1,3	1,5
	France	142 390	21 465	0,4	– 1,3
	Ireland	3 074	3 843	1,6	– 0,8
	Italia	23 611 (^{2})	12 823	0,2	– 0,4
	Luxembourg	6 103	151	– 0,4	1,4
	Nederland	19 474	8 953	0,0	– 0,2
	Portugal (^{4})	340 888	1 951	6,5	2,1
	United Kingdom	8 528	10 934	3,1	– 1,8

3.1.5 *(cont.)*

		1993		% TAV ([1])	
		Mio NC	Mio ECU	$\dfrac{1993}{1985}$	$\dfrac{1993}{1992}$
1	2	3	4	5	6
A — Cereals (excl. rice)	EUR 12 ([4])	–	20 509	– 0,8	– 8,0
	Belgique/België	9 529	235	1,1	4,4
	Danmark	5 979	787	1,7	36,6
	BR Deutschland ([4])	4 946	2 448	– 3,0	– 18,8
	Elláda	168 392	627	3,0	– 4,8
	España	258 600	1 734	– 7,0	6,0
	France	41 234	6 216	0,6	– 8,5
	Ireland	131	164	– 0,6	– 27,6
	Italia	5 375 ([2])	2 919	1,8	1,1
	Luxembourg	365	9	2,6	– 5,3
	Nederland	409	188	4,5	9,3
	Portugal ([4])	28 539	163	– 6,3	– 35,8
	United Kingdom	1 986	2 546	– 2,3	– 11,2
B — Beef/veal, total	EUR 12 ([4])	–	23 623	1,4	– 0,5
	Belgique/België	54 074	1 336	2,1	4,0
	Danmark	3 916	516	– 2,1	– 7,0
	BR Deutschland ([4])	7 919	3 920	– 0,5	– 10,1
	Elláda	72 067	268	– 1,7	– 5,0
	España	237 500	1 593	2,1	– 2,4
	France	42 254	6 370	– 1,1	– 7,8
	Ireland	1 351	1 689	2,3	– 1,4
	Italia	5 881 ([2])	3 194	– 1,5	0,1
	Luxembourg	2 030	50	0,6	– 3,2
	Nederland	3 731	1 715	0,9	– 1,2
	Portugal ([4])	60 201	345	– 0,4	– 10,2
	United Kingdom	1 959	2 512	0,0	– 9,8
C — Milk	EUR 12 ([4])	–	32 798	– 1,9	– 2,0
	Belgique/België	38 926	962	– 1,1	2,0
	Danmark	11 655	1 535	– 1,3	0,2
	BR Deutschland ([4])	13 375	6 620	– 1,9	– 2,6
	Elláda	247 628	922	0,7	2,8
	España	246 400	1 652	– 1,6	– 3,5
	France	50 439	7 603	– 0,9	0,7
	Ireland	1 146	1 433	– 1,0	– 0,5
	Italia	6 766 ([2])	3 675	– 0,1	– 1,5
	Luxembourg	3 410	84	– 1,3	1,9
	Nederland	8 006	3 681	– 2,0	0,5
	Portugal ([4])	84 787	485	4,2	– 2,5
	United Kingdom	3 212	4 118	– 1,1	0,5

3.1.5 *(cont.)*

		1993		% TAV (¹)	
		Mio NC	Mio ECU	$\dfrac{1993}{1985}$	$\dfrac{1993}{1992}$
1	2	3	4	5	6
D — Pigmeat	EUR 12 (⁴)	–	23 145	1,7	2,5
	Belgique/België	51 997	1 285	5,0	5,9
	Danmark	14 715	1 938	4,3	9,3
	BR Deutschland (⁴)	9 926	4 913	0,0	0,0
	Elláda	66 635	248	1,5	1,8
	España	365 200	2 449	3,6	5,8
	France	17 978	2 710	3,9	12,5
	Ireland	204	255	5,7	5,1
	Italia	4 069 (²)	2 210	1,5	0,8
	Luxembourg	553	14	1,6	15,3
	Nederland	4 814	2 213	1,9	6,7
	Portugal (⁴)	86 279	494	1,7	14,9
	United Kingdom	999	1 281	0,8	3,3
E — Eggs and poultrymeat	EUR 12	–	14 212	0,9	0,5
	Belgique/België	17 723	438	4,5	9,4
	Danmark	1 725	227	3,8	5,1
	BR Deutschland	2 896	1 433	1,7	0,5
	Elláda	116 239	433	1,7	– 2,4
	España	244 800	1 642	0,1	1,1
	France	23 721	3 576	3,9	– 0,7
	Ireland	125	156	4,2	– 2,1
	Italia	5 002 (²)	2 717	1,2	– 0,3
	Luxembourg	76	2	– 0,5	– 2,7
	Nederland	2 453	1 128	– 0,7	1,0
	Portugal	72 948	418	5,7	14,6
	United Kingdom	1 452	1 862	1,5	0,3

Source: Eurostat — Agricultural accounts and EC Commission, Directorate-General for Agriculture.

(¹) The changes are calculated on the basis of series at constant 1990 prices for the 1985 to 1993 period. For the Member States, the changes are calculated on the basis of figures in national currency and for EUR 12 on the basis of figures converted into ecus.
(²) In thousand million lire.
(³) At current prices.
(⁴) 1992.

3.1.6   **Final agricultural production, consumption of inputs and gross value-added** (at market prices – 1990): **changes by volume**

(1985 = 100)

1	2	1989	1990	1991	1992	1993
		3	4	5	6	7
Final production	EUR 12	105,5	105,9	107,4	109,2	106,9
	Belgique/België	111,0	108,1	111,9	124,3	128,0
	Danmark	105,9	109,6	108,9	104,3	111,8
	BR Deutschland	103,2	103,0	103,4	105,0	102,0
	Elláda	107,5	94,5	108,9	106,6	106,6
	España	103,6	107,8	106,7	105,9	104,6
	France	105,9	108,9	106,1	111,4	107,0
	Ireland	104,0	111,7	112,6	117,5	114,9
	Italia	105,2	103,1	108,4	109,0	105,8
	Luxembourg	103,1	102,0	96,4	105,2	104,0
	Nederland	109,2	113,6	115,8	118,8	120,0
	Portugal	109,9	119,0	119,2	112,0	102,4
	United Kingdom	101,3	102,9	103,5	103,2	100,2
Consumption of inputs	EUR 12	105,2	105,0	104,3	105,3	103,8
	Belgique/België	114,0	112,5	113,5	124,8	127,0
	Danmark	98,9	101,9	100,8	108,7	108,3
	BR Deutschland	96,8	94,8	93,2	92,7	88,6
	Elláda	102,4	103,5	104,4	109,3	108,5
	España	108,3	110,7	112,5	114,8	111,1
	France	108,5	109,8	107,2	107,2	107,1
	Ireland	109,3	110,9	112,0	112,2	114,6
	Italia	107,8	106,0	107,7	106,2	105,0
	Luxembourg	111,4	114,4	118,7	118,9	118,2
	Nederland	110,3	110,6	112,4	111,8	112,2
	Portugal	115,5	119,0	118,3	108,6	98,3
	United Kingdom	100,2	97,2	94,7	93,6	96,3
Gross value-added	EUR 12	105,1	107,2	110,0	112,4	109,5
	Belgique/België	106,9	101,9	109,7	123,5	129,3
	Danmark	113,3	117,8	117,5	99,7	115,6
	BR Deutschland	112,1	114,4	117,5	122,1	120,6
	Elláda	109,0	91,8	110,2	105,8	106,0
	España	99,8	105,4	102,0	98,7	99,4
	France	103,6	108,1	105,3	114,9	106,8
	Ireland	99,6	112,4	113,1	122,0	115,1
	Italia	104,1	101,9	108,7	110,2	106,2
	Luxembourg	97,5	93,7	81,5	101,0	94,4
	Nederland	108,0	116,8	119,5	126,5	128,5
	Portugal	104,5	119,0	120,2	115,4	106,5
	United Kingdom	102,7	110,2	114,9	112,1	105,3

Source: Eurostat.

3.1.7   **Evolution of the implicit price index of final production:**
   — value/volume (nominal)
   — value/volume, deflated by GDP deflator (real)

(1985 = 100)

1	2	1989	1990	1991	1992	1993
		3	4	5	6	7
Nominal	EUR 12	109,6	109,7	111,7	106,3	103,7
	Belgique/België	104,6	100,7	99,2	92,1	86,9
	Danmark	97,9	93,0	90,3	90,4	80,9
	BR Deutschland	99,0	93,4	92,1	84,2	78,7
	Elláda	156,7	188,5	226,6	231,6	248,0
	España	116,1	117,1	116,9	109,4	113,3
	France	107,1	105,9	103,1	95,2	88,9
	Ireland	117,3	104,3	100,9	103,4	108,6
	Italia	107,9	112,2	117,5	113,7	112,8
	Luxembourg	110,8	110,8	101,6	99,6	98,8
	Nederland	99,8	94,2	94,3	90,2	84,6
	Portugal	135,9	141,0	142,7	134,8	132,3
	United Kingdom	112,0	113,7	112,6	113,3	116,8
Real	EUR 12	89,7	84,8	80,9	73,7	69,3
	Belgique/België	91,9	85,8	82,3	73,8	67,8
	Danmark	82,9	76,8	72,8	71,5	63,3
	BR Deutschland	90,8	83,1	78,9	68,4	61,8
	Elláda	90,3	89,9	91,9	81,7	77,1
	España	88,3	82,9	77,3	67,9	67,7
	France	92,3	88,5	83,7	75,6	67,7
	Ireland	100,5	90,9	87,0	88,3	90,3
	Italia	83,1	80,3	78,3	72,4	69,0
	Luxembourg	98,1	95,3	84,9	79,6	76,5
	Nederland	97,4	89,8	87,4	81,6	75,2
	Portugal	83,3	73,3	64,8	54,0	49,5
	United Kingdom	90,1	85,9	79,9	77,0	77,1

Source: Eurostat.

3.1.8   **Evolution of the implicit price index of intermediate consumption:**
— **value/volume** (nominal)
— **value/volume, deflated by GDP deflator** (real)

(1985 = 100)

1	2	1989	1990	1991	1992	1993
		3	4	5	6	7
Nominal	EUR 12	102,7	103,8	106,4	106,9	108,3
	Belgique/België	94,5	92,3	92,9	92,1	91,8
	Danmark	100,4	96,7	96,5	94,0	93,7
	BR Deutschland	91,8	91,6	93,4	93,4	93,4
	Elláda	159,4	188,1	227,0	245,4	276,1
	España	108,5	110,0	112,9	114,3	117,5
	France	103,5	101,8	101,7	99,8	98,7
	Ireland	99,9	99,3	99,2	99,4	99,3
	Italia	102,2	105,2	106,6	107,9	115,5
	Luxembourg	94,3	95,8	96,6	96,9	93,2
	Nederland	92,0	90,7	92,1	92,6	89,9
	Portugal	134,4	142,1	148,8	147,8	151,4
	United Kingdom	107,7	111,1	116,3	118,8	123,0
Real	EUR 12	85,4	81,8	79,5	76,5	75,0
	Belgique/België	83,0	78,6	77,1	73,9	71,6
	Danmark	85,0	79,8	77,7	74,3	73,2
	BR Deutschland	84,1	81,4	79,9	75,9	73,3
	Elláda	92,0	89,9	92,2	86,8	86,0
	España	82,4	77,8	74,7	71,0	70,2
	France	89,2	85,0	82,5	79,3	76,3
	Ireland	85,7	86,7	85,8	85,0	82,8
	Italia	78,6	75,1	70,9	68,5	70,6
	Luxembourg	83,5	82,4	80,7	77,4	72,2
	Nederland	89,7	86,4	85,4	83,8	80,0
	Portugal	82,5	74,4	67,7	59,3	56,7
	United Kingdom	86,7	84,0	82,6	80,8	81,3

Source : Eurostat and EC Commission, Directorate-General for Agriculture.

3.1.9   The 'cost-price squeeze' (¹) : the ratio of producer prices to input prices

(1985 = 100)

	1989	1990	1991	1992	1193
1	2	3	4	5	6
EUR 12	106,7	105,7	105,0	99,6	95,7
Belgique/België	110,7	109,2	106,8	100,0	94,7
Danmark	97,5	96,1	93,6	96,1	86,3
BR Deutschland	107,8	101,9	98,6	90,1	84,2
Elláda	98,3	100,2	99,9	94,4	89,8
España	107,0	106,4	103,5	95,7	96,4
France	103,5	104,1	101,4	95,3	90,0
Ireland	117,4	105,1	101,6	104,1	109,3
Italia	105,6	106,7	110,3	105,4	97,6
Luxembourg	117,5	115,7	105,2	102,8	106,1
Nederland	108,6	104,0	102,4	97,3	94,1
Portugal	101,1	99,2	95,9	91,2	87,3
United Kingdom	104,0	102,3	96,8	95,3	95,0

Source : Eurostat and EC Commission, Directorate-General for Agriculture.

(¹) The 'cost-price squeeze' is calculated by dividing changes in the deflated index prices of the value of final agricultural production by changes in the deflated index prices of the value of inputs.

3.1.10   Gross fixed capital formation and gross value-added in agriculture at factor cost ([1])

(1985 = 100)

1	2	1989	1990	1991	1992	1993
		3	4	5	6	7
Gross fixed	Belgique/België	116,9	135,1	100,6	111,5	108,8
capital	Danmark	95,4	109,4	90,4	97,0	95,3
formation	BR Deutschland	112,3	122,0	127,2	123,8	0,0
(GFCF)	Elláda	109,4	122,1	125,9	192,7	178,9
	España	0,0	0,0	0,0	0,0	0,0
	France	132,8	133,2	128,7	119,6	110,2
	Ireland	173,4	196,7	155,1	173,7	125,2
	Italia	137,2	133,5	137,0	136,0	135,9
	Luxembourg	143,6	158,3	145,4	141,4	0,0
	Nederland	133,7	143,7	144,5	140,6	127,4
	Portugal	226,1	270,6	277,8	274,4	0,0
	United Kingdom	91,1	91,1	72,4	85,1	83,8
Gross value-	Belgique/België	125,4	117,7	118,3	114,5	112,7
added (GVA)	Danmark	107,9	110,4	104,6	96,8	95,2
	BR Deutschland	121,3	113,5	110,3	115,6	0,0
	Elláda	177,5	181,7	255,8	251,4	274,4
	España	130,6	139,5	137,0	120,1	141,1
	France	115,8	120,1	114,3	113,9	109,8
	Ireland	135,7	135,6	126,5	144,9	148,2
	Italia	116,8	117,5	134,2	133,9	131,2
	Luxembourg	122,1	116,4	108,0	109,8	107,2
	Nederland	118,3	118,6	121,9	115,6	105,4
	Portugal	156,7	179,8	185,7	168,4	0,0
	United Kingdom	120,3	126,8	124,6	131,4	148,3
GFCF/GVA	Belgique/België	14,8	18,2	13,5	15,4	15,3
(%)	Danmark	20,2	22,6	19,7	22,8	22,8
	BR Deutschland	30,8	35,8	38,4	35,7	0,0
	Elláda	8,4	9,2	6,7	10,5	8,9
	España	0,0	0,0	0,0	0,0	0,0
	France	17,8	17,3	17,5	16,3	15,6
	Ireland	21,6	24,6	20,8	20,3	14,3
	Italia	33,9	32,8	29,5	29,3	29,9
	Luxembourg	39,3	45,5	45,1	43,1	0,0
	Nederland	26,7	28,6	28,0	28,7	28,6
	Portugal	19,3	20,1	20,0	21,8	0,0
	United Kingdom	18,1	17,2	13,9	15,5	13,5

Source : Eurostat — Agricultural accounts and EC Commission, Directorate-General for Agriculture.

([1]) At current prices : the series is based on figures exclusive of VAT.

3.1.11   Changes (% TAV) in final production, gross value-added, employment, utilized agricultural area and growth of agricultural productivity '1992' as compared with 1985 (¹)

1	At 1990 prices		Total employment 'agriculture, forestry, hunting, and fisheries'	Utilized agricultural area (UAA) (²)	Labour productivity calculated on the basis of:		Productivity per ha of UAA calculated on the basis of:	
	Final production	Gross value-added			final production	gross value-added	final production	gross value-added
	2	3	4	5	6	7	8	9
EUR 12 (³)	0,3	-0,7	-3,0	-0,8	-2,7	-3,7	-0,5	-1,5
Belgique/België	2,7	2,4	-2,8	-0,6	-0,1	-0,4	2,1	1,8
Danmark	0,9	1,3	-3,4	-0,3	-2,5	-2,1	0,6	1,0
BR Deutschland (³)	0,4	1,7	-3,8	-0,6	-3,4	-2,1	-0,2	1,1
Elláda	0,5	0,3	-4,9	-4,0	-4,4	-4,6	-3,5	-3,7
España	0,5	0,0	-5,0	-0,7	-4,5	-5,0	-0,2	-0,7
France	1,0	1,0	-4,0	-0,5	-3,0	-3,0	0,5	0,5
Ireland	1,7	1,6	-3,0	-0,8	-1,3	-1,4	0,9	0,8
Italia	0,9	1,0	-2,7	-0,7	-1,8	-1,7	0,2	0,3
Luxembourg	0,1	-1,4	-2,2	-0,2	-2,1	-3,6	-0,1	-1,6
Nederland	2,2	3,4	-0,5	-0,2	1,7	2,9	2,0	3,2
Portugal (³)	2,2	2,3	-3,7	-13,6	-1,5	-1,4	-11,4	-11,3
United Kingdom	3,8	10,3	-2,1	-0,9	1,7	8,2	2,9	9,4

Source: Eurostat — Agricultural accounts;
— Social statistics;
— Agricultural statistics.

(¹) The changes are calculated on the basis of series after recording net of VAT.
(²) Estimates for 1993.
(³) '1991'/1985.

3.1.12 Net value-added at factor cost for total manpower per annual work unit (AWU) from 1975 to 1993

('1985' = 100)

	EUR 12	Belgique/ België	Danmark	Deutsch-land	Elláda	España	France	Ireland	Italia	Luxembourg	Nederland	Portugal	United Kingdom
1	2	3	4	5	6	7	8	9	10	11	12	13	14
1975	:	85,8	52,9	118,5	72,4	74,4	94,8	106,9	90,1	64,6	82,8	:	106,0
1976	:	101,1	54,7	123,0	78,5	79,6	93,8	101,6	84,3	57,2	90,2	:	115,8
1977	:	84,1	63,9	117,9	75,4	90,4	91,6	125,0	88,9	71,2	85,9	:	108,0
1978	:	90,7	69,5	113,0	85,3	91,5	95,0	129,5	89,9	70,9	84,6	:	102,1
1979	:	82,1	60,4	101,2	80,8	83,5	97,2	108,0	95,8	73,6	78,0	:	98,8
1980	91,9	86,8	65,3	90,1	92,0	90,3	88,0	88,1	109,2	68,3	75,1	97,4	92,3
1981	93,6	95,2	74,8	90,8	97,4	80,0	91,2	88,4	106,7	77,1	92,2	92,0	98,0
1982	103,4	100,3	90,5	110,6	100,1	92,4	107,6	96,5	107,2	106,8	96,8	102,8	106,3
1983	99,2	108,3	77,5	89,3	90,9	92,2	100,3	100,9	112,3	93,2	93,3	98,6	95,5
1984	102,2	104,4	103,2	102,3	98,8	100,8	99,4	112,1	101,3	96,6	100,7	99,3	114,6
1985	98,4	99,4	96,4	92,6	101,3	102,7	100,0	98,5	101,2	100,0	95,5	97,4	90,5
1986	99,5	96,2	100,4	105,1	99,9	96,4	100,5	90,4	97,5	103,4	103,7	103,3	94,9
1987	96,8	90,6	81,5	87,8	99,4	102,1	101,5	110,7	98,8	105,9	85,8	101,2	95,3
1988	99,5	98,0	84,2	108,8	108,0	116,5	99,6	129,8	93,3	108,8	89,3	85,7	86,5
1989	111,9	122,6	100,5	130,4	121,9	114,6	117,9	133,0	99,4	123,6	105,5	101,6	97,2
1990	109,7	112,1	96,3	114,0	105,8	121,8	125,4	136,7	91,6	116,3	101,3	105,7	99,2
1991	110,5	112,2	88,3	106,2	144,3	127,0	117,1	129,0	98,3	101,1	100,7	95,6	93,8
1992	105,1	102,0	76,9	102,3	119,6	112,1	117,2	154,4	92,3	96,2	87,1	82,3	99,2
1993	103,6	101,3	81,8	79,1	119,5	137,4	113,3	159,4	85,7	90,2	76,9	73,5	114,2
TAV % 1993/1992	-1,4	-5,3	6,5	-22,7	-0,1	22,5	-3,4	3,3	-7,1	-6,2	-11,7	-10,7	15,1

Source: Eurostat.

3.1.13   **Volume of agricultural labour in annual work units (AWUs) from 1974 to 1993**

(1 000)

	EUR 12	Belgique/België	Danmark	BR Deutschland	Elláda	España	France	Ireland*	Italia	Luxembourg	Nederland	Portugal	United Kingdom
1	2	3	4	5	6	7	8	9	10	11	12	13	14
1974	12 990,9	143,3	176,3	1 198,0	1 092,0	2 454,0	2 078,0	333,4	3 336,7	12,2	281,0	1 330,0	556,0
1975	12 492,4	137,2	168,2	1 168,0	1 068,0	2 279,7	2 008,0	324,6	3 209,1	11,5	277,5	1 299,3	541,3
1976	12 220,6	130,5	162,9	1 139,0	1 045,0	2 101,9	1 965,0	318,1	3 207,5	10,8	273,7	1 320,8	545,4
1977	11 774,4	124,9	156,5	1 082,0	1 022,0	1 959,0	1 926,0	312,0	3 094,4	10,6	265,9	1 281,7	539,4
1978	11 543,3	120,8	150,5	1 059,0	999,0	1 898,3	1 895,0	305,4	3 094,5	10,1	259,9	1 212,8	538,0
1979	11 233,8	120,3	144,4	1 007,0	978,0	1 774,9	1 864,0	297,3	3 044,4	9,7	256,5	1 210,7	526,6
1980	10 855,3	115,6	137,6	987,0	956,0	1 634,7	1 817,0	289,6	2 938,8	9,2	254,3	1 202,2	513,3
1981	10 339,0	112,4	131,4	974,0	935,0	1 487,5	1 768,0	283,8	2 751,6	8,6	249,3	1 135,7	501,7
1982	9 987,2	110,2	126,7	951,0	924,0	1 432,5	1 720,0	279,0	2 593,4	8,3	248,0	1 098,1	496,0
1983	9 845,6	109,4	123,8	927,0	917,0	1 415,0	1 671,0	276,1	2 645,8	7,9	248,3	1 012,2	492,1
1984	9 650,3	108,7	120,3	912,0	918,0	1 341,9	1 620,0	275,9	2 598,7	7,5	246,7	1 017,0	483,6
1985	9 444,7	106,1	114,7	904,0	931,0	1 300,4	1 564,0	275,8	2 494,1	7,3	245,4	1 020,7	481,2
1986	9 170,0	104,8	111,8	890,0	898,0	1 252,1	1 509,0	266,0	2 473,4	7,0	242,7	942,0	473,2
1987	8 934,6	101,6	105,1	836,0	849,0	1 218,0	1 455,0	254,5	2 422,9	6,7	240,5	983,2	462,1
1988	8 638,8	98,3	101,0	821,0	828,0	1 191,2	1 401,0	248,0	2 313,2	6,4	237,4	940,7	452,6
1989	8 226,8	96,0	98,5	775,0	770,4	1 137,5	1 335,2	243,0	2 193,6	6,3	237,5	893,5	440,3
1990	7 942,2	93,6	95,2	754,0	752,4	1 070,7	1 272,4	238,0	2 153,2	6,0	236,8	839,2	430,7
1991	7 654,1	90,8	92,7	716,3	683,7	961,5	1 227,9	229,4	2 155,7	5,8	236,9	833,6	419,8
1992	7 404,6	88,0	89,9	684,8	718,0	914,5	1 184,9	223,4	2 060,6	5,9	239,0	782,3	413,4
1993	7 133,5	83,3	88,1	650,0	701,5	870,6	1 125,6	218,9	1 982,0	5,8	235,3	762,0	410,4
% 1993/1992	-3,7	-5,3	-2,0	-5,1	-2,3	-4,8	-5,0	-2,0	-3,8	-1,6	-1,5	-2,6	-0,7

Source : Eurostat.

3.1.14 Percentage variation in 1993 in the consumption of inputs by volume compared with 1992

1	EUR 12	Belgique/België	Danmark	BR Deutschland	Ellada	España	France	Ireland	Italia	Luxembourg	Nederland	Portugal	United Kingdom
	2	3	4	5	6	7	8	9	10	11	12	13	14
Seeds and reproductive material	-1,4	-1,0	-9,9	-4,5	0,0	-3,5	1,0	-12,5	-2,5	-0,8	0,0	:	1,0
Energy and lubricants	-1,5	-1,0	0,0	-4,0	4,0	-3,4	1,0	2,5	-0,5	-0,6	1,5	-18,3	-3,0
Fertilizers and soil improvers	-7,0	-3,0	-10,0	-6,0	-4,0	-21,3	-4,5	4,6	-5,8	-3,2	-4,0	:	-4,7
Crop protection products	-7,0	-5,0	-10,0	-6,0	2,5	-9,2	-10,0	-1,5	-2,2	0,0	-3,0	-29,9	5,8
Animal feed	0,5	5,0	3,0	-4,4	-5,0	-2,6	3,5	4,7	-0,2	-4,2	1,5	0,5	2,8
Farm implements, upkeep and repairs	-1,6	-1,0	0,0	-5,6	0,0	5,4	-2,0	-0,7	:	0,0	-1,0	-18,6	-3,3
Services	-0,7	0,0	0,0	-3,2	-0,2	-5,3	3,0	-1,7	:	:	0,0	-10,7	-1,3
Consumption of inputs	-1,4	1,8	-0,3	-4,5	-0,7	-3,3	0,1	2,1	-1,1	-0,6	0,4	-9,5	0,0

Source : Eurostat.

3.1.15 Main agricultural economic data, by region (1991)

Region	Share of agricult. in whole economy ag. GVA / tot. GVA %	Gross value-added GVA (fc) (Mio ECU)	GVA/ MWU EUR 12 = 100	Share of in- puts/ final production %	Share of other production costs (1)/ fin. production %	Share of main products in final agricultural production (% of total final production)							
						Cereals	Other crops	Fruit, vegetables	Wine	Milk	Cattle	Pigs	Eggs, poultry
1	2	3	4	5	6	7	8	9	10	11	12	13	14
EUR 12 (*)	2,3	126 271,4	100,0	43,3	:	11,7	8,6	15,1	:	15,9	11,3	10,4	:
Belgique/België	1,8	2 997,2	211,9	56,5	18,7	4,4	8,5	17,6	0,0	14,0	18,2	23,0	5,9
— Vlaams gewest + Région bruxelloise/ Brussels gewest	2,1	2 079,7	220,2	59,1	17,1	1,8	5,5	21,7	:	11,0	13,4	30,0	
— Région wallonne	2,1	917,5	196,0	49,4	23,2	11,5	16,8	6,2	:	22,3	31,6	3,7	:
Danmark													
— Danmark	3,1	3 085,3	214,4	50,9	39,2	14,6	8,7	2,5	0,0	23,3	7,9	29,6	:
BR Deutschland	1,0	15 141,1	132,9	53,6	34,2	10,1	8,6	7,4	3,8	23,9	15,2	17,4	5,3
— Schleswig-Holstein	2,3	1 184,3	201,8	56,6	33,8	14,5	8,1	7,4	0,0	28,2	15,3	13,2	2,9
— Hamburg	0,1	81,9	:	46,2	35,5	1,7	0,8	49,8	0,0	2,5	3,0	0,8	0,4
— Niedersachsen	2,7	4 029,7	221,0	50,9	26,4	11,0	10,5	5,3	0,0	21,8	13,7	21,2	9,4
— Bremen	0,1	18,4	:	49,1	34,2	2,8	0,8	34,3	0,0	20,9	13,0	2,2	1,2
— Nordrhein-Westfalen	0,7	2 456,9	166,3	54,4	30,5	10,7	7,5	7,9	0,0	17,9	12,9	27,8	5,5
— Hessen	0,5	810,3	104,0	55,6	36,0	12,3	8,8	11,4	3,1	21,9	14,2	15,6	3,7
— Rheinland-Pfalz	1,5	1 142,2	120,6	45,8	33,1	10,6	7,4	9,8	33,4	13,9	9,8	6,7	2,5
— Baden-Württemberg	0,9	2 123,6	116,8	47,4	38,1	8,3	5,3	11,6	9,6	20,5	14,2	14,4	3,6
— Bayern	1,1	3 213,6	83,9	60,1	43,6	7,7	9,6	3,8	1,4	34,7	21,0	12,9	3,6
— Saarland	0,2	55,2	101,5	61,7	42,3	11,1	5,0	12,6	0,5	25,3	18,8	5,6	5,0
— Berlin	0,1	25,0	31,5	42,5	34,2	0,7	0,1	62,8	0,0	1,6	1,2	3,2	1,7
— Brandenburg	:												
— Mecklenburg-Vorpommern	:												
— Sachsen	:												
— Sachsen-Anhalt	:												
— Thüringen	:												:

Elláda (²)	13,2	7 178,2	59,1	22,0	15,1	11,2	18,3	23,6	1,9	8,6	3,2	3,2	4,6
— Anatoliki Makedonia, Thraki	21,2	566,3	:	23,3	19,8	33,7	29,1	11,4	0,4	6,7	3,3	2,7	1,8
— Kentriki Makedonia	14,6	1 288,5	:	25,3	19,6	18,0	27,0	24,6	0,4	8,2	7,2	1,8	3,1
— Dytiki Makedonia	12,0	209,1	:	32,8	22,1	29,5	13,1	10,2	0,6	16,9	8,5	3,0	1,6
— Thessalia	30,9	1 060,1	83,5	20,7	12,3	11,6	36,3	12,2	0,7	7,9	2,5	2,8	2,2
— Ipeiros	17,4	241,8	34,6	38,0	12,7	6,6	5,5	19,6	0,4	18,1	2,7	6,6	10,4
— Ionia nisia	15,4	155,1	46,4	15,5	8,3	0,7	2,3	21,2	5,0	7,3	3,9	-1,0	2,4
— Dytiki Ellada	22,6	704,2	:	18,4	14,2	7,9	19,2	22,1	3,2	9,3	1,3	2,6	2,2
— Sterea Ellada	16,6	600,4	:	27,1	15,6	10,4	21,5	19,9	3,8	8,0	0,9	9,0	6,6
— Peloponnisos	26,5	870,7	:	16,7	13,2	1,7	4,6	44,9	3,3	6,9	0,7	3,2	3,8
— Attiki	1,8	333,6	:	21,3	9,2	0,8	0,6	19,2	5,1	6,4	2,5	2,0	28,1
— Voreio Aigaio	14,6	121,9	:	22,1	25,9	2,5	2,2	18,3	5,6	16,3	4,1	1,9	7,3
— Notio Aigaio	8,3	132,6	:	21,5	13,2	2,1	6,7	27,5	2,0	11,6	7,9	6,7	5,1
— Kriti	31,4	894,1	74,4	13,3	11,4	0,3	1,8	37,8	2,1	6,5	0,8	2,1	2,5
España (²)	4,2	15 121,8	68,8	42,5	:	10,0	8,8	26,6	4,4	9,1	6,8	11,6	7,6
— Galicia	5,7	1 151,7	26,5	35,2	:	-0,5	8,6	7,9	5,0	31,6	14,4	11,6	12,1
— Principado de Asturias	2,6	263,7	26,7	37,4	:	-4,1	8,1	7,6	0,1	49,3	29,5	1,5	5,6
— Cantabria	4,0	200,7	46,1	32,9	:	0,1	2,8	3,2	0,1	54,4	29,6	2,3	4,2
— País Vasco	1,0	232,6	46,1	46,9	:	5,4	14,6	20,2	7,0	26,2	11,4	2,3	7,0
— Navarra	4,9	313,8	88,7	48,3	:	18,2	3,6	23,8	6,6	9,5	6,8	13,9	7,7
— La Rioja	10,4	284,3	102,5	31,3	:	4,5	12,5	35,4	17,6	2,7	3,5	6,7	6,3
— Aragón	6,7	876,8	80,7	54,6	:	24,0	2,7	15,7	2,9	1,8	5,3	19,0	9,5
— Cataluña	1,7	1 146,1	76,5	65,2	:	6,2	2,0	19,4	2,9	6,6	8,4	28,1	16,4
— Baleares	1,9	146,1	66,1	44,5	:	0,7	6,7	45,3	0,8	11,2	5,7	4,7	5,7
— Castilla-León	7,9	1 737,2	61,3	45,5	:	19,5	14,5	4,8	1,1	14,6	12,3	14,8	6,1
— Madrid	0,3	139,8	76,2	54,8	:	12,9	4,5	21,5	3,9	13,6	5,8	1,4	29,1
— Castilla-La Mancha	11,7	1 574,2	104,2	39,9	:	20,1	8,1	13,0	16,7	5,9	4,0	6,5	6,6
— Comunidad Valenciana	4,2	1 431,6	94,7	38,7	:	1,7	2,4	66,8	2,4	0,8	0,7	7,3	7,5
— Región de Murcia	8,7	738,6	110,8	48,0	:	2,4	4,5	64,0	1,7	1,2	1,2	17,0	1,9
— Extremadura	11,4	804,1	77,0	36,6	:	12,9	16,4	18,1	5,0	5,0	9,4	10,4	1,6
— Andalucía	7,6	3 622,3	92,3	27,8	:	9,0	16,1	32,6	2,6	3,8	2,9	4,0	3,5
— Canarias	3,5	458,3	78,6	34,8	:	0,0	5,5	62,3	4,2	5,0	2,0	1,2	8,8

3.1.15 *(cont.)*

Region	Share of agricult. in whole economy ag. GVA / tot. GVA %	Gross value-added — GVA (fc) (Mio ECU)	GVA/MWU EUR 12 = 100	Share of inputs/final production %	Share of other production costs (¹/ fin. production %	Share of main products in final agricultural production (% of total final production)							
						Cereals	Other crops	Fruit, vegetables	Wine	Milk	Cattle	Pigs	Eggs, poultry
1	2	3	4	5	6	7	8	9	10	11	12	13	14
France	2,6	25 739,9	131,1	44,8	25,6	18,6	9,2	11,5	10,6	15,7	12,9	6,5	7,7
— Ile de France	0,2	578,5	213,7	41,5	33,5	38,8	22,4	15,1	0,1	1,3	1,1	0,4	4,0
— Champagne-Ardennes	9,4	2 088,2	295,8	30,7	24,1	22,0	17,5	1,6	39,5	6,4	3,9	0,9	0,9
— Picardie	4,9	1 199,2	210,7	44,9	26,5	30,0	32,0	8,8	3,0	11,2	6,1	1,6	3,1
— Haute-Normandie	2,2	624,2	146,4	45,6	27,4	26,7	18,7	5,3	0,0	20,9	16,7	2,9	2,8
— Centre	4,5	1 657,0	153,4	43,6	26,7	42,9	14,6	9,9	3,9	5,6	6,5	1,9	5,2
— Basse-Normandie	5,3	1 133,1	117,3	45,2	24,2	11,7	4,3	8,1	0,0	38,7	23,7	5,0	3,4
— Bourgogne	4,8	1 211,0	152,1	40,3	26,9	24,6	10,7	5,3	19,6	7,1	20,2	1,9	3,7
— Nord-Pas-de-Calais	1,8	925,2	142,8	48,9	23,5	20,3	22,2	11,1	0,0	20,5	8,5	7,9	4,6
— Lorraine	1,8	605,6	126,9	45,5	23,9	25,5	11,4	3,4	0,2	30,6	19,4	2,2	2,7
— Alsace	1,8	491,2	128,3	35,8	24,1	22,5	6,6	6,8	31,8	11,8	6,0	2,6	5,3
— Franche-Comté	2,2	407,2	113,3	45,6	21,4	13,2	4,2	3,4	3,3	46,9	19,7	3,2	2,1
— Pays de la Loire	5,1	2 518,2	144,6	46,6	23,2	10,8	2,2	10,4	3,8	21,5	21,6	6,9	15,6
— Bretagne	6,1	2 469,0	137,3	58,6	15,4	6,5	2,6	6,4	0,0	23,2	13,7	27,3	18,0
— Poitou-Charentes	4,5	1 025,3	102,4	52,4	27,7	27,2	11,0	4,1	10,7	13,1	14,0	3,1	6,9
— Aquitaine	3,7	1 668,1	98,6	44,6	33,0	21,1	4,1	15,8	18,6	8,6	10,1	3,5	12,8
— Midi-Pyrénées	4,0	1 548,9	90,6	46,8	25,0	24,6	8,1	10,8	3,4	14,6	15,3	5,1	7,7
— Limousin	2,4	341,0	65,8	60,3	28,8	7,5	1,6	7,0	0,1	10,0	53,5	5,9	2,6
— Rhône-Alpes	1,8	1 686,1	110,3	38,6	22,1	10,3	3,8	16,2	15,8	19,3	11,8	4,0	9,0
— Auvergne	3,3	738,2	90,4	47,3	28,1	14,6	4,1	4,1	0,6	27,9	31,0	5,6	5,1
— Languedoc-Roussillon	4,6	1 356,4	118,9	30,0	43,4	5,4	2,0	31,8	49,3	2,5	2,0	0,6	3,3
— Provence-Alpes-Côte d'Azur	2,0	1 366,2	148,8	34,1	34,1	5,1	2,1	50,0	18,8	0,8	0,7	0,9	1,8
— Corse	2,6	102,1	127,6	37,4	35,7	1,1	2,2	61,3	13,1	8,1	3,4	5,9	3,0

Ireland													
— Ireland	6,5	2 651,9	73,7	44,0	23,5	5,4	3,6	3,4	0,00	31,71	37,15	5,84	4,3
Italia	3,1	32 279,6	99,4	27,1	:	9,7	6,3	27,5	8,55	11,2	7,95	6,07	7,72
— Piemonte	2,3	1 956,5	74,1	35,3	:	18,1	3,6	14,9	11,05	12,36	16,86	6,27	8,5
— Valle d'Aosta	1,3	41,2	48,0	32,1	:	0,1	4,9	7,3	3,81	38,7	30,23	0,71	5,08
— Liguria	1,7	567,0	80,0	15,3	:	0,3	1,7	14,5	1,97	3,73	1,71	0,13	3,48
— Lombardia	1,4	2 872,2	129,4	41,9	:	11,1	4,7	6,0	2,77	27,2	14,5	17,25	10,99
— Trentino-Alto Adige	4,4	800,0	109,8	21,7	:	0,1	0,7	51,5	14,96	17,53	7,33	1,14	3,74
— Veneto	3,3	2 871,8	105,4	33,1	:	8,9	11,7	18,8	11,92	11,2	10,06	4,49	17,97
— Friuli-Venezia Giulia	2,1	511,9	81,5	34,2	:	18,6	13,6	8,8	11,57	14,72	9,16	6,27	7,36
— Emilia-Romagna	4,2	3 514,3	138,0	33,6	:	9,7	6,4	26,0	6,23	15,78	7,98	11,69	12,22
— Toscana	1,9	1 344,4	73,1	27,2	:	16,0	6,5	12,8	14,6	5,33	4,83	6,5	6,08
— Umbria	3,6	585,5	84,5	31,4	:	20,8	20,6	7,0	6,4	4,28	7,00	12,99	11,38
— Marche	3,4	860,4	85,1	29,4	:	21,2	10,8	19,6	9,79	3,41	6,99	6,31	9,85
— Lazio	1,7	1 777,3	86,3	24,6	:	8,9	3,8	36,4	8,93	12,74	6,26	2,59	4,41
— Campania	3,9	2 969,7	83,2	17,6	:	4,0	13,0	42,7	3,75	5,99	4,96	1,91	4,33
— Abruzzi	4,9	968,3	87,0	22,5	:	7,0	7,1	31,9	19,29	5,77	4,54	2,56	7,1
— Molise	5,1	241,5	61,1	31,1	:	25,3	9,3	11,0	7,14	10,88	7,42	5,59	15,06
— Puglia	6,9	3 661,5	124,5	15,4	:	7,2	4,7	38,9	10,14	3,08	2,05	0,34	1,54
— Basilicata	6,1	477,9	63,4	25,2	:	27,0	2,9	31,4	3,69	6,31	6,3	5,76	2,34
— Calabria	7,4	1 571,1	89,0	15,2	:	2,7	2,6	37,0	3,01	5,01	5,46	2,84	2,62
— Sicilia	6,2	3 762,2	129,8	14,8	:	5,6	2,1	52,7	12,43	3,61	4,51	0,77	2,69
— Sardegna	3,8	924,9	79,8	31,4	:	7,2	2,5	24,1	5,33	20,87	11,02	7,73	3,48
Luxembourg													
— Luxembourg	1,2	118,3	130,5	47,5	33,1	5,9	2,7	1,4	5,8	47,4	25,5	9,1	1,0
Nederland	3,6	8 111,8	229,5	48,2	:	1,2	6,5	14,1	:	21,1	9,9	16,7	6,6
— Noord-Nederland	4,4	1 095,6	212,6	47,6	:	2,5	17,3	3,2	:	43,3	13,2	5,0	6,3
— Oost-Nederland	4,3	1 756,8	175,6	56,5	:	0,7	5,6	6,2	:	28,2	17,3	23,4	7,9
— West-Nederland	3,0	3 378,2	291,2	35,3	:	1,4	5,4	22,8	:	11,9	3,7	3,3	1,2
— Zuid-Nederland	3,9	1 881,0	218,8	56,3	:	0,8	3,4	16,1	:	15,2	8,7	32,0	12,1

3.1.15 (cont.)

Region	Share of agricult. in whole economy ag. GVA tot. GVA %	Gross value-added — GVA (fc) (Mio ECU)	GVA/MW EUR 12 = 100	Share of inputs/ final production %	Share of other production costs (¹)/ fin. production %	Share of main products in final agricultural production (% of total final production)							
						Cereals	Other crops	Fruit, vegetables	Wine	Milk	Cattle	Pigs	Eggs, poultry
1	2	3	4	5	6	7	8	9	10	11	12	13	14
Portugal													
— Portugal	3,5	2 241,0	17,7	51,1	26,7	8,3	9,5	20,0	7,9	11,8	10,0	9,8	9,4
United Kingdom	1,1	9 854,9	146,8	54,3	33,0	17,0	9,1	10,4	:	21,1	14,7	7,2	9,9
— North	0,9	466,5	142,0	64,3	26,4	14,1	0,2	1,6	:	36,2	22,8	3,5	:
— Yorkshire-Humberside	1,3	863,0	170,6	56,4	29,4	23,2	1,0	8,7	:	9,9	11,0	18,0	:
— East Midlands	2,1	1 184,3	209,6	48,9	29,2	26,1	1,6	13,4	:	9,9	10,5	5,3	:
— East Anglia	4,0	1 194,4	263,8	44,9	29,3	24,7	1,6	16,6	:	3,1	3,8	11,8	:
— South-East	0,5	1 534,1	163,3	45,5	43,6	23,4	1,7	22,7	:	11,4	6,8	5,8	:
— South-West	1,5	1 102,1	111,0	61,0	35,2	12,0	0,6	5,0	:	41,3	14,2	6,4	:
— West Midlands	1,1	845,8	162,4	55,1	31,5	14,6	0,8	9,0	:	25,7	12,8	4,7	:
— North-West	0,6	535,6	154,1	51,3	29,7	4,5	0,3	15,7	:	37,7	11,5	6,1	:
— Wales	1,1	562,2	85,6	65,8	31,0	1,9	0,0	1,2	:	37,5	27,7	1,2	:
— Scotland	1,2	1 015,8	126,0	57,9	39,0	17,2	0,2	3,7	:	17,8	26,7	4,1	:
— Northern Ireland	2,7	550,9	92,0	60,4	25,6	1,7	0,0	3,7	:	28,2	33,7	9,4	:

Source : Eurostat.

(¹) Other production costs = depreciation + wages + rent + interest.

(²) 1989.

NB : The negative values reflect 'negative' final production (for example, fodder cereals which are products of one region but are consumed on holdings in another region), i.e. production which is not entered as such in the accounts but which incurs costs.

3.2.1   **The farm accountancy data network — Explanatory note**

The farm accountancy data network (FADN) collects accountancy data from a sample of agricultural holdings in the Community. The FADN field of survey relates to 'commercial' farms, i.e. farms which market the bulk of their production and which exceed a minimum level of economic activity defined in terms of economic size (see the definition of the European size unit below).

In the most recent accounting years there were almost 58 000 holdings (Community of Twelve) representative of commercial farms in the FADN sample.

The terms used in the tables relate to the following definitions.

BASIC FADN TERMS

Accounting year

The accounting year is a 12-month period starting between 1 January and 1 July, the exact date varying from one Member State to another.

Economic size and European size unit (ESU)

The European size unit (ESU) is a unit of measurement of the economic size of the agricultural holding. A farm has an economic size of 1 ESU if its total standard gross margin is ECU 1 200 of 1988 SGM. The standard gross margin for each enterprise corresponds to the average value, over a three-year period and in a given region, of production minus certain variable costs (Decision 85/377/EEC).

Type of farming (TF)

The type of farming (TF) of a holding is determined by the relative share in the holding's total standard gross margin of each of the enterprises of the holding. A description is given in Table 3.2.2. The results given in the following tables relate to nine groups aggregated from the 17 principal types of farming in the Community farm typology (Decision 85/377/EEC).

Weighting and number of holdings represented

The holdings in the FADN sample are selected in such a way as to be representative, for each division, of the holdings belonging to each cell formed by the combination of TF and economic size class. The populations to be represented are derived from the Community farm structure surveys.

The results presented are weighted averages. Each holding in the FADN sample is attributed a weight proportional to the number of holdings belonging to the same type of farming and the same economic size class in the division.

The number of holdings represented is the sum of the weights of the holdings in the sample. Some cells (division — TF — economic size class) may have no holdings in the sample, either because very high selection rates would be necessary or because there are technical difficulties in selecting holdings.

STRUCTURAL DATA

UAA: utilized agricultural area (in hectares).

3.2.1   *(continued)*

Annual work unit (AWU and FWU)

This represents the agricultural work done by one full-time worker in one year. Part-time and seasonal work are fractions of an AWU.

An FWU is an AWU of unpaid (family) labour.

AVERAGE RESULTS PER HOLDING

Total output

This is the value of total production during the accounting year. Included are off-farm sales, home-grown feed and seed, farmhouse consumption and benefits in kind, as well as changes in the value of livestock and stocks of crop products.

Intermediate consumption

This corresponds to all the fixed and variable costs that are necessary for agricultural activity and includes home-grown feed and seed but excludes financial charges, labour costs, rent and depreciation.

Depreciation

This is the annual provision designed to replace the fixed components of working capital at the end of their life (buildings, machinery, equipment, etc.). It is calculated on the basis of replacement value.

Farm net value-added (FNVA)

Total output less intermediate consumption and depreciation, adjusted to take account of taxes, grants and subsidies linked to production.

This is an indicator of the economic performance of the holding. It renumerates family and hired labour, own and borrowed capital and the management of the holding.

Family farm income (FFI)

This corresponds to farm net value-added, less other real costs in the accounting year: interest and financial charges, wages and social security costs paid and rent.

This indicator represents the return on the labour of farmer and family, and on owned capital.

3.2.2   Description of the types of farming in Table 3.2.3

The nine types of farming: shares of each enterprise in total output

Type of farming (principal types) EUR 12**	TF codes	Enterprise output as % of total output												
		Cereals	Other field crops	Vegetables and flowers	Fruits	Wine and grapes	Olives and olive oil	Dairying	Beef and veal	Sheep and goats (¹)	Pigmeat	Poultry and eggs	Other	Total
1	2	3	4	5	6	7	8	9	10	11	12	13	14	15
All farms		14,7	6,9	10,1	4,6	6,8	1,8	19,5	11,2	3,2	9,8	2,6	8,7	100
A — Cereals	11	76,0	6,9	0,7	0,2	0,8	0,5	0,5	2,3	1,2	0,8	1,9	8,2	100
B — General cropping	12 + 60	30,3	26,0	13,0	2,2	4,8	1,8	1,4	3,9	0,8	4,1	1,5	10,1	100
C — Horticulture	20	0,5	0,5	93,0	0,6	0,1	0,2	0,1	0,0	0,0	0,0	0,0	5,0	100
D — Vineyards	31	1,4	0,5	0,4	0,8	90,2	1,0	0,1	0,3	0,0	0,0	0,1	5,2	100
E — Fruit (and other permanent crops)	32 + 33 + 34	2,1	1,0	1,9	52,0	10,0	17,3	0,4	0,4	0,7	0,4	0,2	13,5	100
F — Dairying	41	4,3	0,9	0,1	0,1	0,1	0,0	64,7	19,4	0,5	2,1	0,4	7,4	100
G — Dry stock	42 + 43 + 44	6,7	0,9	0,1	0,1	0,2	0,2	16,2	36,9	23,7	1,9	0,5	12,5	100
H — Pigs and/or poultry	50	2,8	0,5	0,2	0,1	0,0	0,0	0,3	0,7	0,1	67,8	22,4	5,0	100
I — Mixed	71 + 72 + 81 + 82	17,4	5,3	1,0	0,5	1,1	0,4	19,0	14,6	3,4	24,6	4,1	8,6	100

Source: FADN results for 1992/93 (weighted with the 1990 Farm Structure Survey using »1988« standard gross margins).
(¹) Including milk.

3.2.3 Results by type of farming, 1991/92 and 1992/93

Type of farming	Number of holdings				Size of holdings			
	In the FADN field of observation		In the sample (^{1})		UAA (ha)		Labour input (AWU)	
	91/92	92/93	91/92	92/93	91/92	92/93	91/92	92/93
1	2	3	4	5	6	7	8	9
All types of farming								
EUR 12	4 169 729	4 157 089	57 600	58 210	23,5	24,2	1,50	1,50
Belgique/België	51 340	50 500	1 193	1 191	26,4	27,6	1,67	1,66
Danmark	76 692	77 182	2 277	2 268	35,6	35,7	1,18	1,16
BR Deutschland	331 670	330 850	4 984	5 072	31,3	31,9	1,59	1,59
Elláda	561 103	561 343	5 799	6 008	5,9	5,8	1,68	1,68
España	656 112	638 952	7 666	6 959	19,9	22,5	1,14	1,18
France	516 910	516 535	7 458	7 757	47,8	49,5	1,67	1,67
Ireland	134 670	134 970	1 267	1 293	39,4	39,3	1,26	1,22
Italia	1 199 827	1 204 425	19 176	19 515	10,7	10,6	1,38	1,37
Luxembourg	2 201	2 280	295	288	51,2	51,6	1,72	1,70
Nederland	91 784	92 452	1 493	1 513	22,1	22,4	2,16	2,17
Portugal	411 500	409 570	2 912	3 146	12,3	12,6	1,62	1,56
United Kingdom	135 920	138 030	3 080	3 200	125,2	126,6	2,36	2,35
A. Cereals								
EUR 12	281 054	278 394	3 732	3 712	39,7	40,3	1,05	1,00
Belgique/België	27	95	1	2	:	:	:	
Danmark	14 620	14 620	222	276	26,4	26,3	0,52	0,51
BR Deutschland	5 389	6 111	80	100	35,9	37,4	1,03	1,07
Elláda	19 320	19 320	348	292	14,2	15,3	1,32	1,28
España	90 001	88 821	1 019	852	51,3	51,8	0,87	0,92
France	33 940	32 890	537	601	71,3	74,3	1,32	1,31
Ireland	1 272	2 097	22	27	76,8	67,6	1,24	1,00
Italia	91 940	94 460	1 228	1 305	15,9	15,7	1,01	1,04
Luxembourg	0	0	0	0	:	:	:	
Nederland	46	119	1	3	:	:	:	
Portugal	13 588	9 291	88	76	29,0	43,3	1,26	1,19
United Kingdom	10 910	10 570	186	178	119,7	118,0	1,95	1,82
B. General cropping								
EUR 12	1 082 853	1 077 234	14 261	14 550	22,1	22,6	1,52	1,51
Belgique/België	5 673	5 605	132	127	40,3	42,2	1,41	1,37
Danmark	22 180	22 180	519	460	38,9	38,4	0,85	0,87
BR Deutschland	51 731	51 019	977	975	43,8	43,2	1,50	1,51
Elláda	198 960	199 060	2 566	2 740	7,3	6,9	1,82	1,82
España	129 971	130 590	1 747	1 876	22,7	25,1	1,12	1,07
France	107 270	107 520	1 579	1 649	62,7	66,1	1,57	1,61
Ireland	2 908	2 453	41	35	56,8	55,5	1,67	1,70
Italia	386 951	382 811	5 449	5 382	9,6	9,6	1,39	1,37
Luxembourg	1	0	1	0	:	:	:	
Nederland	13 304	13 231	318	314	44,4	43,8	1,58	1,43
Portugal	142 025	140 145	470	539	11,0	10,5	1,71	1,63
United Kingdom	21 880	22 620	462	453	154,1	153,0	2,93	3,06
C. Horticulture								
EUR 12	142 518	137 597	3 205	3 478	3,6	3,8	2,44	2,50
Belgique/België	5 080	5 080	128	127	2,9	3,2	2,57	2,63
Danmark	1 670	1 670	206	205	5,9	6,4	4,45	4,07
BR Deutschland	7 820	8 330	187	212	3,1	2,6	3,63	3,69
Elláda	9 350	9 320	86	93	2,7	2,6	1,90	1,88
España	41 730	30 640	363	339	2,8	3,5	1,52	1,60
France	16 332	17 804	275	295	7,3	7,8	2,84	2,75
Ireland	70	70	2	1	:	:	:	
Italia	28 795	31 168	1 196	1 310	1,8	2,0	2,00	2,09
Luxembourg	0	0	0	0	:	:	:	
Nederland	15 164	15 192	339	346	4,8	4,5	5,06	5,10
Portugal	13 437	14 253	383	424	2,8	2,4	1,63	1,51
United Kingdom	3 070	4 070	40	126	10,5	7,3	4,81	5,17

Average results per holding in 1 000 ECU (current)

Total output		Intermediate consumption		Depreciation		Farm net value-added		Farm net value-added per AWU		Family farm income per unit unpaid labour	
91/92	92/93	91/92	92/93	91/92	92/93	91/92	92/93	91/92	92/93	91/92	92/93
10	11	12	13	14	15	16	17	18	19	20	21
43,6	43,6	21,7	22,3	5,5	5,7	17,5	17,4	11,6	11,6	9,4	9,2
106,6	106,3	54,8	55,9	9,6	10,8	44,0	41,3	26,4	24,8	23,1	20,3
95,6	87,1	53,7	54,6	9,8	10,0	32,1	24,2	27,2	20,9	10,3	0,6
83,3	84,6	47,7	49,3	13,5	14,2	24,4	23,7	15,3	14,9	11,3	10,3
14,5	13,0	4,2	4,4	1,4	1,5	10,3	8,4	6,1	5,0	5,9	4,8
19,9	23,2	9,2	9,7	2,4	2,2	8,8	12,7	7,7	10,7	7,2	10,9
84,0	83,9	42,9	43,6	11,0	11,5	30,8	32,1	18,5	19,2	13,8	14,3
32,7	35,5	19,0	20,0	3,2	3,2	13,4	16,2	10,6	13,2	8,7	11,3
27,5	26,0	10,4	10,5	3,2	3,3	14,3	12,8	10,4	9,3	9,8	8,7
92,2	98,8	50,7	50,8	18,0	19,8	33,2	35,2	19,3	20,7	17,5	18,5
175,2	181,0	95,7	106,2	23,8	23,3	55,8	52,1	25,8	24,0	18,2	12,8
9,9	9,3	5,0	5,2	1,4	1,6	4,2	3,4	2,6	2,2	2,5	1,7
138,9	134,9	81,0	77,5	16,3	15,1	50,0	52,6	21,2	22,4	16,8	20,3
33,7	31,5	16,5	15,6	5,1	5,1	12,1	12,4	11,5	11,7	8,2	8,5
:	:	:	:	:	:	:	:	:	:	:	:
30,5	24,1	17,9	16,9	4,9	4,9	7,2	2,9	13,8	5,7	– 8,0	– 18,0
45,1	44,4	25,5	25,6	10,3	10,9	9,2	10,2	8,9	9,5	2,6	3,1
14,1	12,4	5,7	5,8	1,9	1,8	8,2	6,4	6,2	5,0	5,0	3,8
20,5	20,0	10,7	9,3	2,6	2,2	6,9	9,9	7,9	10,7	6,4	9,7
89,5	81,9	41,6	41,8	14,0	15,2	29,8	27,6	22,5	21,0	15,3	13,2
92,7	76,2	43,5	36,4	10,4	6,4	40,0	34,5	32,2	34,7	31,1	32,5
19,4	17,8	8,5	8,0	3,3	3,3	8,6	7,5	8,5	7,3	7,6	6,3
:	:	:	:	:	:	:	:	:	:	:	:
16,3	15,5	10,4	12,1	2,2	2,8	6,2	4,6	4,9	3,9	3,5	1,5
136,3	135,8	69,9	63,9	19,7	18,9	48,3	57,4	24,8	31,5	19,7	32,5
36,4	34,1	16,1	16,2	4,8	5,0	15,7	14,3	10,3	9,5	8,2	7,1
94,8	93,4	45,0	46,0	7,7	8,3	43,1	40,6	30,6	29,6	26,3	23,6
57,7	52,3	32,3	33,0	7,8	8,0	17,3	13,4	20,3	15,5	– 4,4	– 9,1
91,7	89,0	49,7	49,5	15,0	15,9	28,5	26,0	19,0	17,2	13,8	11,2
16,1	14,6	4,5	4,7	1,7	1,7	10,9	8,8	6,0	4,9	5,5	4,4
21,3	20,7	8,9	7,9	2,3	2,3	10,1	11,6	9,0	10,8	8,4	11,0
95,3	88,9	46,4	46,4	13,8	14,2	32,7	32,8	20,8	20,4	14,7	14,0
85,2	72,1	46,1	43,8	7,1	6,6	35,2	25,2	21,0	14,8	16,8	10,6
22,6	20,1	7,6	7,4	2,9	3,0	12,4	10,3	8,9	7,5	8,3	6,8
:	:	:	:	:	:	:	:	:	:	:	:
128,3	119,3	62,7	68,9	16,7	18,4	48,9	31,3	30,9	21,9	19,0	4,7
8,2	7,3	3,6	3,6	1,2	1,3	3,9	3,1	2,3	1,9	2,2	1,5
220,9	213,3	114,0	116,3	27,3	26,8	83,7	82,0	28,5	26,8	26,7	26,9
92,5	99,1	43,2	46,3	10,5	12,3	38,6	40,8	15,8	16,0	14,7	13,4
116,9	119,0	49,3	52,2	15,5	16,8	51,7	49,6	20,1	18,9	20,8	17,7
267,5	211,5	149,2	105,5	20,8	18,1	96,8	88,2	21,7	21,7	9,7	6,8
172,3	190,4	86,6	106,5	18,2	20,1	66,7	63,9	18,4	17,3	21,0	14,6
22,1	22,7	6,5	7,5	3,1	3,2	13,1	12,4	6,9	6,6	6,9	6,8
23,9	27,7	9,0	8,1	0,8	1,2	13,9	18,4	9,2	11,5	9,8	13,5
114,8	106,7	52,2	47,0	14,4	15,9	48,2	43,9	17,0	16,0	16,0	15,1
:	:	:	:	:	:	:	:	:	:	:	:
47,1	52,2	14,8	17,3	4,4	4,6	27,7	30,1	13,8	14,4	13,6	14,3
:	:	:	:	:	:	:	:	:	:	:	:
347,7	363,0	169,1	187,4	47,8	52,9	130,2	125,6	25,7	24,6	29,4	18,6
10,4	8,9	3,8	3,9	2,6	3,3	4,8	1,7	2,9	1,1	3,6	0,7
303,6	210,7	210,6	109,0	17,9	16,9	74,8	85,3	15,6	16,5	14,5	19,1

3.2.3 *(cont.)*

Type of farming	Number of holdings				Size of holdings			
	In the FADN field of observation		In the sample (¹)		UAA (ha)		Labour input (AWU)	
	91/92	92/93	91/92	92/93	91/92	92/93	91/92	92/93
1	2	3	4	5	6	7	8	9
D. Vineyards								
EUR 12	248 230	243 123	2 961	2 992	9,0	9,1	1,49	1,46
Belgique/België	0	0	0	0	:	:	:	:
Danmark	0	0	0	0	:	:	:	:
BR Deutschland	17 320	17 320	169	161	7,7	7,7	1,85	1,80
Elláda	11 542	11 543	148	151	3,5	3,6	1,54	1,55
España	22 680	19 412	214	164	13,7	17,8	0,93	0,94
France	54 511	54 860	799	874	17,1	16,6	2,05	2,05
Ireland	0	0	0	0	:	:	:	:
Italia	107 587	104 918	1 427	1 414	5,7	5,8	1,27	1,24
Luxembourg	250	250	15	13	6,6	6,4	2,40	2,43
Nederland	0	0	0	0	:	:	:	:
Portugal	34 340	34 820	189	215	6,1	4,8	1,44	1,27
United Kingdom	0	0	0	0	:	:	:	:
E. Fruit (and other permanent crops)								
EUR 12	811 389	814 030	7 031	7 292	6,7	7,4	1,36	1,37
Belgique/België	1 760	1 760	49	48	8,7	8,9	2,56	2,62
Danmark	820	820	74	80	14,0	13,8	2,48	2,65
BR Deutschland	4 360	4 520	88	85	10,7	10,1	2,98	3,15
Elláda	240 191	240 190	1 511	1 600	4,4	4,4	1,54	1,55
España	136 530	135 389	1 328	964	11,4	15,7	1,18	1,20
France	13 872	14 082	270	284	19,0	20,4	2,85	3,06
Ireland	0	0	0	0	:	:	:	:
Italia	357 556	360 149	3 171	3 572	5,7	5,5	1,19	1,18
Luxembourg	0	0	0	0	:	:	:	:
Nederland	3 960	4 410	53	92	6,5	7,2	2,17	2,80
Portugal	51 180	51 180	464	507	8,0	9,0	1,27	1,30
United Kingdom	1 160	1 530	23	60	25,9	25,0	9,04	7,59
F. Dairy								
EUR 12	571 803	561 572	10 018	9 477	29,6	30,7	1,58	1,61
Belgique/België	15 900	14 930	287	295	28,7	31,8	1,56	1,57
Danmark	15 780	15 780	471	490	39,3	40,5	1,57	1,52
BR Deutschland	138 850	137 350	1 663	1 611	29,8	31,0	1,53	1,51
Elláda	2 178	2 652	19	17	6,4	5,1	1,91	1,91
España	84 780	80 420	1 508	1 260	7,1	7,4	1,24	1,47
France	111 900	111 860	1 388	1 339	41,2	42,2	1,53	1,50
Ireland	41 598	40 784	447	428	39,0	39,3	1,51	1,53
Italia	64 573	65 650	2 296	2 120	16,8	17,4	1,86	1,84
Luxembourg	1 520	1 520	226	220	56,8	58,5	1,70	1,68
Nederland	36 100	36 100	507	484	28,7	29,1	1,57	1,58
Portugal	23 573	19 646	356	373	9,3	10,3	1,74	1,76
United Kingdom	35 050	34 880	850	840	70,0	70,6	2,29	2,25
G. Drystock (excl. milk)								
EUR 12	498 533	505 074	7 294	7 503	45,7	46,6	1,45	1,41
Belgique/België	7 500	7 460	147	155	35,5	35,2	1,50	1,46
Danmark	740	970	15	18	28,7	26,6	1,01	0,91
BR Deutschland	15 910	15 660	193	200	35,0	36,2	1,47	1,46
Elláda	42 908	43 524	662	672	4,4	4,3	1,83	1,81
España	94 920	97 040	1 063	909	20,3	22,0	1,20	1,18
France	92 453	92 583	1 198	1 245	56,4	57,9	1,49	1,46
Ireland	84 207	85 076	656	705	37,3	36,9	1,08	1,03
Italia	66 744	64 019	1 774	1 951	33,3	32,1	1,74	1,72
Luxembourg	210	210	20	21	52,6	56,1	1,30	1,33
Nederland	6 520	6 520	38	34	15,2	18,9	1,55	1,45
Portugal	39 212	44 322	475	533	27,0	27,1	1,58	1,48
United Kingdom	47 210	47 690	1 053	1 060	171,4	177,9	1,69	1,63

Average results per holding in 1 000 ECU (current)

Total output		Intermediate consumption		Depreciation		Farm net value-added		Farm net value-added per AWU		Family farm income per unit unpaid labour	
91/92	92/93	91/92	92/93	91/92	92/93	91/92	92/93	91/92	92/93	91/92	92/93
10	11	12	13	14	15	16	17	18	19	20	21
38,1	39,2	11,4	12,0	5,4	5,7	20,9	21,5	14,1	14,7	12,3	12,4
:	:	:	:	:	:	:	:	:	:	:	:
:	:	:	:	:	:	:	:	:	:	:	:
54,1	55,8	23,0	22,1	9,3	9,3	21,5	25,0	11,6	13,9	8,3	10,9
11,0	11,1	1,9	2,0	1,5	1,8	8,9	9,5	5,8	6,1	5,8	6,4
13,0	13,4	3,4	4,1	3,5	4,7	6,1	4,5	6,6	4,8	5,8	3,0
100,4	105,6	31,7	33,9	11,4	11,9	55,7	58,8	27,2	28,6	26,0	27,1
:	:	:	:	:	:	:	:	:	:	:	:
21,7	20,4	4,9	4,9	3,9	3,9	12,8	11,7	10,1	9,4	9,6	8,8
90,3	137,3	30,4	37,2	15,0	17,0	47,1	87,3	19,6	36,0	18,8	39,3
:	:	:	:	:	:	:	:	:	:	:	:
7,9	6,0	2,1	1,7	1,2	1,1	4,6	3,3	3,2	2,6	2,4	2,0
:	:	:	:	:	:	:	:	:	:	:	:
18,4	16,7	4,8	5,2	2,7	2,6	11,8	9,9	8,7	7,3	8,4	6,4
107,5	113,3	32,0	38,8	11,3	12,3	64,1	61,5	25,1	23,5	30,9	27,2
108,5	98,0	47,0	41,2	7,1	7,7	53,5	48,6	21,6	18,3	13,5	2,5
131,3	115,1	53,5	49,8	15,5	16,2	63,0	49,9	21,1	15,9	25,5	12,4
11,2	9,2	2,1	2,3	1,2	1,3	9,2	7,0	6,0	4,5	6,0	4,6
15,4	17,2	4,3	4,4	4,0	2,9	7,8	11,7	6,6	9,8	6,2	11,0
114,3	101,4	38,2	44,3	13,4	14,2	63,2	46,7	22,1	15,3	25,6	11,3
:	:	:	:	:	:	:	:	:	:	:	:
18,0	15,1	4,0	4,1	2,5	2,5	12,2	9,0	10,3	7,6	9,9	6,7
:	:	:	:	:	:	:	:	:	:	:	:
110,3	131,7	37,4	56,4	14,9	21,0	57,2	54,5	26,3	19,5	28,3	16,5
8,4	7,5	2,8	2,9	1,6	1,8	4,2	3,1	3,3	2,4	3,5	1,9
355,6	268,6	173,0	153,5	25,1	20,3	157,3	94,9	17,4	12,5	22,0	− 0,2
66,7	71,8	36,6	38,2	8,5	9,0	23,1	26,1	14,6	16,3	11,1	12,7
79,8	91,9	38,0	40,7	8,8	10,7	35,9	42,8	23,1	27,3	18,6	21,9
115,8	119,7	62,6	68,0	11,5	11,9	42,5	41,4	27,1	27,2	14,3	12,4
66,9	72,9	38,1	40,1	12,6	13,3	19,1	22,3	12,5	14,7	8,9	10,9
37,2	41,8	24,4	25,4	2,3	2,2	40,0	14,9	20,9	7,8	20,4	7,3
20,0	23,7	10,8	10,6	1,6	1,7	7,6	11,5	6,1	7,8	5,9	7,6
67,5	70,5	38,2	38,2	8,5	8,8	22,0	24,7	14,4	16,5	10,8	12,7
52,9	60,5	28,4	30,7	4,8	4,8	21,1	27,2	14,0	17,8	11,9	16,0
61,9	62,3	31,9	31,6	5,2	5,4	25,2	25,5	13,6	13,9	12,9	13,7
95,7	99,6	54,2	54,2	19,3	21,9	32,8	30,5	19,3	18,2	17,7	15,4
124,5	132,5	66,7	73,8	15,6	17,0	42,8	42,3	27,2	26,8	15,9	13,9
18,1	22,6	11,4	14,6	1,7	2,1	5,5	6,8	3,2	3,8	3,1	3,3
147,0	148,9	80,7	78,0	15,6	14,3	54,5	60,2	23,8	26,8	20,8	25,8
33,3	34,3	19,6	19,5	4,1	4,1	14,1	16,1	9,7	11,4	7,9	9,7
68,7	73,8	39,4	38,8	7,7	7,7	25,2	31,8	16,8	21,7	12,3	16,9
63,7	43,8	39,1	32,8	7,5	6,9	17,9	5,1	17,7	5,6	− 2,2	− 15,0
75,3	80,6	46,9	48,7	13,0	14,2	19,1	22,1	13,0	15,1	8,9	10,6
17,6	17,3	8,1	8,3	0,9	0,9	11,7	11,5	6,4	6,4	6,3	6,2
18,2	22,7	9,9	10,0	1,2	1,1	8,9	14,5	7,4	12,3	7,1	12,3
47,6	49,3	28,2	28,1	7,3	7,4	19,0	21,2	12,8	14,6	9,4	11,0
14,8	15,2	9,3	9,3	1,8	1,8	7,1	8,6	6,6	8,4	5,5	7,3
52,5	53,7	26,5	27,7	3,8	4,2	23,2	23,8	13,3	13,8	12,5	13,2
60,5	63,4	39,3	40,0	13,2	13,9	20,1	20,4	15,5	15,3	12,4	12,2
56,9	75,3	46,5	52,5	9,9	10,9	1,8	13,6	1,1	9,4	− 9,2	− 0,2
10,0	9,4	6,0	6,3	1,5	1,8	4,4	3,3	2,8	2,2	3,7	2,2
51,1	48,4	36,3	34,2	9,2	8,0	22,5	23,9	13,3	14,7	8,4	10,8

3.2.3 *(cont.)*

Type of farming	Number of holdings				Size of holdings			
	In the FADN field of observation		In the sample (¹)		UAA (ha)		Labour input (AWU)	
	91/92	92/93	91/92	92/93	91/92	92/93	91/92	92/93
1	2	3	4	5	6	7	8	9
H. *Granivores*								
EUR 12	51 423	56 971	1 109	1 202	12,2	11,3	1,68	1,70
Belgique/België	3 940	3 980	136	116	5,3	7,4	1,33	1,35
Danmark	3 892	3 932	201	199	35,1	35,9	1,88	1,87
BR Deutschland	4 890	4 940	70	82	22,0	21,7	1,65	1,45
Elláda	1 090	1 010	26	27	1,1	1,1	2,16	2,23
España	9 330	12 680	153	222	3,9	2,6	1,11	1,28
France	7 840	8 220	146	158	19,3	18,2	1,64	1,65
Ireland	350	355	6	5	:	:	:	:
Italia	4 303	4 823	64	51	9,8	9,8	2,09	2,09
Luxembourg	0	0	0	0	:	:	:	:
Nederland	9 250	9 250	173	166	4,9	4,8	1,46	1,46
Portugal	2 870	3 720	54	62	10,3	7,4	1,69	1,82
United Kingdom	3 668	4 061	80	114	15,6	15,0	3,30	3,27
I. *Mixed (crops + livestock)*								
EUR 12	481 926	483 094	7 989	8 004	29,6	30,8	1,65	1,65
Belgique/België	11 460	11 590	313	321	30,5	30,6	1,64	1,61
Danmark	16 990	17 210	569	540	40,1	40,3	1,27	1,25
BR Deutschland	85 400	85 600	1 557	1 646	34,2	34,8	1,51	1,50
Elláda	35 564	34 724	433	416	7,6	7,5	1,79	1,80
España	46 170	43 960	271	373	20,2	26,6	1,10	1,25
France	78 792	76 716	1 266	1 312	53,9	57,7	1,67	1,64
Ireland	4 265	4 136	93	92	65,5	68,9	1,88	1,91
Italia	91 378	96 427	2 571	2 410	17,4	17,3	1,76	1,76
Luxembourg	220	300	33	34	61,9	51,0	1,52	1,46
Nederland	7 440	7 630	64	74	20,9	21,1	1,52	1,53
Portugal	91 275	92 192	433	417	12,5	12,8	1,79	1,75
United Kingdom	12 972	12 609	386	369	129,0	133,8	2,96	2,70

Source : EC Commission, Directorate-General for Agriculture, FADN—Weighting by farm structure survey 1987, classification as Decision 85/377/EEC, standard gross margins » 1984 «.

(¹) Results for groups of less than 10 holdings are not considered representative and are therefore not included in the table, although they are included in totals.

Average results per holding in 1 000 ECU (current)

Total output		Intermediate consumption		Depreciation		Farm net value-added		Farm net value-added per AWU		Family farm income per unit unpaid labour	
91/92	92/93	91/92	92/93	91/92	92/93	91/92	92/93	91/92	92/93	91/92	92/93
10	11	12	13	14	15	16	17	18	19	20	21
205,7	207,2	137,8	144,1	19,2	13,7	48,6	49,7	28,9	29,2	26,2	26,9
252,2	200,8	156,1	142,6	12,5	14,7	84,6	43,6	63,5	32,3	58,9	24,8
287,1	261,5	159,8	170,2	21,7	23,2	104,8	70,0	55,8	37,5	50,4	18,8
162,8	132,2	98,5	92,6	24,6	17,0	38,6	24,7	23,4	17,0	19,1	10,4
125,2	125,2	75,3	82,8	4,3	4,6	44,3	36,9	20,6	16,6	20,3	18,5
76,2	117,7	56,0	76,1	3,4	4,3	16,1	36,5	14,5	28,6	15,7	34,5
221,7	237,9	158,1	165,2	17,2	17,8	46,9	56,3	28,6	33,3	22,3	27,9
:	:	:	:	:	:	:	:	:	:	:	:
240,9	298,9	159,2	179,9	9,3	9,6	72,0	109,2	34,4	52,1	34,1	53,7
:	:	:	:	:	:	:	:	:	:	:	:
275,8	270,7	183,1	204,8	47,5	22,6	45,2	42,9	30,9	29,3	20,5	15,7
76,0	77,3	63,3	56,8	2,0	2,7	11,3	17,1	6,7	9,4	6,8	12,1
294,3	309,9	217,9	229,9	19,3	16,4	57,8	64,1	17,5	19,6	18,2	23,8
62,0	60,9	36,3	37,0	7,2	7,5	19,8	18,5	12,0	11,2	9,1	8,1
119,6	113,2	64,3	66,2	9,2	10,0	47,6	38,6	29,1	24,0	24,0	18,1
122,4	105,7	71,4	72,5	11,3	11,5	39,6	23,8	31,2	19,1	15,6	− 1,6
99,7	95,4	62,0	63,3	14,1	14,9	26,2	20,1	17,4	13,4	13,3	8,3
18,1	16,9	8,1	7,8	1,2	1,2	10,8	10,0	6,0	5,6	5,7	5,3
20,2	29,8	10,7	14,3	1,7	1,9	8,7	15,1	7,9	12,1	7,4	12,0
95,5	98,7	57,7	59,9	11,9	12,6	27,0	30,1	16,2	18,3	11,2	12,9
89,8	104,2	54,1	63,5	7,4	9,2	32,4	36,9	17,2	19,3	14,4	16,3
41,4	39,5	20,0	19,3	4,3	4,3	17,6	16,6	10,0	9,4	9,3	8,7
99,8	87,1	60,4	52,4	17,1	15,2	32,7	26,1	21,5	17,9	18,5	15,4
167,1	166,1	111,8	119,4	15,2	16,4	39,9	30,8	26,2	20,2	13,8	7,4
8,7	8,5	5,1	5,3	1,1	1,4	3,3	2,7	1,8	1,6	1,8	1,3
198,9	185,0	120,6	112,6	20,8	18,3	64,8	64,5	21,9	23,9	16,1	20,8

3.2.4 Results by income class, 1991/92 and 1992/93

Class of income in 1 000 ECU		Number of holdings in the FADN field of observation		Area (ha UAA)		Average results per holding in 1 000 ECU											
						Total output		Intermediate consumption		Depreciation		Farm net value-added		Farm net value-added per AWU		Family farm income per unit of unpaid labour	
		91/92	92/93	91/92	92/93	91/92	92/93	91/92	92/93	91/92	92/93	91/92	92/93	91/92	92/93	91/92	92/93
1	2	3	4	5	6	7	8	9	10	11	12	13	14	15	16	17	18
EUR 12 (p)																	
	<0-5	1 869 052	1 908 299	15,6	15,2	19,5	21,0	11,7	13,3	3,9	4,1	4,5	4,5	3,1	3,1	0,8	0,4
	5-10	937 298	906 525	18,7	18,1	28,5	29,6	13,9	14,7	3,7	4,0	12,1	12,5	8,6	8,7	7,3	7,3
	10-20	823 402	814 036	29,3	30,8	54,1	53,5	26,4	26,4	6,3	6,4	22,9	23,2	15,8	16,2	14,1	14,2
	20-30	273 040	261 505	41,8	45,9	94,5	95,4	44,9	46,0	10,1	10,4	41,3	42,4	24,8	25,7	24,2	24,3
	>30	266 937	266 723	59,0	68,0	180,4	171,4	79,8	76,6	16,4	15,8	86,5	83,8	39,3	39,7	52,2	52,4
	All holdings	4 169 729	4 157 089	23,5	24,2	43,6	43,6	21,7	22,3	5,5	5,7	17,5	17,4	11,6	11,6	9,4	9,2
Belgique/België																	
	<0-5	5 610	5 695	25,3	17,2	48,2	68,5	31,9	47,5	8,3	10,9	8,8	10,4	5,5	6,3	-1,2	-2,0
	5-10	7 495	7 225	22,6	23,3	48,8	67,3	26,6	41,0	6,2	7,6	17,7	19,5	11,1	12,5	7,7	7,9
	10-20	15 948	16 494	25,3	25,7	70,6	77,8	36,9	40,7	6,9	8,2	28,3	30,3	18,1	19,2	14,8	15,0
	20-30	9 551	10 634	30,2	30,2	115,0	122,1	58,3	61,2	11,1	12,3	48,4	50,7	28,1	29,2	24,0	24,4
	>30	12 736	10 451	27,5	36,3	205,1	182,6	101,5	89,6	14,6	15,6	91,5	81,1	50,2	44,7	51,7	43,8
	All holdings	51 340	50 500	26,4	27,6	106,6	106,3	54,8	55,9	9,6	10,8	44,0	41,3	26,4	24,8	23,1	20,3
Danmark																	
	<0-5	39 015	51 088	28,7	31,5	47,1	56,4	31,1	41,0	6,4	7,9	9,4	9,0	11,5	9,7	-13,6	-15,6
	5-10	8 904	7 061	32,1	33,7	69,8	86,1	40,6	51,0	7,5	8,8	21,7	27,6	19,7	21,7	7,6	7,5
	10-20	12 990	10 101	37,7	39,5	104,3	112,8	56,5	65,0	10,0	11,1	38,0	38,7	26,8	27,0	14,8	14,7
	20-30	7 713	4 619	41,3	46,8	137,0	165,0	71,6	90,2	12,5	15,9	53,0	60,8	34,7	35,9	24,5	24,2
	>30	8 070	4 314	64,2	68,7	305,3	309,6	155,7	158,3	25,4	28,2	123,3	126,5	54,2	49,9	57,1	51,6
	All holdings	76 692	77 182	35,6	35,7	95,6	87,1	53,7	54,6	9,8	10,0	32,1	24,2	27,2	20,9	10,3	0,6
BR Deutschland																	
	<0-5	113 070	113 260	25,0	26,4	51,6	61,8	35,6	44,2	12,2	13,0	4,9	6,1	3,4	4,1	-1,9	-2,6
	5-10	64 799	69 588	27,7	27,2	62,8	67,2	37,1	40,1	11,0	12,2	17,2	17,4	10,9	11,1	7,5	7,5
	10-20	86 137	89 812	32,7	33,1	88,1	87,7	48,7	48,0	13,4	14,0	28,6	29,0	17,5	17,9	14,4	14,5
	20-30	36 605	34 574	38,6	41,9	122,4	123,4	65,5	64,6	16,1	17,3	43,9	45,4	26,0	26,5	24,2	24,5
	>30	31 059	23 616	49,7	52,1	182,4	176,1	90,2	84,2	20,6	22,2	75,4	74,5	39,3	39,3	45,6	45,1
	All holdings	331 670	330 850	31,3	31,9	83,3	84,6	47,7	49,3	13,5	14,2	24,4	23,7	15,3	14,9	11,3	10,3
Elláda																	
	<0-5	280 144	341 290	4,8	5,0	9,3	8,7	3,4	3,5	1,2	1,4	5,4	4,7	3,1	2,8	2,8	2,5
	5-10	193 349	158 513	6,0	6,2	15,3	16,1	4,1	4,9	1,4	1,5	11,3	11,5	6,9	6,9	6,8	6,9
	10-20	75 063	54 937	8,4	7,9	24,9	24,6	6,1	7,0	2,0	2,0	19,3	18,2	12,1	11,9	13,1	13,1
	20-30	8 874	4 850	14,7	16,6	45,5	48,0	11,0	12,2	3,2	3,3	34,8	36,7	18,2	19,1	23,7	24,1
	>30	3 673	1 752	16,3	14,1	71,8	94,2	14,6	35,0	3,5	5,0	83,1	58,5	42,9	29,8	57,7	42,6
	All holdings	561 103	561 343	5,9	5,8	14,5	13,0	4,2	4,4	1,4	1,5	10,3	8,4	6,1	5,0	5,9	4,8
España																	
	<0-5	313 597	169 391	16,9	25,0	13,3	15,6	7,3	9,8	2,9	3,9	3,0	2,4	2,6	2,1	1,0	-0,2
	5-10	141 901	177 725	20,2	14,3	18,0	16,9	8,2	6,6	1,6	1,4	8,5	9,7	7,5	7,4	7,1	7,2
	10-20	123 975	188 134	21,7	17,3	25,5	21,7	10,8	7,6	1,8	1,4	13,5	14,0	12,7	13,2	13,8	13,9
	20-30	42 140	47 087	24,4	33,3	35,6	35,8	14,2	12,8	2,0	1,7	20,9	23,8	18,0	19,7	24,0	24,0
	>30	34 499	56 615	34,6	49,2	49,1	60,4	18,5	23,5	2,4	2,5	30,4	39,2	24,5	29,0	45,1	50,9
	All holdings	656 112	638 952	19,9	22,5	19,9	23,2	9,2	9,7	2,4	2,2	8,8	12,7	7,7	10,7	7,2	10,9

France																
<0-5	135 200	122 721	38,7	39,6	47,7	50,3	31,2	32,9	9,1	10,1	8,6	10,3	5,4	6,1	-0,2	-1,0
5-10	110 615	101 796	43,2	42,1	58,9	55,9	33,7	32,6	8,4	8,4	18,6	18,4	11,5	11,5	7,7	7,5
10-20	155 767	167 212	50,0	51,3	83,3	112,7	44,3	42,3	10,6	10,4	29,7	30,0	18,5	18,8	14,3	14,4
20-30	60 929	67 231	57,8	59,6	113,7	185,2	54,9	55,2	13,5	13,9	45,3	47,2	28,0	28,7	24,2	24,2
>30	54 400	57 575	61,9	66,2	193,7		73,3	75,9	19,8	20,5	97,9	91,6	44,9	45,8	53,9	51,3
All holdings	516 910	516 535	47,8	49,5	84,0	83,9	42,9	43,6	11,0	11,5	30,8	32,1	18,5	19,2	13,8	14,3
Ireland																
<0-5	65 854	46 194	30,3	27,2	15,5	17,6	11,3	13,2	1,9	1,9	4,3	5,1	3,7	4,4	1,8	2,3
5-10	32 295	35 841	37,6	35,7	22,5	18,5	13,2	10,8	2,2	1,9	10,0	9,5	8,5	8,4	7,2	7,4
10-20	24 316	35 130	49,1	44,3	46,4	35,7	25,3	19,2	4,4	3,2	20,6	18,3	15,6	15,4	14,1	13,9
20-30	6 664	8 850	61,6	55,2	84,8	76,8	44,2	39,1	7,5	6,1	37,7	36,5	24,2	25,9	23,5	24,6
>30	5 542	8 956	89,8	80,6	173,9	154,8	85,5	76,3	13,4	11,6	78,9	71,6	35,8	37,2	50,1	48,9
All holdings	134 670	134 970	39,4	39,3	32,7	35,5	19,0	20,0	3,2	3,2	13,4	16,2	10,6	13,2	8,7	11,3
Italia																
<0-5	507 184	628 941	7,0	7,1	11,5	10,8	5,2	5,1	2,6	2,6	3,9	3,6	2,9	2,8	2,3	2,0
5-10	317 010	289 842	8,6	9,6	19,3	21,6	7,2	8,5	2,6	3,0	9,9	10,6	7,8	8,0	7,4	7,4
10-20	252 859	189 952	13,2	14,8	34,4	37,8	12,6	14,2	3,5	3,9	19,0	20,5	14,3	14,2	14,0	14,0
20-30	62 256	48 204	21,5	24,9	70,3	79,2	25,4	30,9	5,8	6,6	39,9	43,1	23,5	23,6	24,5	24,4
>30	60 518	47 486	30,5	31,4	131,3	153,4	45,7	60,0	8,0	9,0	79,6	86,9	37,9	39,4	51,9	53,1
All holdings	1 199 827	1 204 425	10,7	10,6	27,5	26,0	10,4	10,5	3,2	3,3	14,3	12,8	10,4	9,3	9,8	8,7
Luxembourg																
<0-5	333	573	50,6	50,2	69,6	63,6	50,0	43,3	18,3	17,4	10,4	7,5	6,4	5,0	-1,3	-2,0
5-10	301	201	48,2	55,8	77,8	100,5	46,1	60,9	17,9	22,4	22,4	24,7	12,9	13,9	7,8	7,8
10-20	749	636	47,1	57,8	82,6	93,2	45,4	50,6	15,2	19,8	31,8	31,6	17,9	19,3	15,1	15,0
20-30	462	428	57,8	50,3	108,3	101,1	57,3	47,1	20,1	19,3	43,4	41,8	25,7	24,5	24,6	23,2
>30	356	442	54,3	43,6	124,7	149,3	58,0	59,9	20,9	22,1	53,6	74,9	30,6	37,3	41,0	45,7
All holdings	2 201	2 280	51,2	51,6	92,2	98,8	50,7	50,8	18,0	19,8	33,2	35,2	19,3	20,7	17,5	18,5
Nederland																
<0-5	22 236	32 930	21,8	22,0	116,0	139,9	86,4	104,4	30,3	23,5	-2,0	11,4	-1,0	5,3	-23,8	-18,8
5-10	7 724	9 235	20,7	18,9	114,1	122,1	67,3	75,9	15,1	14,9	31,2	31,6	17,5	16,9	7,7	7,8
10-20	22 117	16 676	19,7	22,3	121,1	166,0	68,1	95,9	13,6	20,4	39,5	50,1	22,7	24,9	14,6	15,3
20-30	14 760	13 128	22,5	20,0	159,3	162,4	80,0	87,4	18,5	18,2	61,3	57,7	30,6	32,1	24,4	24,6
>30	24 947	20 483	24,6	26,2	304,2	297,6	146,6	143,3	32,8	32,7	126,1	124,7	44,2	46,0	53,9	52,1
All holdings	91 784	92 452	22,1	22,4	175,2	181,0	95,7	106,2	23,8	23,3	55,8	52,1	25,8	24,0	18,2	12,8
Portugal																
<0-5	344 411	357 698	9,5	11,0	7,1	6,8	4,0	4,1	1,2	1,4	2,4	1,9	1,5	1,2	1,0	0,6
5-10	33 207	30 130	16,8	15,7	16,9	17,1	7,4	8,1	1,6	2,2	9,1	8,3	6,0	5,8	6,9	6,8
10-20	22 346	13 843	30,7	25,4	22,8	28,9	9,5	14,1	2,4	2,7	12,2	14,5	8,8	10,1	13,6	13,9
20-30	5 127	4 343	37,2	37,3	33,5	37,4	13,7	19,4	3,7	3,1	20,0	17,6	9,5	11,1	24,4	23,8
>30	6 409	3 557	57,5	58,2	59,1	85,6	27,2	40,5	5,0	5,6	31,1	46,7	12,2	16,9	45,8	58,4
All holdings	411 500	409 570	12,3	12,6	9,9	9,3	5,0	5,2	1,4	1,6	4,2	3,4	2,6	2,2	2,5	1,7
United Kingdom																
<0-5	42 398	38 519	90,0	80,1	79,0	84,1	55,8	62,7	12,7	12,8	17,3	16,1	8,6	7,4	-5,6	-7,5
5-10	19 697	19 368	91,3	83,7	68,1	64,9	43,9	41,1	9,3	8,6	22,6	22,9	12,8	12,5	7,5	7,6
10-20	31 137	31 110	116,6	125,7	108,9	93,4	64,8	57,3	13,2	10,8	40,2	36,3	18,5	18,3	14,5	14,8
20-30	17 960	17 558	133,6	126,6	156,4	130,0	87,7	72,5	17,8	14,1	60,2	55,8	25,3	25,5	24,3	24,3
>30	24 728	31 476	217,1	210,6	323,3	283,9	169,3	140,5	30,9	26,6	132,6	130,1	36,2	38,9	59,5	64,2
All holdings	135 920	138 030	125,2	126,6	138,9	134,9	81,0	77,5	16,3	15,1	50,0	52,6	21,2	22,4	16,8	20,3

Source : EC Commission, Directorate-General for Agriculture, FADN—Weighting by farm structure survey 1990, classification as Decision 94/376/EEC, standard gross margins « 1988 ». Results for groups of less than 10 holdings are not considered representative and are therefore not included in the table, although they are included in totals.

(¹) This indicator represents the farmer's income and that of his family.

3.3.1 Agricultural prices and amounts of Community aid (beginning of marketing year)

	Category of price or amount in ECU/tonne except as stated	1991/92	1992/93	1993/94	1994/95	% TAV 1992/93 / 1991/92	% TAV 1993/94 / 1992/93	% TAV 1994/95 / 1993/94
1	2	3	4	5	6	7	8	9
Cereals								
Compensatory payments (arable scheme):								
1. All cereals (¹)		—	—	25,00	35,00	x	x	40,0
2. Supplement durum wheat/ha		—	—	297,00	297,00	x	x	0,0
3. Protein products (¹)		—	—	65,00	65,00	x	x	0,0
4. Set-aside (¹)		—	—	45,00	57,00	x	x	26,7
5. Seed flax (¹)		—	—	85,00	87,00	x	x	2,4
6. Oilseeds (²)		—	—	359,00	359,00	x	x	0,0
Marketing year: July-June								
Beginning of single market: 1967/1968								
1. Durum wheat	Target price	277,21	269,10	128,32	118,45	-2,9	-52,3	-7,7
	Single/basic intervention price	227,70	220,87	115,49	106,60	-3,0	-47,7	-7,7
	Threshold price	272,62	264,31	172,74	162,87	-3,0	-34,6	-5,7
	Production aid/ha	181,88/146,34	181,88	—	—	0,0	x	x
2. Common wheat	Target price	233,26	226,47	128,32	118,45	-2,9	-43,3	-7,7
	Single/basic intervention price	168,55	163,49	115,49	106,60	-3,0	-29,4	-7,7
	Threshold price	228,67	221,68	172,74	162,87	-3,1	-22,1	-5,7
3. Barley	Target price	212,33	206,16	128,32	118,45	-2,9	-37,8	-7,7
	Single/basic intervention price	160,13	155,33	115,49	106,60	-3,0	-25,6	-7,7
	Threshold price	207,74	201,37	172,74	162,87	-3,1	-14,2	-5,7
4. Rye	Target price	212,33	206,16	128,32	118,45	-2,9	-37,8	-7,7
	Single/basic intervention price	160,13	155,33	115,49	106,60	-3,0	-25,6	-7,7
	Threshold price	207,74	201,37	172,74	162,87	-3,1	-14,2	-5,7
5. Maize	Target price	212,33	206,16	128,32	118,45	-2,9	-37,8	-7,7
	Single/basic intervention price	168,55	163,49	115,49	106,60	-3,0	-29,4	-7,7
	Threshold price	207,74	201,37	172,74	162,87	-3,1	-14,2	-5,7
Rice								
Marketing year: September-August								
Beginning of single market: 1967/1968								
1. Paddy rice	Target price	313,65	313,65	309,60	309,60	0,0	-1,3	0,0
2. Husked rice	Target price	546,13	545,52	530,60	530,60	-0,1	-2,7	0,0
Round-grain	Threshold price	540,05	539,44	523,88	523,88	-0,1	-2,9	0,0
Long-grain	Threshold price	540,05	539,44	523,88	523,88	-0,1	-2,9	0,0

3. Wholly milled							
Round-grain	Threshold price	718,65	717,86	697,78	697,78	-0,1	-2,8	0,0
Long-grain	Threshold price	789,52	788,64	766,09	766,09	-0,1	-2,9	0,0
4. Broken rice	Threshold price	292,49	281,91	241,83	276,88	-3,6	-14,2	14,5

Sugar and isoglucose

Marketing year : July-June
Beginning of single market :
1968/1969 : sugar
1977/1978 : isoglucose

1. Beet							
Basic price							
Community	40,00	40,00	39,48	39,48	0,0	-1,3	0,0
Italia	40,00	40,00	39,48	39,48	0,0	-1,3	0,0
Ireland	40,00	40,00	39,48	39,48	0,0	-1,3	0,0
United Kingdom	40,00	40,00	39,48	39,48	0,0	-1,3	0,0
Minimum price for 'A' sugarbeet							
Community	39,20	39,20	38,69	38,69	0,0	-1,3	0,0
Italia	41,72	41,72	41,21	41,21	0,0	-1,2	0,0
Ireland	40,77	40,77	40,26	40,26	0,0	-1,3	0,0
United Kingdom	40,77	40,77	40,26	40,26	0,0	-1,3	0,0
España/Portugal	46,04/42,03	41,02/40,77	40,51/40,26	40,51/40,26	-10,9/-3,0	-1,2/-1,3	0,0/0,0
Minimum price for 'B' sugarbeet							
Community	27,20	27,20	26,85	26,85	0,0	-1,3	0,0
Italia	29,72	29,72	29,37	29,37	0,0	-1,2	0,0
Ireland	28,77	28,77	28,42	28,42	0,0	-1,2	0,0
United Kingdom	28,77	28,77	28,42	28,42	0,0	-1,2	0,0
España/Portugal	34,04/30,03	29,02/28,77	28,67/28,42	28,67/28,42	-14,7/-4,2	-1,2/-1,2	0,0/0,0
2. Raw sugar							
Threshold price	546,00	546,00	539,90	539,90	0,0	-1,1	0,0
3. White sugar							
Target price	557,90	557,90	550,70	550,70	0,0	-1,3	0,0
Intervention price							
Community	530,10	530,01	523,30	523,30	-0,0	-1,3	0,0
Italia	549,50	549,50	542,70	542,70	0,0	-1,2	0,0
French OD	530,10	530,01	523,30	523,30	-0,0	-1,3	0,0
Ireland	542,20	542,20	535,40	535,40	0,0	-1,3	0,0
United Kingdom	542,20	542,20	535,40	535,40	0,0	-1,3	0,0
España/Portugal	612,90/533,50	544,10/542,20	537,30/535,40	537,30/535,40	-11,2/1,6	-1,2/-1,3	0,0/0,0
Threshold price	639,00	639,00	631,80	631,80	0,0	-1,1	0,0
4. Molasses							
Threshold price	68,90	68,00	68,00	68,00	0,0	-1,3	0,0

Olive oil

Marketing year : November-October
Beginning of single market : 1966/1967

Target production price	3 220,10	3 211,60	3 178,20	3 178,20	-0,3	-1,0	0,0
Intervention price	2 158,70	2 018,40	1 919,80	1 624,00 (³)	-6,5	-4,9	-15,4
Production aid	708,30	841,10	881,80	1 177,60	18,7	4,8	33,5

3.3.1 (cont.)

1	Category of price or amount in ECU/tonne except as stated	1991/92	1992/93	1993/94	1994/95	% TAV 1992/93 / 1991/92	% TAV 1993/94 / 1992/93	% TAV 1994/95 / 1993/94
	2	3	4	5	6	7	8	9
Oilseeds								
Marketing year:								
Rapeseed: July-June								
Sunflower: October-September								
From 1972/73: September-August								
Soya: November-October								
Flax seeds: August-July								
Castor beans: October-September								
Beginning of single market: 1967/68								
Soya: 1974/75								
Flax seeds: 1976/77								
Castor beans: 1978/79								
1. Rape	Target price	442,70	442,70	–	–	0,0	x	x
	Basic intervention price	400,80	400,80	–	–	0,0	x	x
2. Sunflower	Target price	573,80	573,80	–	–	0,0	x	x
	Basic intervention price	525,80	525,80	–	–	0,0	x	x
3. Soya	Target price	549,10	549,10	–	–	0,0	x	x
	Minimum price	481,30	481,30	–	–	0,0	x	x
4. Flax seeds	Target price	544,90	544,90	–	–	0,0	x	
Dried fodder								
Marketing year:								
Dehydrated lucerne: April-March								
Beginning of single market: 1974/75								
1. Dehydrated lucerne	Target price	178,61	176,37	176,29	176,29	–1,3	–0,0	0,0
Cotton (natural)								
Marketing year: September-August								
Beginning of single market: 1981/82								
	Target price ([*])	958,60	1 027,90	1 014,60	1 014,60	7,2	–1,3	0,0
	Minimum price ([*])	910,70	976,50	963,90	963,90	7,2	–1,3	0,0
Flax and hemp — ECU/ha								
Marketing year: August-July								
Beginning of single market: 1970/71								
1. Flax	Community aid	374,36	374,36	774,86	774,86	0,0	107,0	0,0
2. Hemp	Flat-rate aid	339,42	339,42	641,60	641,60	0,0	89,0	0,0

Seeds (⁵)

Marketing year: July-June
Beginning of single market: 1972/1973
(Fibre flax: 1973/1974,
Monoecious hemp: 1975/1976 and
Seed flax: 1977/1978)

Item	Type							
1. Monoecious hemp (²)	Aid	172,00	172,00	170,00	170,00	0,0	-1,2	0,0
2. Fibre flax (²)	Aid	238,00	238,00	235,00	235,00	0,0	-1,3	0,0
3. Seed flax (²)	Aid	188,00	188,00	186,00	186,00	0,0	-1,1	0,0
4. Grasses (²)	Aid	161,00 à 701,00	161,00 à 701,00	159,00 à 692,00	159,00 à 692,00	0,00,0	-1,2/-1,3	0,00,0
5. Legumes (²)	Aid	168,00 à 594,00	168,00 à 630,00	166,00 à 622,00	166,00 à 622,00	0,06,1	-1,2/-1,3	0,00,0

Wine — ECU/degree-hl orhl (according to type)

Marketing year: September-August
Beginning of single market: 1969/1970

Item	Type							
A — 1. Type R I	Guide price	3,21/3,01	3,21/3,01	3,17	3,17	0,00,0	-1,2/-5,3	0,0
2. Type R II	Guide price	3,21/3,01	3,21/3,01	3,17	3,17	0,00,0	-1,2/-1,2	0,0
3. Type R III	Guide price	52,14/48,81	52,14/48,81	51,47	51,47	0,00,0	-1,3/-5,4	0,0
4. Type A I	Guide price	3,21/3,01	3,21/3,01	3,17	3,17	0,00,0	-1,2/-5,3	0,0
5. Type A II	Guide price	69,48/65,04	69,48/65,04	68,58	68,58	0,00,0	-1,3/-5,4	0,0
6. Type A III	Guide price	79,35/74,28	79,35/74,28	78,32	78,32	0,00,0	-1,3/-5,4	0,0
B — 1. Red wine	Reference price	4,37	4,37	4,31	4,31	0,0	-1,4	0,0
2. White wine	Reference price	4,37	4,37	4,31	4,31	0,0	-1,4	0,0
3. Liqueur wine	Reference price	69,00/75,20	69,00/75,20	59,22/98,02	59,22/98,02	0,00,0	-14,2/30,3	0,00,0
4. Liqueur wine (processed)	Reference price	60,60/86,70	60,60/86,70	59,82/85,58	59,82/85,58	0,00,0	-1,3/-1,3	0,00,0
5. Wine (fortified for distillation)	Reference price	2,59	2,59	2,56	2,56	0,0	-1,2	0,0
6. Grape must	Reference price	2,78	2,78	2,74	2,74	0,0	-1,4	0,0
7. White wine (Riesling-Sylvaner)	Reference price	88,76	88,76	87,61	87,61	0,0	-1,3	0,0
C — Grape juice								
1. White	Reference price	3,98	3,98	3,93	3,93	0,0	-1,3	0,0
2. Other	Reference price	3,98	3,98	3,93	3,93	0,0	-1,3	0,0

Leaf tobacco — ECU/kg (⁶)

Item	Type							
I — Flue cured	Premium	–	–	2,244	2,244	x	x	0,0
Virgin D, Virginia and hybrids thereof (⁷)	Supplementary amount	–	–	0,321	0,321	x	x	0,0
II — Light air cured	Premium	–	–	1,795	1,795	x	x	0,0
Badischer Burley E and hybrids thereof (⁷)	Supplementary amount	–	–	0,562	0,562	x	x	0,0

3.3.1 *(cont.)*

1	2 Category of price or amount in Ecus/tonne except as stated	3 1991/92	4 1992/93	5 1993/94	6 1994/95	7 % TAV 1992/93 / 1991/92	8 % TAV 1993/94 / 1992/93	9 % TAV 1994/95 / 1993/94
III — Dark air cured Badischer Geudertheimer, Pereg, Korso (7)	Premium	–	–	1,795	1,795	x	x	0,0
	Supplementary amount	–	–	0,351	0,351	x	x	0,0
Paraguay and hybrids thereof, Dragon vert and hybrids thereof, Philippin, Petit Grammont (Flobecq), Semois, Appelterre (7)								
Nijkerk (7)	Supplementary amount	–	–	0,262	0,262	x	x	0,0
	Supplementary amount	–	–	0,153	0,153	x	x	0,0
Misionero and hybrids thereof, Rio Grande and hybrids thereof (7)								
	Supplementary amount	–	–	0,167	0,167	x	x	0,0
IV — Fire cured	Premium	–	–	1,974	1,974	x	x	0,0
V — Sun cured	Premium	–	–	1,795	1,795	x	x	0,0
VI — Basmas	Premium	–	–	2,961	3,109	x	x	5,0
VII — Katerini and similar varieties	Premium	–	–	2,512	2,638	x	x	5,0
VIII — Kaba Koulak classique	Premium	–	–	1,795	1,885	x	x	5,0
Fruit and vegetables — ECU/100 kg Marketing year: differs according to product Beginning of single marketing year: 1966/1967								
1. Cauliflowers	Basic price	30,91	30,91	30,91	30,56	0,0	0,0	-1,1
	Buying-in price	13,45	13,45	13,45	13,30	0,0	0,0	-1,1
2. Tomatoes (open grown)	Reference price	81,39	81,39	81,39	81,39	0,0	0,0	0,0
	Basic price	28,41	28,41	28,15	28,09	0,0	-0,9	-0,2
	Buying-in price	10,80	10,80	10,70	10,68	0,0	-0,9	-0,2
3. Oranges (Group 1)	Basic price	34,45	31,39	29,12	33,04	-8,9	-7,2	13,5
	Buying-in price	21,78	19,85	18,42	20,88	-8,9	-7,2	13,4
4. Mandarins	Basic price	36,95	30,03	36,48	32,93	-18,7	21,5	-9,7
	Buying-in price	23,64	19,27	23,34	21,09	-18,5	21,1	-9,6
5. Lemons	Reference price	51,18	51,18	51,18	51,18	0,0	0,0	0,0
	Basic price	40,18	42,47	40,85	34,45	5,7	-3,8	-15,7
	Buying-in price	23,59	24,95	23,99	20,19	5,8	-3,8	-15,8

6. Table grapes	Reference price	49,42	49,42	49,42	49,42	0,0	0,0	0,0
	Basic price	36,25	36,25	35,83	35,82	0,0	-1,2	-0,0
	Buying-in price	23,31	23,31	23,05	23,04	0,0	-1,1	-0,0
7. Apples (Group 1)	Reference price	48,71	48,71	48,71	48,71	0,0	0,0	0,0
	Basic price	26,16	26,46	24,62	24,04	1,1	-7,0	-2,4
	Buying-in price	13,33	13,48	12,54	12,25	1,1	-7,0	-2,3
8. Pears	Reference price	46,98	46,98	46,98	46,98	0,0	4,1	0,0
	Basic price	28,62	28,62	28,27	28,27	0,0	-1,2	0,0
	Buying-in price	14,73	14,73	14,54	14,54	0,0	-1,3	0,0
9. Peaches	Reference price	64,30	64,30	64,30	64,30	0,0	0,0	0,0
	Basic price	42,26	42,70	36,17	37,45	1,0	-15,3	3,5
	Buying-in price	23,47	23,71	20,06	20,78	1,0	-15,4	3,6
10. Cherries	Reference price	119,32	119,32	119,32	119,32	0,0	0,0	0,0
11. Plums (Group 1)	Reference price	66,21	66,21	66,21	66,21	0,0	0,0	0,0
12. Cucumbers	Reference price	76,47	76,47	76,47	76,47	0,0	0,0	0,0

Products processed from fruit and vegetables
— ECU/100 kg

Marketing years: varies according to product
Beginning of single market:
Tomato concentrates: 1975/1976
Preserved pineapple: 1976/1977
Other: 1978/1979

1. Preserved pineapple	Aid	112,615	104,729	111,090	125,565	-7,0	6,1	13,0
	Minimum price	31,586	31,586	31,178	31,178	0,0	-1,3	0,0
2. Peaches in syrup	Production aid	11.543/10,894	8,428	6,794	7,174	-27,0/-22,2	-19,4	5,6
	Minimum producer price	26,738/26,089	26,738	22,962	23,832	0,0/2,5	-14,1	3,8
3. Prunes	Production aid	66,357/61,340	66,570	64,751	61,094	0,3/8,5	-2,7	-5,6
	Minimum producer price	158.403/153,122	158,403	160,266	160,266	0,0/3,4	1,2	0,0

Milk products
Marketing year: April-March
Beginning of single market: 1968/1969

1. Milk (3.7% FC)	Target price	268,10	268,10	260,60	256,60	0,0	-2,8	-1,5
2. Butter	Intervention price	2 927,80	2 927,80	2 803,30	2 718,00	0,0	-4,3	-3,0
3. Cheese								
— Grana Padano (30-60 days)	Intervention price	3 796,70	3 796,70	3 672,40	–	0,0	-3,3	×
— Grana Padano (6 months)	Intervention price	4 704,30	4 704,30	4 565,30	–	0,0	-3,0	×
— Parmigiano Reggiano (6 months)	Intervention price	5 192,10	5 192,10	5 047,60	–	0,0	-2,8	×
4. Skimmed-milk powder	Intervention price	1 724,30	1 724,30	1 702,00	1 702,00	0,0	-1,3	0,0

3.3.1 *(cont.)*

	Category of price or amount in ECU/tonne except as stated	1991/92	1992/93	1993/94	1994/95	% TAV		
						1992/93 / 1991/92	1993/94 / 1992/93	1994/95 / 1993/94
1	2	3	4	5	6	7	8	9
5. Pilot products								
	Serum powder — Threshold price	572,10	572,10	561,30	554,10	0,0	-1,9	-1,3
	Milk powder (15%) — Threshold price	1 937,60	1 937,60	1 912,00	1 912,50	0,0	-1,3	0,0
	Milk powder (26%) — Threshold price	2 687,20	2 687,20	2 620,30	2 588,70	0,0	-2,5	-1,2
	Condensed milk (unsweetened) — Threshold price	1 002,20	1 002,20	980,60	971,80	0,0	-2,2	-0,9
	Condensed milk (sweetened) — Threshold price	1 316,60	1 316,60	1 289,00	1 278,70	0,0	-2,1	-0,8
	Butter — Threshold price	3 284,30	3 284,30	3 145,60	3 048,50	0,0	-4,2	-3,1
	Emmental — Threshold price	3 817,60	3 817,60	3 711,60	3 655,60	0,0	-2,8	-1,5
	Blue-veined cheese — Threshold price	3 181,40	3 181,40	3 105,10	3 070,30	0,0	-2,4	-1,1
	Parmigiano Reggiano — Threshold price	5 961,70	5 961,70	5 846,60	5 804,40	0,0	-1,9	-0,7
	Cheddar — Threshold price	3 441,10	3 441,10	3 352,40	3 310,10	0,0	-2,6	-1,3
	Gouda and other — Threshold price	3 170,01	3 170,01	3 091,90	3 055,70	0,0	-2,5	-1,2
	Lactose — Threshold price	947,20	947,20	930,40	920,60	0,0	-1,8	-1,1
Beef/veal								
Marketing year: April-March								
Beginning of single market: 1968/69								
1. Beef animals (live)	Guide price	2 000,00	2 000,00	1 974,20	1 974,20	0,0	-1,3	0,0
	Intervention price	3 430,00	3 430,00	3 216,40	3 047,10	0,0	-6,2	-5,3
Pigmeat								
Marketing year: July-June								
Beginning of single market: 1967/68								
Pig carcasses	Basic price	1 897,00	1 897,00	1 872,00	1 300,00	0,0	-1,3	-30,6
Eggs								
Marketing year: August-July								
Beginning of single market: 1967/68								
Eggs in shell	Sluice-gate price	826,00	836,40	830,40	829,50	1,3	-0,7	-0,1
Poultrymeat								
Marketing year: August-July								
Beginning of single market: 1967/68								
1. 70% chickens	Sluice-gate price	1 107,30	1 111,60	1 101,90	1 111,20	0,4	-0,9	0,8
2. 70% ducks	Sluice-gate price	1 244,60	1 258,90	1 250,60	1 249,40	1,1	-0,7	-0,1
3. 75% geese	Sluice-gate price	1 522,00	1 537,00	1 528,30	1 527,10	1,0	-0,6	-0,1
4. 80% turkeys	Sluice-gate price	1 465,50	1 475,20	1 469,60	1 468,70	0,7	-0,4	-0,1
5. 70% guinea-fowl	Sluice-gate price	1 695,90	1 709,50	1 701,60	1 700,50	0,8	-0,5	-0,1

Silkworms — ECU/box of seed

Marketing year: April-March							
Beginning of single market: 1972/73							
Aid	111,81	111,81	110,41	110,36	0,0	-1,3	-0,0

Peas, beans and field beans

Marketing year: July-June							
Beginning of single market: 1978/79							
Activating price	440,10	440,10	–	–	0,0	x	x
Minimum purchase price (peas)	253,40	253,40	–	–	0,0	x	x
Minimum purchase price (beans)	234,70	234,70	–	–	0,0	x	x

Sheepmeat and goatmeat —
ECU/100 kg

Marketing year: April-March							
Beginning of single market: 1980/81							
Basic price	432,32	422,95	418,53	417,45	-2,2	-1,0	-0,3
Intervention price	367,47	–	–	–	x	x	x
Derived intervention price (Ireland)	347,66	–	–	–	x	x	x
Reference price	432,32	–	–	–	x	x	x

Source: EC Commission, Directorate-General for Agriculture.

(1) Per tonne of cereals (regionalization plan).
(2) Per tonne of oilseeds (regionalization plan).
(3) To be reduced as a result of maximum guaranteed quantity for 1992/93 marketing year having been exceeded.
(4) To be reduced as a result of maximum guaranteed quantity having been exceeded.
(5) Seed subsidies 1993 (ECU/100 kg).

1. Gramineae:

Festuca ovina L.	36,1
Festuva pratensis Huds	31,9
Poa pratensis L.	32,2
Poa trivialis L.	28,9
Lolium perenne L. (var. haute persi)	21,5
Lolium perenne L. (nouv. var. &x a.)	15,9
Lolium multiflorum lam	17,5
Festuca ovina L.	36,1
Lolium x Boucheanum	17,5
Arrhenatherum elatius L.	55,6
Festuca arundianacea Schr.	48,8
Poa nemoralis L.	32,2
Festololium	26,8

2. Leguminosae:

Pisum sativum L. partim	69,2
Vicia faba L. partim	42,2
Vicia sativa L.	36,1
Vicia villosa Roth	43,7
Trifolium pratense L.	62,9
Trifolium repens L.	62,9
Trifolium repens L. gigan.	62,9
Trifolium alexandrinum L.	62,9
Trifolium hybridum L.	62,9

Trofolium incarnatum L.	0,0	37,9
Trifolium resupinatum L.	0,0	37,9
Medicago sativa (ecotypes)	25,4	18,3
Medicago sativa (variétés)	19,9	30,3
Medicago lupolina L.	44,3	26,4
Onobrichis viciifolia Se.	62,2	16,6
Hedysarum coronarium L.	58,6	30,2

3. Ceres:

Triticum spelta L.	37,9	11,9
Oryza sativa L.	38,0	13,8
— type indica		14,3
— type japonica		12,3

4. Oléagineux:

Linum usatissimum (textile)	23,5
Linum usatissimum (oléagineux)	18,6
Cannabis sativa L. (mono)	17,0

(6) The new COM for tobacco entered into force from the 1993 harvest onwards (Regulation (EEC) No 2075/92. OJ L 215, 30.7.1992). This Regulation provides that the amount of the premiums for the eight groups of varieties is to be paid in its entirety to tobacco producers in addition to the tobacco buying-in price. By contrast, the previous COM (Regulation (EEC) No 727/70) provided for payment of premiums to first buyers (processors) of tobacco calculated individually for 34 varieties. As a result, the new premium system cannot be compared with the system that applied to previous harvests.
(7) Germany, Belgium, France.

3.3.2 **Producer prices for agricultural products in the Community** (excluding VAT)   **EUR 12**

1	Nominal index 1985 = 100			% TAV		Real index 1985 = 100			% TAV	
	1991	1992	1993	1992/1991	1993/1992	1991	1992	1993	1992/1991	1993/1992
	2	3	4	5	6	7	8	9	10	11
Total	117,9	113,8	114,5	− 3,5	0,6	85,3	78,3	75,0	− 8,2	− 4,2
Crop products	133,3	121,3	122,6	− 9,0	1,1	92,3	78,9	75,5	− 14,5	− 4,3
Cereals and rice	101,2	98,5	98,6	− 2,7	0,1	73,8	68,4	65,5	− 7,3	− 4,2
Common wheat	99,9	98,0	95,9	− 1,9	− 2,1	75,2	70,8	66,6	− 5,9	− 5,9
Durum wheat	104,2	100,5	106,7	− 3,6	6,2	66,5	59,6	60,6	− 10,4	1,7
Fodder barley	94,5	94,2	92,3	− 0,3	− 2,0	71,4	68,4	64,8	− 4,2	− 5,3
Barley for brewing	100,9	102,2	101,7	1,3	− 0,5	74,5	72,1	69,0	− 3,2	− 4,3
Oats	99,0	109,6	110,6	10,7	0,9	73,4	78,7	76,1	7,2	− 3,3
Grain maize	111,9	99,5	102,1	− 11,1	2,6	77,1	63,4	61,2	− 17,8	− 3,5
Paddy rice	105,7	112,8	133,8	6,7	18,6	68,2	68,5	78,0	0,4	13,9
Other	82,7	82,6	76,3	− 0,1	− 7,6	66,5	64,1	56,8	− 3,6	− 11,4
Roots and brassicas	147,9	120,3	124,9	− 18,7	3,8	109,0	84,7	83,0	− 22,3	− 2,0
Ware potatoes	216,0	137,2	146,9	− 36,5	7,1	152,7	90,5	89,7	− 40,7	− 0,9
Sugarbeet	107,6	111,6	113,6	3,7	1,8	82,6	81,7	79,8	− 1,1	− 2,3
Other	91,6	86,2	81,4	− 5,9	− 5,6	80,1	73,3	67,3	− 8,5	− 8,2
Fresh vegetables	147,6	139,7	145,8	− 5,4	4,4	99,9	87,5	87,2	− 12,4	− 0,3
Fruits	160,9	131,2	128,0	− 18,5	− 2,4	105,3	79,2	71,8	− 24,8	− 9,3
Fresh fruits	163,1	129,4	123,0	− 20,7	− 4,9	109,5	81,0	72,1	− 26,0	− 11,0
Dried fruits	141,7	146,5	171,0	3,4	16,7	68,7	63,6	69,6	− 7,4	9,4
Wine/must	145,8	136,6	130,1	− 6,3	− 4,8	106,6	95,1	86,0	− 10,8	− 9,6
Olives and olive oil	187,1	160,0	169,6	− 14,5	6,0	104,4	81,8	80,9	− 21,6	− 1,1
Seeds	122,0	119,9	120,6	− 1,7	0,6	94,0	88,6	87,2	− 5,7	− 1,6
Flowers and plants	113,5	110,1	113,2	− 3,0	2,8	91,9	84,8	84,1	− 7,7	− 0,8
Other crop products	123,9	115,6	121,2	− 6,7	4,8	72,7	59,0	58,0	− 18,8	− 1,7
Animals and livestock products	104,7	107,4	107,6	2,6	0,2	79,4	77,8	74,6	− 2,0	− 4,1
Animals (for slaughter and export)	100,0	103,8	101,6	3,8	− 2,1	75,3	74,6	69,8	− 0,9	− 6,4
Beef animals	95,4	99,6	108,3	4,4	8,7	74,4	74,9	78,5	0,7	4,8
Calves	114,3	121,6	129,9	6,4	6,8	85,4	86,6	88,1	1,4	1,7
Pigs	97,7	101,9	81,3	4,3	− 20,2	76,2	75,6	57,5	− 0,8	− 23,9
Sheep and lambs	101,8	111,8	118,9	9,8	6,4	65,5	67,7	69,4	3,4	2,5
Poultry	102,4	100,3	105,4	− 2,1	5,1	73,4	68,7	69,2	− 6,4	0,7
Other animals	121,1	118,0	119,5	− 2,6	1,3	78,4	70,5	67,2	− 10,1	− 4,7
Milk	114,3	116,3	119,8	1,7	3,0	88,5	86,4	85,3	− 2,4	− 1,3
Eggs	102,6	97,1	105,3	− 5,4	8,4	74,6	66,7	69,4	− 10,6	4,0
Other livestock production	101,0	108,1	107,3	7,0	− 0,7	63,0	62,6	57,1	− 0,6	− 8,8

Source: Eurostat.

3.3.3 Producer price indices (excl. VAT)

(1985 = 100)

	Nominal indices			% TAV		Indices in real terms (deflated)			% TAV	
	1991	1992	1993	1992/1991	1993/1992	1991	1992	1993	1992/1991	1993/1992
1	2	3	4	5	6	7	8	9	10	11
Crop products:										
EUR 12	133,3	121,3	122,6	-9,0	1,1	92,3	78,9	75,5	-14,5	-4,3
Belgique/België	103,8	91,9	85,4	-11,5	-7,1	90,6	78,3	70,8	-13,6	-9,6
Danmark	91,9	91,3	78,9	-0,7	-13,6	74,0	72,0	61,5	-2,7	-14,6
BR Deutschland	100,3	88,2	83,8	-12,1	-5,0	90,6	76,6	69,9	-15,5	-8,7
Elláda	251,5	254,3	269,6	1,1	6,0	94,6	82,5	76,5	-12,8	-7,3
España	131,8	116,9	125,8	-11,3	7,6	90,9	76,2	78,4	-16,2	2,9
France	111,3	95,7	90,4	-14,0	-5,5	92,6	77,7	72,0	-16,1	-7,3
Ireland	108,7	104,9	108,8	-3,5	3,7	89,6	83,9	85,7	-6,4	2,1
Italia	134,8	120,6	121,9	-10,5	1,1	96,3	81,9	79,3	-15,0	-3,2
Luxembourg	114,8	88,2	88,7	-23,2	0,6	102,1	76,1	73,9	-25,5	-2,9
Nederland	107,1	95,3	92,3	-11,0	-3,1	98,9	85,3	80,5	-13,8	-5,6
Portugal	179,1	160,4	165,6	-10,4	3,2	94,5	77,6	75,3	-17,9	-3,0
United Kingdom	115,4	108,5	109,0	-6,0	0,5	81,8	74,1	73,3	-9,4	-1,1
Livestock products:										
EUR 12	104,7	107,4	107,6	2,6	0,2	79,4	77,8	74,6	-2,0	-4,1
Belgique/België	90,2	92,8	86,2	2,9	-7,1	78,8	79,1	71,5	0,4	-9,6
Danmark	92,1	92,5	81,6	0,4	-11,8	74,2	73,0	63,6	-1,6	-12,9
BR Deutschland	91,5	93,5	85,8	2,2	-8,2	82,7	81,2	71,6	-1,8	-11,8
Elláda	208,0	233,7	253,5	12,4	8,5	78,2	75,8	71,9	-3,1	-5,1
España	100,1	100,4	104,0	0,3	3,6	69,0	65,4	64,8	-5,2	-0,9
France	100,1	101,3	98,1	1,2	-3,2	83,3	82,4	78,1	-1,1	-5,2
Ireland	102,5	106,4	113,8	3,8	7,0	84,5	85,1	89,7	0,7	5,4
Italia	111,1	114	119,8	2,6	5,1	79,3	77,4	77,9	-2,4	0,6
Luxembourg	102,4	101,8	99,6	-0,6	-2,2	91,1	87,8	82,9	-3,6	-5,6
Nederland	93,3	92,4	85,0	-1,0	-8,0	86,0	82,8	74,1	-3,7	-10,5
Portugal	121,9	116,3	118,8	-4,6	2,1	64,3	56,3	54,0	-12,4	-4,1
United Kingdom	110,8	117,5	126,5	6,0	7,7	78,5	80,2	85,1	2,2	6,1

Total:

	117,9	113,8	114,5	-3,5	0,6	85,3	78,3	75,0	-8,2	-4,2
EUR 12										
Belgique/België	94,7	92,5	85,9	-2,3	-7,1	82,6	78,8	71,3	-4,6	-9,5
Danmark	92,0	92,1	80,7	0,1	-12,4	74,1	72,7	62,9	-1,9	-13,5
BR Deutschland	94,0	92,0	85,3	-2,1	-7,3	84,9	79,9	71,1	-5,9	-11,0
Elláda	238,4	248,1	264,8	4,1	6,7	89,6	80,5	75,1	-10,2	-6,7
España	117,1	109,2	115,7	-6,7	6,0	80,7	71,1	72,0	-11,9	1,3
France	105,8	98,5	94,2	-6,9	-4,4	88,0	80,0	75,0	-9,1	-6,3
Ireland	103,2	106,3	113,2	3,0	6,5	85,1	84,9	89,3	-0,2	5,2
Italia	124,9	117,9	121,0	-5,6	2,6	89,2	80,0	78,7	-10,3	-1,6
Luxembourg	104,5	99,6	97,8	-4,7	-1,8	93,0	85,9	81,4	-7,6	-5,2
Nederland	98,0	93,4	87,5	-4,7	-6,3	90,5	83,6	76,3	-7,6	-8,7
Portugal	147,1	135,7	139,4	-7,7	2,7	77,6	65,7	63,4	-15,3	-3,5
United Kingdom	112,5	114,1	119,9	1,4	5,1	79,7	77,9	80,7	-2,3	3,6

Source: Eurostat.

3.3.4   Annual rate of change of: (a) consumer prices for foodstuffs and beverages; (b) producer prices for agricultural products

	% TAV		% trend compared with preceding year					% trend compared with the corresponding month of preceding year	
1	1992/1985	1993/1985	1985	1990	1991	1992	1993	III 1994	VI 1994
	2	3	4	5	6	7	8	9	10
Consumer prices for foodstuffs and beverages									
EUR 12	5,0	4,5	6,0	5,9	5,7	3,1	1,0	2,1	3,1
Belgique/België	1,3	1,0	3,2	3,4	1,6	-0,5	-0,6	0,6	3,0
Danmark	2,0	1,7	4,2	0,3	0,7	1,7	-0,3	1,9	3,6
BR Deutschland	1,5	1,4	0,2	3,6	3,1	2,4	0,6	0,9	2,0
Elláda	28,0	29,2	19,5	20,4	20,0	14,1	12,7	11,8	12,8
España	6,5	5,7	9,6	6,4	3,3	3,0	0,1	5,4	6,0
France	2,8	2,4	4,9	3,8	2,8	0,6	-0,2	-0,8	1,2
Ireland	2,7	2,4	3,1	1,4	0,7	1,2	-0,2	3,9	4,7
Italia	6,9	6,3	8,8	5,9	10,3	5,7	1,3	3,4	3,0
Luxembourg	1,8	1,5	3,5	4,0	3,0	0,4	-0,7	0,7	1,5
Nederland	0,8	0,7	0,5	2,2	3,1	2,1	-0,3	0,8	2,9
Portugal	12,0	10,7	17,4	12,5	11,2	4,3	0,9	3,8	5,6
United Kingdom	5,0	4,7	3,2	8,0	5,1	2,2	1,7	0,2	1,3

Producer prices for agricultural products	1,9	1,7	1,6	1,6	2,6	-3,5	0,6	2,3	6,9
EUR 12	1,9	1,7	1,6	1,6	2,6	-3,5	0,6	2,3	6,9
Belgique/België	-1,1	-1,9	2,6	-5,6	-1,2	-2,3	-7,1	1,5	14,8
Danmark	-1,2	-2,6	-2,0	-7,5	-1,9	0,1	-12,4	-4,9	2,5
BR Deutschland	-1,2	-2,0	-4,4	-5,0	-0,9	-2,1	-7,3	-2,5	2,4
Elláda	13,9	12,9	18,0	21,2	18,0	4,1	6,7	20,0	17,9
España	1,3	1,8	3,1	0,6	-0,2	-6,7	5,9	11,1	13,8
France	-0,2	-0,7	1,2	-0,2	0,2	-6,9	-4,3	-2,4	3,1
Ireland	0,9	1,6	-2,5	-11,4	-3,1	3,0	6,6	5,2	3,8
Italia	2,4	2,4	6,4	4,7	5,6	-5,6	2,7	-6,8	5,2
Luxembourg	-0,1	-0,3	3,9	-2,0	-7,1	-4,7	-1,8	-0,3	-3,4
Nederland	-1,0	-1,7	-2,1	-5,6	2,9	-4,6	-6,4	0,3	0,5
Portugal	4,5	4,2	15,4	4,1	-5,4	-7,7	2,7	11,0	11,1
United Kingdom	1,9	2,3	-5,2	1,3	-0,7	1,4	5,1	-3,8	0,9

Source: Eurostat.

3.3.5 Input prices (excl. VAT)

		Belgique/België	Danmark	Deutschland (¹)	Elláda	España	France (¹)	Ireland	Italia	Luxembourg (¹)	Nederland	Portugal	United Kingdom
A — Animal feed													
Barley ECU/100 kg	1991	21,29	17,71	16,74	21,81	20,04	21,08	..	20,35	19,24	17,96	..	22,55
	1992	20,85	17,73	16,67	19,67	19,20	20,79	..	19,31	18,27	17,58	..	21,10
	1993	19,97	17,00	15,88∞	19,97	17,31	19,98	..	19,06	17,61	16,55	..	21,28
Oats ECU/100 kg	1991	21,89	..	15,42	30,25	20,50	18,73	..	22,26	19,19	0,00	..	20,81
	1992	23,72	..	16,40	27,26	20,65	22,76	..	26,78	18,75	–	..	20,92
	1993	24,89	..	16,41	27,20	18,83	21,31	..	27,08	19,28	–		23,98
Maize ECU/100 kg	1991	26,68	..	22,21	27,08	24,79	26,71	..	23,42	25,33	23,54	..	31,30
	1992	22,91	..	20,95	24,42	22,43	24,38	..	21,42	23,17	19,87	..	30,49
	1993	22,90	..	17,16∞	23,93	20,82	23,87	..	20,38	23,01	19,31	..	29,32
Toasted extracted soya bean meal ECU/100 kg	1991	20,64	19,47	23,39	..	..	- 22,27	..	19,11		17,18	..	26,13
	1992	20,76	19,18	20,58	..	..	22,60	..	18,96		17,06	..	25,54
	1993	23,50	19,51	22,33	..	..	24,66	..	21,25		19,22	..	28,08
Fish meal ECU/100 kg	1991	47,38	55,91	46,23	..	42,57	..	..	58,46	..	43,23	..	49,57
	1992	47,63	58,88	48,41	..	43,58	..	..	57,25	..	42,86	..	49,12
	1993	45,74	54,71	45,50	..	40,34	..	..	50,01	..	35,81	..	46,85
Dried sugarbeet pulp ECU/100 kg	1991	17,69	..	15,18	10,29	19,38	15,59	15,95	19,24	..	18,91	..	20,77
	1992	15,29	..	14,94	10,58	19,03	11,78	15,85	18,70	..	17,01	..	20,08
	1993	15,33	..	13,93∞	11,17	16,81	10,84	15,05	16,35	..	16,87	..	18,77
B — Compound feedingstuffs													
Supplementary feed for breeding calves ECU/100 kg	1991	26,17	19,83	23,53	25,82	30,40	..	23,72	28,44	26,36	20,86	..	..
	1992	27,31	19,70	20,21	25,08	29,36	..	0,00	28,26	27,29	21,23	..	..
	1993	27,50	19,24	20,31∞	26,18	26,70	..	25,45	26,14	28,25	21,61	..	..
Supplementary feed for dairy cattle (stall-fed) (bags) ECU/100 kg	1991	22,41	..	15,73	25,41	28,00	21,60	21,52	28,46	23,01	17,57	..	..
	1992	23,73	..	16,64	24,45	26,98	21,53	21,86	26,98	23,36	18,29	..	..
	1993	23,76	..	16,77	24,27	24,32	21,84	20,95	24,46	23,86	18,53	..	..
Supplementary feed for dairy cattle (grass-fed) ECU/100 kg	1991	21,18	..		..	..	20,78	–	27,29	21,01	17,57	..	..
	1992	21,77	..		..	..	21,05	–	26,53	21,03	–	..	..
	1993	21,92	..		..	..	21,58	20,68	24,62	21,23	–	..	..
Complete feed for breeding piglets (bags) ECU/100 kg	1991	31,33	20,27	24,51	35,00	39,43	27,36	26,06	29,79	31,72	28,65	..	..
	1992	32,23	20,40	25,63	34,55	39,22	28,18	26,34	31,40	32,28	29,54	..	..
	1993	31,97	20,82	26,02∞	36,10	35,50	27,66	24,65	30,62	33,08	29,84	..	..
Complete feed for fattening pigs (bags) ECU/100 kg	1991	24,33	20,24	20,63	28,95	28,84	21,04	25,33	28,57	25,26	20,12	..	..
	1992	24,66	20,25	20,81	27,69	27,66	21,26	25,76	27,31	25,83	20,88	..	..
	1993	24,08	20,53	20,96∞	28,30	24,98	20,79	24,49	24,93	26,26	20,92	..	..

Item		1	2	3	4	5	6	7	8	9	10	11
Complete feed for broilers (bags) ECU/100 kg	1991	31.83	23.94	27.00	33.11	..	..	34.00	35.41	26.01	..	..
	1992	32.15	24.65	26.00	31.57	..	..	33.00	37.52	26.64	..	..
	1993	31.99	24.87	30.00	28.73	..	..	30.00	39.23	27.22	..	..
Complete feed for 'battery' laying hens (bags) ECU/100 kg	1991	27.45	22.53	29.11	30.79	..	..	32.39	31.74	22.67	..	..
	1992	28.05	23.49	28.04	29.63	..	..	30.85	32.82	22.73	..	..
	1993	28.13	23.13	28.31	26.76	..	..	27.70	33.67	22.57	..	..
C — Fertilizers (²):												
Nitrate of ammonia (³) (26% in bags) ECU/100 kg	1991	55.98	61.60	29.87	66.01	54.52	56.43	49.72	56.23	52.40	84.89	48.73
	1992	51.87	57.63	33.77	58.86	51.81	54.63	47.78	53.35	48.71	93.26	45.35
	1993	49.16	56.89	31.05	49.60	50.09	46.90	49.39	51.47	46.34	65.19	39.04
Superphosphate ECU/100 kg	1991	65.79	..	36.88	74.38	64.38	..	70.48	..	69.19	77.88	..
	1992	66.78	..	68.70	74.88	62.92	..	65.10	..	70.55	87.20	..
	1993	68.64	..	63.18	66.39	63.07	..	54.19	..	72.59	74.86	..
Potassium chloride ECU/100 kg	1991	32.58	35.21	..	30.06	25.63	–	26.33	26.17	33.36	36.58	26.58
	1992	32.70	36.39	..	28.69	26.30	–	27.17	26.25	34.24	40.40	24.83
	1993	31.83	37.80	..	26.01	25.63	30.91	24.21	28.66	35.22	75.01	23.58
D — Compound fertilizers (²):												
Fertilizers containing nutrients N-P-K 20-10-10 ECU/100 kg	1991	..	..	..	37.41	20.95	..	18.29	..	19.56	..	16.35
	1992	..	..	..	35.61	19.99	..	17.57	..	17.45	..	14.74
	1993	..	..	..	32.69	19.63	..	–	..	18.34	..	13.22
Fertilizers containing nutrients N-P-K 17-17-17 (bags) ECU/100 kg	1991	20.33	24.65	15.34	23.87	..	21.70	23.37	23.28	21.81	32.30	20.03
	1992	20.38	23.47	19.71	22.30	..	21.64	–	22.00	19.39	33.02	17.79
	1993	20.18	22.40	18.13	18.96	..	19.41	–	21.22	20.60	23.94	16.19
Fertilizers containing nutrients N-P-K 9-9-18 ECU/100 kg	1991	..	16.55	..	17.44	..	..	15.35	..	14.37	..	12.50
	1992	..	15.93	..	16.01	..	..	14.76	..	13.05	..	11.19
	1993	..	15.67	..	13.75	..	..	–	..	13.82	..	10.15
E — Motor fuels and other fuels:												
Diesel fuel for tractors: ECU/100 kg	1991	18.40	26.20	34.63	29.63	53.85	..	23.64	21.93	26.51	57.67	25.39
	1992	16.23	25.91	36.66	29.96	–	..	24.87	20.09	24.88	57.24	23.43
	1993	17.74	26.98	48.42	29.89	–	..	31.34	21.71	–	55.74	26.90
Heating fuel ECU/100 kg	1991	18.40	22.07	34.63	10.70	33.26	32.52	71.60	..	24.12	..	30.61
	1992	16.23	20.59	36.66	9.84	29.72	29.79	69.38	..	23.60	..	30.60
	1993	17.74	21.75	35.18	9.85	30.98	29.76	61.50	..	25.91	..	–

Source: Eurostat.

(¹) Bulk price: Germany, for all products for which the reference is normally in bags, excluding nitrate of ammonia and fertilizers containing nutrients N-P-K 17-17-17 (in bags); France, for complete feed for fattening pigs.

(²) Price for 100 kg of pure nutrient content, except for fertilizers containing nutrient: price per 100 kg of product.

(³) Nitrate of ammonia (33%) in bags for Greece, France and the United Kingdom.

3.3.6 Agricultural wages, input prices (¹) and producer prices (excl. VAT)

(1985 = 100)

1	1989	1990	1991	1992	1993	% TAV 1993/1985	% TAV 1992/1991	% TAV 1993/1992
	2	3	4	5	6	7	8	9
Farm wages								
EUR 12	:	:	:	:	:	×	×	×
Belgique/België	112,9	115,4	119,1	127,0	132,0	4,1	6,6	3,9
Danmark	123,1	126,0	132,0	136,2	139,4	4,9	3,2	2,4
BR Deutschland ∞	109,6	111,6	116,4	121,9	125,8	3,3	4,7	3,2
Elláda	173,4	202,6	231,7	254,1	262,7	14,8	9,7	3,4
España	133,2	148,7	162,4	178,3	188,2	9,5	9,8	5,6
France (⁴)	116,1	121,1	127,0	132,0	135,4	4,4	3,9	2,6
Ireland	117,7	121,8	124,0	128,7	:	×	3,8	×
Italia	123,6	129,3	138,0	152,1	159,9	6,9	10,2	5,1
Luxembourg	113,5	116,2	122,3	134,7	143,9	5,3	10,1	6,8
Nederland	109,0	113,5	120,2	128,5	131,5*	4,0	6,9	2,3
Portugal	:	:	:	:	:	×	×	×
United Kingdom	124,8	138,7	151,9	157,4	:	×	3,6	×
Inputs (²)								
EUR 12	103,8	104,9	108,3	110,5	113,3	1,6	2,0	2,6
Belgique/België	94,1	90,1	90,6	90,5	89,3	− 1,4	0,1	− 1,4
Danmark	99,2	97,4	96,6	96,3	96,6	− 0,4	− 0,3	0,3
BR Deutschland	93,1	91,5	93,5	95,3	93,6	− 0,8	1,8	− 1,7
Elláda	153,3	183,4	227,8	257,7	292,7	14,4	13,1	13,6
España	108,1	109,0	111,3	111,6	114,7	1,7	0,3	2,8
France	100,3	99,7	100,6	100,6	100,5	0,1	0,1	− 0,1
Ireland	99,3	99,5	99,8	100,0	100,2	0,0	0,2	0,2
Italia	108,3	109,2	110,8	113,5	121,8	2,5	2,4	7,3
Luxembourg	95,4	96,6	98,7	98,3	98,4	− 0,2	− 0,4	0,1
Nederland	89,3	86,4	86,9	87,5	85,5	− 1,9	0,7	− 2,3
Portugal	124,4	130,6	137,3	136,5	133,9	3,7	− 0,6	− 1,9
United Kingdom	109,3	113,4	117,7	121,3	127,5	3,1	3,1	5,1
Producer prices (³)								
EUR 12	113,1	114,9	117,9	113,8	114,5	1,7	− 3,5	0,6
Belgique/België	101,5	95,8	94,7	92,5	85,9	− 1,9	− 2,3	− 7,1
Danmark	101,4	93,8	92,0	92,1	80,7	− 2,6	0,1	− 12,4
BR Deutschland	99,8	94,8	94,0	92,0	85,3	− 2,0	− 2,1	− 7,3
Elláda	166,4	202,1	238,4	248,1	264,8	12,9	4,1	6,7
España	116,6	117,3	117,1	109,2	115,7	1,8	− 6,7	5,9
France	105,9	105,6	105,8	98,5	94,2	− 0,7	− 6,9	− 4,3
Ireland	120,2	106,5	103,2	106,3	113,2	1,6	3,0	6,6
Italia	112,9	118,2	124,9	117,9	121,0	2,4	− 5,6	2,7
Luxembourg	114,8	112,5	104,5	99,6	97,8	− 0,3	− 4,7	− 1,8
Nederland	100,8	95,2	98,0	93,4	87,5	− 1,7	− 4,6	− 6,4
Portugal	149,4	155,5	147,1	135,7	139,4	4,2	− 7,7	2,7
United Kingdom	111,8	113,3	112,5	114,1	119,9	2,3	1,4	5,1

Source: Eurostat ('Purchase price of inputs' and 'Producer prices for agricultural products' are harmonized indices, whereas 'Farm wages' remain heterogeneous national indices).

(¹) The EC index of farm input prices is a Laspeyres index, whereas the deflated price series (see Table 3.1.8) is a Paasche index. The discrepancies between the figures in the two tables are mainly a matter of the differing index formulae.
(²) Indices of the prices of goods and services of current agricultural consumption.
(³) Annual indices include fruit and vegetables.
(⁴) *Source:* SCEES.

3.3.7 EC price indices for feedingstuffs, fertilizers and soil improvement, fuels and lubricants, and investments in machinery (excl. VAT)

(1985 = 100)

1	1989	1990	1991	1992	1993	% TAV 1993/1985	% TAV 1992/1991	% TAV 1993/1992
	2	3	4	5	6	7	8	9
Feedingstuffs								
EUR 12	104,7	101,0	101,8	103,0	106,0	0,7	1,2	2,9
Belgique/België	93,8	85,6	84,5	85,3	82,1	-2,4	1,0	-3,8
Danmark	97,4	89,2	83,4	82,9	81,8	-2,5	-0,6	-1,3
BR Deutschland	89,8	82,2	80,7	81,9	79,0	-2,9	1,5	-3,5
Elláda	160,8	183,6	219,0	231,0	262,3	12,8	5,3	13,5
España	107,7	106,0	105,5	104,7	106,5	0,8	-0,7	1,7
France	104,5	96,9	95,2	97,1	95,7	-0,5	2,1	-1,4
Ireland	105,2	103,2	100,0	100,2	101,2	0,1	0,2	1,0
Italia	109,6	107,4	109,2	110,4	121,1	2,4	1,1	9,7
Luxembourg	92,5	90,8	89,3	88,5	87,3	-1,7	-0,9	-1,4
Nederland	88,9	79,6	78,9	79,8	75,9	-3,4	1,0	-4,9
Portugal	117,2	117,2	120,2	116,5	117,1	2,0	-3,1	0,5
United Kingdom	112,7	113,6	115,8	119,0	127,7	3,1	2,8	7,3
Fertilizers and soil improvement								
EUR 12	89,8	90,8	93,1	92,2	88,6	-1,5	-1,0	-3,9
Belgique/België	83,9	85,4	86,8	81,2	78,0	-3,1	-6,4	-3,9
Danmark	72,1	74,5	74,2	69,8	67,9	-4,7	-5,9	-2,7
BR Deutschland	81,7	82,0	84,9	81,0	77,4	-3,2	-4,6	-4,4
Elláda	166,7	193,1	252,6	329,8	330,1	16,1	30,6	0,1
España	85,5	84,2	85,9	82,8	80,0	-2,8	-3,6	-3,4
France	86,3	85,8	86,1	83,2	79,8	-2,8	-3,3	-4,1
Ireland	84,3	84,1	85,6	84,4	79,8	-2,8	-1,5	-5,5
Italia	99,7	99,5	99,8	101,0	103,5	0,4	1,3	2,5
Luxembourg	83,0	80,8	86,2	80,8	76,9	-3,2	-6,3	-4,8
Nederland	76,1	75,5	78,4	72,5	67,7	-4,8	-7,6	-6,6
Portugal	121,9	133,7	142,9	146,7	122,6	2,6	2,7	-16,4
United Kingdom	92,4	94,1	91,1	86,3	81,6	-2,5	-5,2	-5,4

Fuels and lubricants

EUR 12	81,5	92,7	100,7	103,4	112,8	1,5	2,8	9,1
Belgique/België	60,7	63,3	62,9	59,0	62,2	−5,8	−6,2	5,4
Danmark	81,3	82,6	82,4	76,5	76,2	−3,3	−7,2	−0,4
BR Deutschland	72,5	77,5	83,2	82,6	83,1	−2,3	−0,8	0,6
Elláda	118,1	159,1	220,6	262,8	331,3	16,2	19,1	26,1
España	96,8	102,4	111,8	117,8	126,0	2,9	5,3	7,0
France	73,6	78,6	79,1	73,0	73,7	−3,7	−7,7	1,0
Ireland	85,3	88,1	89,4	84,4	86,3	−1,8	−5,6	2,3
Italia	75,1	101,9	94,3	103,8	129,4	3,3	10,1	24,7
Luxembourg	78,5	82,7	82,2	79,3	81,2	−2,6	−3,5	2,4
Nederland	67,8	85,0	85,5	82,9	82,3	−2,4	−3,0	−0,7
Portugal	120,0	136,3	155,7	153,4	162,3	6,2	−1,5	5,8
United Kingdom	81,9	90,7	95,8	96,0	100,6	0,1	0,2	4,8

Investment in machinery

EUR 12	122,5	129,8	137,0	143,9	150,3	5,2	5,0	4,4
Belgique/België	114,3	118,5	122,8	125,4	131,2	3,5	2,1	4,6
Danmark	118,6	124,7	128,8	131,1	134,5	3,8	1,8	2,6
BR Deutschland	109,3	113,6	118,9	124,4	128,2	3,2	4,6	3,1
Elláda	178,7	204,7	242,2	269,5	307,8	15,1	11,3	14,2
España	123,1	129,5	131,2	133,4	138,4	4,1	1,6	3,7
France	117,9	122,1	126,4	130,6	133,8	3,7	3,3	2,5
Ireland	119,0	123,5	126,7	128,9	131,6	3,5	1,7	2,1
Italia	123,6	131,9	139,5	148,0	155,1	5,6	6,1	4,8
Luxembourg	115,7	126,1	131,1	136,2	140,7	4,4	3,9	3,3
Nederland	112,3	117,9	124,9	128,9	131,0	3,4	3,2	1,6
Portugal	171,8	179,2	186,1	197,8	205,5	9,4	6,3	3,9
United Kingdom	124,3	131,4	139,3	146,7	149,7	5,2	5,3	2,0

Source : Eurostat.

3.3.8   Market value of agricultural land (parcels)

1	2	ECU/ha (1)			% TAV (real) (2)		
		1991	1992	1993	1993/1979	1992/1991	1993/1992
		3	4	5	6	7	8
Belgique/België (3)	Arable land	11 308	11 696	11 944	−4,6	−1,5	−3,5
	Meadow	9 332	9 488	9 527	−5,0	−3,2	−5,1
Danmark (4)	Agricultural land	6 537	6 274	5 987	−5,0	−7,0	−9,2
BR Deutschland	Agricultural land	15 922	14 937	15 380	−3,4	−12,3	−5,2
Deutschland (10)	Agricultural land	:	11 811	11 423	x	x	−10,9
Elláda	Irrigated land	12 344	11 900	11 546	x	−8,0	−8,3
	Non-irrigated land	5 886	5 635	5 597	x	−8,6	−6,2
España	Irrigated land	14 369	11 591	9 871	−4,9	−21,9	−8,0
	Non-irrigated land	3 464	3 019	2 709	−3,9	−15,6	−3,1
France	Arable land	3 141	3 095	3 060	−5,7	−5,2	−7,9
	Natural meadow	2 266	2 234	2 216	−6,9	−5,2	−7,6
Ireland (12)	Agricultural land	4 733	4 819	:	−6,9 (8)	−0,2	x
Italia (11)	Agricultural land	4 686	4 460	:	−3,7 (8)	−5,4	x
Luxembourg (9)	Agricultural land	55 420	:	:	x	x	x
Nederland (5)	Arable land	16 616	17 100	:	2,9 (8)	−1,2	x
	Meadow	18 780	21 495	:	−1,1 (8)	9,9	x
United Kingdom							
— England (6)	Agricultural land	5 989	4 981	:	−5,7 (8)	−16,2	x
— Wales (6)	Agricultural land	4 830	3 326	:	−6,7 (8)	−30,6	x
— Scotland (7)	Agricultural land	4 198	3 261	:	−5,5	−21,7	x
— Northern Ireland (6)	Agricultural land	4 933	4 899	3 022	−6,0 (8)	0,1	−6,1

Source: Eurostat.

(1) Converted at current exchange rates.
(2) In national currencies, deflated (GDP deflator).
(3) Weighted average of public and private sales.
(4) Agricultural holdings with buildings (10-100 ha).
(5) Land with vacant possession.
(6) Sales of all agricultural land with vacant possession of more than 5 ha (2 ha in Northern Ireland).
(7) Price of farms (land and buildings) of more than 5 ha.
(8) 1992/1979.
(9) Sales of all utilizable agricultural land whether for agricultural or non-agricultural purposes (industrial estates, road building, building plots).
(10) Ex-German Democratic Republic included.
(11) *Source:* INEA.
(12) *Source:* ESRI.

3.3.9 Rents for agricultural land

	ECU/ha (¹)			% TAV (real) (²)			Ratio rent/ market value %
	1991	1992	1993	1993 / 1979	1992 / 1991	1993 / 1992	1993
1	2	3	4	5	6	7	8
Belgique/België							
— Arable land	141,51	146,36	155,25	− 2,1	− 1,5	0,3	1,30
— Meadow	137,22	140,35	149,04	− 1,3	− 2,6	0,4	1,56
BR Deutschland (³) (⁴)							
— Total rents	207,24	:	222,06*	− 0,7	×	×	1,44*
— New rents	238,45	:	263,38*	×	×	×	×
Elláda (⁵)							
— Arable land	393,55	389,27	445,06	− 1,8	5,9	8,0	3,85
France (⁶)							
— Arable land	106,12	110,09	115,77	− 1,7	− 0,2	− 2,0	3,78
Luxembourg	136,65	144,13	142,32	×	− 0,6	− 5,7	×
Nederland (⁷)							
— Arable land	253,14	259,35	266,64	− 0,2	− 1,6	− 5,3	1,51 (⁹)
— Meadow	214,20	211,00	255,15	1,1	− 5,4	11,4	0,98 (⁹)
United Kingdom (⁸)							
England	142,37	136,03*	129,49	− 0,4	− 3,7	− 3,6	2,73 (⁹)*
Wales	77,47	75,93	74,36	0,4	− 1,2	− 0,8	2,28 (⁹)
Scotland	97,00	92,20	89,75	0,4	− 4,2	− 1,4	2,97

Source : Eurostat.

(¹) Converted at current exchange rates.
(²) In national currencies, deflated (GDP deflator).
(³) Biannual surveys.
(⁴) Data for the Federal Republic of Germany, including West Berlin, as constituted prior to 3 October 1990.
(⁵) Most of this land is irrigated.
(⁶) Series based on surveys in 1969, 1980 and 1992, updated using the rent index for wheat production.
(⁷) Weighted by area across agricultural regions.
(⁸) Prices for all kinds of land.
(⁹) 1992.

3.3.11 Value-added tax (VAT) rates: producer prices ([1])
at 1 January 1994

(%)

1	2	Scheme Normal	Scheme Flat-rate ([2])
		3	4
Belgique/België	Most products (excl. flowers)	6,0	6,0
	Flowers	20,5	20,5 ([3])
Danmark	All products	25,0	–
BR Deutschland	Most products	7,0	9,0
	Wine must, beverages, services	15,0	15,0
Elláda	All products	8,0	8,0
España	Products used for animal feed, excluding wine:		
	— Not processed on the holding	6,0	4,0
	— Processed on the holding	6,0	4,0
	Wine	15,0	–
	All products not used for human or animal consumption:		
	— Not processed on the holding	15,0	4,0
	— Processed on the holding	15,0	–
France	All plant products except wine and horticultural products	5,5	3,05
	Wine	18,6	–
	Horticultural products	–	3,05
	All livestock products except animals for meat	–	3,05
	Animals for meat	–	4,0
	Products sold through a producers' group:		
	— fruit, vegetables and wine	–	4,0
	— pigs, eggs and poultry	–	4,0
Ireland	Horses, live cattle, sheep, pigs, goats and cervidae	2,5	2,5
	Other livestock including poultry and fish, carcasses, raw wool, horsehair, bristles, feathers, hides and skins, non-edible horticultural produce	21,0	2,5
	Other agricultural products excluding live animals	0,0	2,5
Italia	Cereals, paddy rice, fresh and dried vegetables, potatoes, fresh and dried fruit, oilseeds for edible oil, olive oil, butter, cheese	4,0	4,0
	Wine and wine must	12,0	9,0-4,0 m.
	Eggs	9,0	9,0
	Cattle	19,0	9,0
	Pigs	19,0	9,0
	Raw milk	19,0	9,0
	All other products	19,0	4,0
Luxembourg	Most products and services	8,0	8,0
	Wine and must	12,0	12,0
Nederland	Most products	6,0	5,932
Portugal	Fresh vegetables, fresh fruit	5,0	–
	Ordinary table wine	5,0	–
	Flowers	16,0	–
	Dried fruit, honey, table wines	16,0	–
	All other agricultural products	0,0	–
United Kingdom	Products generally used for human and animal consumption (including seeds, seedlings and animals)	0,0	4,0
	Other products and services	17,5	4,0

Source: Eurostat.

([1]) The figures are for agriculture in the strict sense, excluding forestry. The most important products are given only as examples.

([2]) The flat-rate schemes applicable to agriculture are all designed to offset on a general sales-related basis the VAT paid on purchases of agricultural inputs.

([3]) VAT on flowers sold by auction is invoiced at 19,5 %. Growers covered by the flat-rate scheme receive only the normal flat-rate of 6 %, the remaining 13,5 % being payable to the central tax authority by the purchaser.

3.3.12   Value-added tax (VAT) rates: input prices
at 1 January 1994

(%)

Belgique/België	Purchase and tenancy of land	(¹)
	Animal feedingstuffs, seeds, fertilizers, liming, agricultural services, veterinary services	6,0
	Coal (solid fuel)	12,0
	Construction and maintenance of farm buildings	20,5
	Farm equipment, pesticides	20,5
	Road diesel fuel, petrol, liquefied petroleum gas for non-agricultural purposes	20,5
	Diesel fuel for agricultural purposes, light fuel oil, natural gas, liquefied petroleum gas, electricity	20,5
Danmark	Purchase of land and buildings	0,0
	All products	25,0
BR Deutschland	Purchase of farmland	(¹)
	Inputs of agricultural origin (animal feedingstuffs, seeds and propagating material, breeding stock)	7,0
	Inputs of industrial origin (fertilizers, pesticides, fuel and power, buildings and machinery, building materials and accessories), non-agricultural services	15,0
Elláda	Purchase and tenancy of land, manual workers' wages, insurance premiums	0,0
	Seed animal feedingstuffs, breeding stock, fertilizers, pesticides, phytopharmaceutical products	8,0
	Most farm equipment, maintenance and repair of machinery, installations and buildings, electricity, lubricants, liquefied gases, asbestos cement piping, wire fencing	18,0
	Motor fuels	18,0
España	Purchase and tenancy of agricultural land	(¹)
	Inputs of agricultural origin: medicines	6,0
	Inputs of industrial origin	15,0
	Most services	15,0
France (²)	Non-processed agricultural products (including breeding stock), work under contract	5,5
	Fertilizers, animal feedingstuffs, pesticides	5,5
	Motor fuel, certain building work and services provided by persons eligible for the special deduction, purchase and maintenance of farm equipment, construction and maintenance of farm buildings	18,6 (³)
Ireland	Animal feedingstuffs, fertilizers (put up in quantities of 10 kg or more), cereals, beet, hay, cake, etc., seeds and propagating material of products used for food, veterinary products for oral administration	0,0
	Concrete and blocks of concrete	12,5
	Electricity, solid fuels, diesel fuel for heating, diesel fuel for tractors, gas for heating and lighting	12,5
	Most services	12,5
	Machinery repairs	12,5
	Fertilizers (quantities less than 10 kg), pesticides, disinfectants and detergents, veterinary products for injection and veterinary equipment, farm equipment including tractors, building materials, second-hand goods, petrol and lubricants, motor vehicle and motorcycles, other services (transport, storage, hiring of equipment)	21,0
Italia	Agricultural loans, rural leases, veterinary services	(¹)
	Animal feedingstuffs of vegetable origin, fertilizers	4,0
	Agricultural work under contract	9,0
	Animal feedingstuffs of animal origin, seeds, breeding stock, pesticides	
	Products of mineral and chemical origin and additives for animal feed	9,0
	Fuels and lubricants, pharmaceuticals	9,0
	Equipment and machinery, gas and electricity, lubricants, building materials, most services	19,0
Luxembourg	Water supplied by public enterprises, disposal of real property.	(¹)
	Inputs: seeds and propagating material, livestock and livestock products, animal feedingstuffs, fertilizers, plant protection products, pharmaceuticals, agricultural servicses, e.g. artificial insemination.	3,0
	Services rendered by professional personnel (veterinary medicine), solid mineral fuel, mineral oils and timber to be used as fuel, unleaded petrol.	12,0
	Farm machinery and equipment, construction and maintenance of farm buildings, motor fuel other than unleaded petrol, certain services (transport).	15,0
Nederland	Telecommunications, indemnity insurance, purchase, renting and tenancy of immovable property (except sale by builder)	(¹)
	Seeds, fertilizers, fuel for hothouses, animal feedingstuffs, breeding stock, some services, pesticides, pharmaceuticals, work under contract, equipment	6,0
	Veterinary services, motor fuels and other fuels, structural work, maintenance and repair of farm buildings, machinery, tractors and equipment, small items of equipment and accessories, transport services, petrol, electricity	17,5
Portugal	Fertilizers and crop protection products, animal feedingstuffs and seeds, live animals, machinery, equipment and tractors, veterinary services	5,0
	Electricity, fuels and gas	5,0
	Maintenance and repair of machinery, petrol, coal	16,0
United Kingdom	Interest relief grants on purchase and renting of land, insurance, financial costs	(¹)
	Most products generally used for human consumption and animal consumption, including seeds, propagating material and animals reared for the purpose. Construction of farm buildings and most civil engineering work (excluding repair and maintenance). Power fuels and other fuels (except road diesel fuel and petrol), electricity and water	(¹)
	Road diesel fuel, lubricants, petrol, fertilizers, chemicals, purchase and maintenance of agricultural machinery, other goods and services not specified	17,5
	Purchase of motor vehicles	17,5

Source: Eurostat.

(¹) Exempt.
(²) Reimbursement at a subsequent stage.
(³) 50% deductible from 1.1.1986.

3.3.13   Producer prices in the Member States in 1993

(ECU/100 kg)

Products	Belgique/ België	Danmark	BR Deutsch-land	Ellada	España	France	Ireland	Italia	Neder-land	Portu-gal	United Kingdom
1	2	3	4	5	6	7	8	9	10	11	12
1. Crop products :											
Common wheat	13,64	14,55	15,04	15,98	17,95	13,48	13,70	19,47	13,68	18,32	14,58
Durum wheat	–	–	–	21,70	20,75	19,62	–	22,22	–	23,14	–
Rye	12,00	13,46	14,36	–	15,31	14,34	–	12,11	13,03	16,28	–
Barley	12,68	14,35	13,30	15,51	14,64	12,25	11,84	17,25	14,80	16,28	14,53
Oats	15,08	16,25	15,64	19,55	15,83	16,16	14,39	21,83	13,40	18,70	15,60
Maize	–	–	13,60	16,86	19,10	13,06	–	18,37	–	17,13	–
Potatoes	3,22	11,78	10,41	23,83	13,89	5,12	–	18,21	4,87	12,32	8,01
Sugarbeet	–	42,89	45,50	42,42	50,52	–	55,39	–	–	41,55	45,36
Dessert apples (1)	19,77	34,69	32,73	31,71	25,03	36,78	–	24,51	21,61	40,70	47,22
Dessert pears (1)	40,92	31,71	37,38	55,39	31,21	51,25	–	39,36	37,24	36,86	48,85
Table grapes (1)	–	–	–	49,63	35,08	82,76	–	45,40	–	54,14	–
Oranges (1)	–	–	–	32,03	12,41	–	–	28,08	–	29,13	–
Cauliflowers (2)	34,34	–	31,63	43,13	28,39	–	39,63	45,16	50,11	46,92	38,06
Lettuces (2)	–	–	38,57	22,56	21,01	–	–	54,83	32,64	55,05	–
Asparagus (2)	320,92	–	–	–	133,75	–	–	201,28	261,12	–	366,26
Tomatoes (2)	–	–	43,74	31,57	32,24	–	–	43,77	–	32,40	–
Carrots (2)	–	–	22,53	22,31	13,86	–	21,66	38,68	24,83	24,55	18,74
Onions (2)	13,10	13,56	13,75	22,63	18,60	41,15	13,13	34,27	–	25,29	17,63
Dry peas	–	–	137,23	124,90	67,92	–	–	67,89	–	54,95	63,96
Dry beans	118,78	–	83,14	88,93	11,83	–	–	97,24	62,98	88,95	–
2. Livestock products :											
Calves (3)	508,98	–	–	–	–	446,81	–	501,31	–	387,54	–
Pigs (3)	125,27	109,70	–	141,68	143,83	–	–	162,45	–	136,32	109,08
Poultry carcasses (class A)	–	–	153,90	–	127,45	113,51	–	156,01	–	–	126,00
Whole drinking milk	–	–	51,02	76,16	40,24	–	71,70	74,49	–	49,81	32,30
Cream	–	–	–	198,13	150,21	–	–	–	–	202,63	514,33
Butter	318,87	325,93	329,13	431,58	309,14	419,53	–	326,13	318,13	411,77	302,03

Source : Eurostat.
(1) All varieties.
(2) All qualities.
(3) Carcass weight.

3.3.14   **Institutional prices in national currency, expressed as indices in real terms for all agricultural products**

(1985/86 = 100)

	1988/89	1989/90	1990/91	1991/92	1992/93 (¹)	1993/94	1994/95
1	2	3	4	5	6	7	8
Belgique/België	94,6	91,3	86,0	83,1	80,0	76,9	67,2
Danmark	89,6	87,0	82,3	80,0	78,0	75,8	67,8
BR Deutschland	91,4	88,0	83,1	79,5	75,6	71,7	63,6
Elláda	96,0	101,9	101,3	90,8	81,5	83,7	79,5
España •	78,7	73,2	66,8	62,9	57,9	66,6	62,5
France	96,0	94,6	91,7	88,1	85,2	82,2	76,8
Ireland	104,8	103,3	100,8	96,9	95,3	90,6	83,8
Italia	87,8	84,8	81,6	76,0	72,1	84,6	82,0
Luxembourg	99,3	96,6	92,4	89,2	85,1	81,5	76,3
Nederland	93,9	91,9	87,7	85,2	83,0	80,6	70,9
Portugal	86,3	83,5	80,5	70,5	58,0	57,9	54,0
United Kingdom	94,2	92,1	93,9	89,1	84,2	95,9	85,4
EUR 12	92,4	89,9	86,5	82,4	78,1	81,4	75,2

Source : EC Commission, Directorate-General for Agriculture.

(¹) Since the beginning of the 1992/93 marketing year, certain CMOs have been reformed, and farmers have been compensated for the fall in institutional prices by direct aid.

3.4.1 Budgetary expenditure on the common agricultural policy

1	Unit	1991	1992	1993	1994 (²)	1995 (³)
	2	3	4	5	6	7
EC budget	Mio ECU	53 823,1 (¹)	58 857,0 (¹)	65 268,5 (¹)	68 354,6	76 526,1
1. EAGGF-Guarantee						
— Plant products	Mio ECU	32 385,9	32 107,5	34 748,1	34 786,9	36 972,5
— Animal products (⁴)	Mio ECU	18 030,5	19 043,2	20 989,0	27 677,0	23 506,5
— Ancillary expenditure	Mio ECU	12 194,3	10 542,7	11 657,0	10 493,0	11 069,0
— Set-aside and income aid (⁸)(⁹)	Mio ECU	1 286,9	1 574,5	1 417,9	1 134,9	983,5
— Accompanying measures	Mio ECU	76,9 (⁵)	147,6 (⁵)	462,6	30,0	44,5
— Monetary reserve	Mio ECU			221,7	675,0	1 369,0
— Reserves and provisions	Mio ECU	(1 000)	(1 000)	(1 000)	(1 000)	(500)
— Depreciation of stocks and disposal of butter	Mio ECU	797,3	799,5	..	..	..
2. EAGGF-Guidance (⁷)	Mio ECU	2 011,0	2 715,4	3 386,0	2 619,0 (¹⁰)	2 827,0 (¹⁰)
3. OTHER AGRICULTURAL EXPENDITURE	Mio ECU	144,8	177,1	203,7	126,5	147,4
4. TOTAL AGRICULTURAL EXPENDITURE	Mio ECU	34 541,7	35 000,0	38 337,8	37 532,4	39 946,9
Charges under the common agricultural policy:	Mio ECU	2 763,0	2 209,2	2 144,3	2 304,5	2 182,0
— ordinary levies	Mio ECU	1 621,2	1 206,8	1 029,1	922,4	946,2
— sugar levies	Mio ECU	1 141,8	1 002,4	1 115,2	1 382,1	1 235,8
Net cost of the CAP:	Mio ECU	31 778,7	32 790,8	36 193,5	35 227,9	37 764,9
— as % of GDP	%	0,6 (⁷)	0,6 (⁷)	0,7 (⁷)	0,6 (⁷)	0,6
— per head in the EEC	ECU	92,1 (⁷)	94,7 (⁷)	104,6 (⁷)	101,3 (⁷)	108,2

Source: EC Commission, Directorate-General for Agriculture.
(¹) Financial Report of the European Communities.
(²) Provisional data.
(³) 1995 budget: EUR 12.
(⁴) Including the common organization of the market in fishery products (Chapter B1-26, formerly Chapter B2-90).
(⁵) Does not include either EAGGF Guidance Section share of set-aside or income aid, not covered in this year by EAGGF Guarantee Section appropriations.
(⁶) Including the EAGGF Guidance Section's share of set-aside (including up to 1992) but not including payments in respect of Regulation (EEC) No 1852/78 (Fisheries).
(⁷) Provisional data.
(⁸) From 1993 onwards expenditure on set-aside (Guidance Section share) is financed by the EAGGF Guarantee Section.
(⁹) From 1994 the 'Set-aside' Chapter B1.40 will become item B1.106 and will be entered in Chapter B1.10 'Arable crops'.
(¹⁰) Not including the amounts for Community initiative programmes.

3.4.2 EAGGF Guarantee and Guidance expenditure, by Member State

(Mio ECU)

	EAGGF Guarantee expenditure (1)					EAGGF Guidance expenditure (2)				
1	1989	1990	1991	1992	1993	1989	1990	1991	1992	1993
	2	3	4	5	6	7	8	9	10	11
EUR 12	25 872,9	26 453,5	32 385,9	32 107,5	34 748,2	1 468,0	1 968,0	2 408,2	2 874,8	3 093,4
Belgique/België	585,8	873,7	1 468,5	1 378,2	1 298,7	31,6	23,1	30,5	28,2	41,7
Danmark	1 015,1	1 113,7	1 220,3	1 166,8	1 334,7	17,2	16,9	18,0	23,5	20,0
BR Deutschland	4 188,7	4 355,2	5 234,5	4 830,5	4 976,2	133,0	204,1	200,2	253,8	348,7
Elláda	1 650,9	1 949,7	2 211,2	2 231,4	2 715,0	235,3	270,2	274,2	392,2	402,9
España	1 903,2	2 120,8	3 314,3	3 578,1	4 175,7	203,9	301,8	514,2	633,6	412,9
France	4 810,5	5 142,2	6 394,4	6 916,5	8 184,8	179,8	383,8	425,5	554,3	633,5
Ireland	1 241,2	1 668,4	1 731,1	1 452,8	1 649,9	121,9	125,0	168,5	194,5	165,8
Italia	4 621,8	4 150,3	5 353,4	5 141,5	4 765,4	263,6	282,7	326,5	375,9	625,0
Luxembourg	1,8	5,2	2,8	1,1	7,3	3,6	4,6	6,7	6,3	9,0
Nederland	3 749,9	2 868,7	2 679,3	2 389,8	2 328,1	20,7	11,4	20,5	21,9	19,5
Portugal	174,4	214,2	315,6	423,8	478,1	179,4	241,6	313,4	289,8	313,9
United Kingdom	1 917,0	1 975,9	2 391,3	2 451,1	2 737,9	78,0	102,8	110,2	100,8	99,5
Community (3)	12,6	15,5	69,2	145,9	96,4	:	:	:	:	1,0 (4)

Source: EC Commission, Directorate-General for Agriculture.
(1) Adjusted for the financial consequences of the clearance of accounts + ECU 29,2 million in 1988, – ECU 202,7 million in 1989, ECU – 377,9 million in 1990, – ECU 437,8 million in 1991, ECU 78,9 million in 1992 and ECU – 384,8 million in 1993).
(2) Expenditure from appropriations for commitment.
(3) Payments direct to recipients made by the Commission for the EAGGF Guarantee Section and 'multi-State' expenditure for the EAGGF Guidance Section.
(4) Financing under Art. 8 for 25 beneficiaries across the Member States.

3.4.3   EAGGF Guarantee expenditure, by product (1991-93)

Product	1991 (¹) Mio ECU	%	1992 (¹) Mio ECU	%	1993 (¹) Mio ECU	%
1	2	3	4	5	6	7
Cereals	5 077,4	15,7	5 456,9	17,0	6 560,3	18,9
Refunds	3 601,5		3 139,7		2 788,8	
Intervention, of which:	1 475,9		2 317,2		3 771,5	
— production refund	419,1		360,7		415,7	
— aid for durum wheat	516,1		456,4		425,9	
— storage	1 419,4		2 497,4		2 723,8	
— co-responsibility levy	-924,3		-1 098,8		89,6	
— small producer aid	38,7		30,8		60,1	
Rice	111,9	0,3	87,3	0,3	69,5	0,2
Refunds	77,8		91,8		59,3	
Intervention	34,1		-4,5		10,2	
Sugar	1 814,9	5,6	1 937,4	6,0	2 188,6	6,3
Refunds	1 251,2		1 305,6		1 531,4	
Intervention, of which:	563,7		631,8		657,2	
— refund of storage costs	460,9		496,1		501,7	
Olive oil	1 874,2	5,8	1 754,3	5,5	2 468,2	7,1
Refunds	111,8		48,4		68,8	
Intervention	1 762,4		1 705,9		2 399,4	
Oils and fats	3 549,5	11,0	4 132,0	12,9	3 063,4	8,8
Refunds	0,5		0,1		0,0	
Intervention	3 549,0		4 131,9		3 063,4	
Protein products	959,0	3,0	862,0	2,7	1 083,8	3,1
Refunds						
Intervention, of which	959,0		862,0		1 083,8	
— peas, field beans	550,8		480,3		558,7	
— dried fodder	403,9		380,2		523,7	
Textile plants and silkworms, of which:	521,8	1,6	771,3	2,4	860,6	2,5
— flax and hemp	33,6		29,0		29,6	
— cotton	487,9		742,1		830,8	
Fruit and vegetables	1 106,5	3,4	1 261,7	3,9	1 672,2	4,8
Refunds	94,8		116,7		187,5	
— fresh	76,9		91,6		156,4	
— processed	17,9		25,1		31,1	
Intervention	1 011,7		1 145,0		1 484,7	
— fresh	412,0		516,0		919,5	
— processed	599,7		629,0		565,2	
Wine	1 047,8	3,2	1 087,2	3,4	1 509,6	4,3
Refunds	55,5		77,3		100,2	
Intervention, of which:	992,3		1 009,9		1 409,4	
— aid for private storage	41,1		40,4		57,5	
— distillation	367,2		320,7		464,3	
— compulsory distillation of the by-products of wine-making	72,3		89,2		76,7	
Tobacco	1 329,6	4,1	1 233,0	3,8	1 165,1	3,4
Refunds	65,3		71,9		36,2	
Intervention	1 264,3		1 161,1		1 128,9	
Other sectors or agricultural products, of which:	67,6	0,2	302,7 (*)	1,0	190,0	0,5
— seeds	66,7		81,5		70,4	
— hops	0,9		9,9		24,5	

		%		%		%
Milk products	5 636,6	17,4	4 006,8	12,5	5 211,3	15,0
Refunds	2 249,0		2 056,1		2 287,5	
Intervention, of which:	3 387,6		1 950,7		2 923,8	
— aids for skimmed milk	1 052,7		1 086,5		857,0	
— skimmed milk storage	270,5		−432,9		−44,6	
— butter storage	661,4		88,3		161,6	
— butter disposal	669,6		508,5		684,9	
— contribution milk producers	−352,4		−368,0		−299,1	
— extension of the markets	248,3		301,8		421,7	
Beef/veal	4 295,0	13,2	4 413,8	13,7	3 986,3	11,5
Refunds	1 282,4		1 332,5		1 711,2	
Intervention, of which:	3 012,6		3 081,3		2 275,1	
— public and private storage	2 302,8		2 190,6		1 383,1	
— premiums for suckler cows	366,9		436,7		558,2	
— special premium	334,9		453,5		318,7	
Sheepmeat and goatmeat	1 790,4	5,5	1 749,2	5,4	1 800,4	5,2
Refunds						
Intervention	1 790,4		1 749,2		1 800,4	
Pigmeat	252,2	0,8	141,6	0,4	200,9	0,6
Refunds	199,5		130,4		193,5	
Intervention	52,7		11,2		7,4	
Eggs and poultrymeat	169,2	0,5	193,2	0,6	290,9	0,8
Refunds	169,2		193,2		290,9	
— eggs	35,7		32,8		40,7	
— poultrymeat	133,5		160,4		250,2	
Intervention						
Other measures for livestock products	p.m.	p.m.	6,0	p.m.	134,8	0,4
Non-Annex II products	704,1	2,2	699,6	2,2	743,5	2,1
Refunds	704,1		699,6		743,5	
Fishery products	26,2	0,1	32,1	0,1	32,4	0,1
Refunds	p.m.		1,9		0,1	
Intervention	26,2		31,2		32,3	
Total market organizations	30 333,9	93,6	30 128,0	93,8	33 231,8	95,6
Accession compensatory amounts (ACAs) in intra-Community trade	28,3	0,1	28,2	0,1	7,1	p.m.
Monetary compensatory amounts (MCAs)	126,2	0,4	−3,0	p.m.	136,4	0,4
— intra-Community trade	126,2		−3,6		118,9	
— extra-Community trade (granted on exports)	p.m.		0,6		17,5	
Total market organizations ACAs+MCAs	30 488,4	94,1	30 153,2	93,9	33 375,3	96,0
Other (5)	−183,5	−0,6	323,7	1,0	−74,2	−0,2
Food aid refunds	217,0	0,7	221,6	0,7	160,4	0,5
Rural development schemes linked to market operation	388,2	1,2	304,4	1,0	444,7	1,3
Depreciation and disposal of stocks	797,3	2,5	799,5	2,5	–	–
Set-aside (6)	76,9	0,2	138,0	0,4	426,8	1,2
Income aid	–	–	9,7	p.m.	35,8	0,1
Accompanying measures	–	–	–	–	221,7	0,6
Expenditure from appropriations carried over from past years	601,6	1,9	157,4	0,5	157,7	0,5
Grand total	32 385,9 (7)	100,0	32 107,5 (3) (7)	100,0	34 748,1 (4)	100,0

Source: EC Commission, Directorate-General for Agriculture.

(1) The expenditure items are taken from the returns made by the Member States under the advance payments system and are charged to a given financial year under Article 100 of the Financial Regulation.
(2) Including ECU −437,8 million from clearance of the 1987 accounts.
(3) Including ECU +79,0 million from clearance of the 1988 accounts.
(4) Including ECU −384,8 million from clearance of the 1989 accounts.
(5) Of which clearance of accounts.
(6) Amounts credited to the Guarantee Section, i.e. 50% of the amount entered in the specific chapter of the budget. In 1993 all expenditure under this chapter falls to the Guarantee Section.
(7) Not including either the Guidance Section's share in payment for set-aside of land from agricultural production or income aid not charged to the Guarantee Section but the Community's contribution to which is paid using Guarantee Section procedures.
(8) Including ECU 202 million for food aid to certain Central and East European countries.

3.4.3.1   EAGGF Guarantee expenditure, by product (1994 and 1995)

Product	1994 ([1])		1995 ([2])	
	Mio ECU	%	Mio ECU	%
1	2	3	4	5
Arable crops ([3])	12 840	36,9	14 779	38,9
Refunds	1 633		1 263	
Intervention, of which:	11 207		13 516	
— production refund	382		303	
— direct aid	8 824		10 861	
— storage of cereals	284		245	
— set-aside	1 705		2 105	
Sugar	2 170	6,2	1 947	5,1
Refunds	1 463		1 399	
Intervention, of which:	707		548	
— refund of storage costs	563		407	
Olive oil	2 060	5,9	893	2,4
Refunds	55		73	
Intervention	2 005		820	
Dried fodder and dried vegetables	389	1,1	292	0,8
Refunds	–		–	
Intervention, of which:	389		292	
— dried fodder	376		245	
Textile plants and silkworms, of which	867	2,5	808	2,2
— flax and hemp	45		84	
— cotton	821		723	
Fruit and vegetables ([4])	1 665	4,8	1 833	4,8
Refunds	216		166	
— fresh	186		140	
— processed	30		26	
Intervention	1 449		1 667	
— fresh	863		651	
— processed	586		1 016	
Wine	1 179	3,4	1 515	4,0
Refunds	81		70	
Intervention, of which:	1 098		1 445	
— aid for private storage	57		41	
— distillation	304		352	
— compulsory distillation of the by-products of wine-making	60		86	
Tobacco	1 087	3,1	1 119	3,0
Refunds	50		21	
Intervention	1 037		1 098	
Other sectors or agricultural products, of which:	420	1,2	321	0,9
— rice	27		45	
— seeds	88		80	
— hops	15		16	
Milk products	4 344	12,5	4 059	10,6
Refunds	1 926		1 942	
Intervention, of which:	2 418		2 117	
— aids for skimmed milk	821		731	
— skimmed milk storage	71		– 15	
— butter storage	85		180	
— butter disposal	685		630	
— contribution milk producers	1		p.m.	
— extension of the markets	197		125	

3.4.3.1   *(cont.)*

Product	1994 (^{1})		1995 (^{2})	
	Mio ECU	%	Mio ECU	%
1	2	3	4	5
Beef/veal	3 569	10,3	5 255	13,9
Refunds	1 730		1 351	
Intervention, of which:	1 839		3 904	
— public and private storage	− 187		960	
— cow premiums	872		1 217	
— special premium	660		1 221	
Sheepmeat and goatmeat	1 740	5,0	1 264	3,3
Refunds	p.m.		p.m.	
Intervention	1 740		1 264	
Pigmeat	423	1,2	159	0,4
Refunds	270		98	
Intervention	153		61	
Eggs and poultrymeat	251	0,7	134	0,4
Refunds	251		134	
— eggs	29		20	
— poultrymeat	222		114	
Intervention	−		−	
Other measures for livestock products	129	0,4	151	0,4
Non-Annex II products	637	1,8	535	1,4
Refunds	637		535	
Fishery products	44	0,1	47	0,1
Refunds	p.m.		p.m.	
Intervention	44		47	
Total market organizations	33 807	97,2	35 110	92,6
Accession compensatory amounts (ACAs)	1	p.m.	p.m.	:
Monetary compensatory amounts (MCAs)	5	p.m.	p.m.	:
Total market organizations ACAs+MCAs	33 813	97,2	35 110	92,6
Food aid refunds	88	0,2	140	0,4
Rural development schemes linked to market operation	404	1,2	471	1,2
Income aid	30	0,1	45	0,1
Accompanying measures	675	1,9	1 369	3,6
Enlargement	−		950	2,5
Others	− 223	− 0,6	− 162	− 0,4
Grand total	34 787 (^{5})	100,0	36 973 (^{6})	100,0

Source: EC Commission, Directorate-General for Agriculture.

(^{1}) Supplementary and amending budget No 2/94.

(^{2}) 1995 budget: EUR 12.

(^{3}) From 1994, following the new budget nomenclature, appropriations relating to cereals, oilseeds, peas and field beans and set-aside will be brought together under Chapter BI.10 'Arable crops'.

(^{4}) From 1994, aid for dried vegetables will be included in Chapter BI.13 'Dried fodder and dried vegetables'.

(^{5}) Not including the appropriations entered for the monetary reserve (ECU 1 billion).

(^{6}) Not including the appropriations entered for the monetary reserve (ECU 500 million), and in payment appropriations (ECU 37 925,5 million in commitment appropriations).

3.4.4   Breakdown of appropriations by sector according to the economic nature of the measures — financial year 1993 — financial year 1994

	1993 — Mio ECU [1] [2]						
		Breakdown by economic nature of the measures					
Budget nomenclature 1994	Appropriations	Export refunds	Storage	Withdrawals from the market + similar operations	Price subsidies	Guidance premiums	Total
					Interventions		
1	2 = 3 + 8	3	4	5	6	7	8 = 4+5+6+7
A – Arable crops [4]	10 610,7	2 788,8	2 723,8	–	5 098,1	–	7 821,9
Sugar	2 188,6	1 531,4	501,7	–	155,5	–	657,2
Olive oil	2 468,2	68,8	177,3	–	2 218,1	–	2 395,4
Dried fodder and dried vegetables	530,0	–	–	–	530,0	–	530,0
Textile plants, of which:	860,6	–	–	–	860,6	–	860,6
– flax and hemp	(29,6)	–	–	–	(29,6)	–	(29,6)
– cotton	(830,8)	–	–	–	(830,8)	–	(830,8)
Fruit and vegetables	1 663,9	187,4	1,8	605,0	747,8	113,6	1 468,2
Wine	1 509,6	100,2	290,6	541,0	174,0	403,8	1 409,4
Tobacco	1 165,1	36,2	– 17,7	–	1 118,8	26,6	1 127,7
Other sectors or agricultural products, of which:	259,4	59,3	– 3,0	–	203,1	–	200,1
– rice	(69,5)	(59,3)	(– 3,0)	–	(13,1)	–	(10,1)
– seeds	(70,4)	–	–	–	(70,4)	–	(70,4)
– hops	(24,5)	–	–	–	(24,5)	–	(24,5)
Milk and milk products, [4]	5 211,2	2 287,5	293,8	–	1 961,7	668,2	2 923,8
of which: – skimmed milk	(1 005,2)	(192,9)	(– 44,6)	–	(857,0)	–	(812,3)
– butter [5]	(1 276,6)	(430,1)	(161,6)	–	(684,9)	–	(846,5)
Beef/veal	3 986,3	1 711,1	1 383,1	–	892,2	–	2 275,2
Sheepmeat and goatmeat	1 800,4	–	3,9	–	1 796,5	–	1 800,4
Pigmeat	200,9	193,5	2,4	–	5,0	–	7,4
Eggs and poultrymeat	290,9	290,9	–	–	–	–	–
Other measures in favour of animal products	134,8	–	–	–	88,8	–	88,8
Non-Annex II products	743,5	743,5	–	–	–	–	–
Fishery products	32,4	0,1	0,5	13,8	18,0	–	32,3
Total A	33 658,5	9 998,7	5 358,2	1 159,8	15 870,2	1 212,2	23 600,4
B – Accession compensatory amounts in trade	7,1	–	–	–	7,1	–	7,1
C – Monetary compensatory amounts							
– in intra-Community trade	118,9	–	–	–	118,9	–	118,9
– in extra-Community trade	17,5	0,1	–	–	17,4	–	17,4
D – Food aid refunds	160,4	160,4	–	–	–	–	–
E – Differentiation of the agricultural market mechanisms	444,7	–	–	–	444,7	–	444,7
F – Total A + B + C + D + E	34 407,1	10 159,2	5 358,2	1 159,8	16 458,3	1 212,2	24 188,5
G – Income aid	35,8	–	–	–	–	–	–
H – Accompanying measures	221,7	–	–	–	–	–	–
I – Other	83,5 [5]	–	–	–	–	–	–
Total A + B + C + D + E + F	34 748,1	10 159,2	5 358,2	1 159,8	16 458,3	1 212,2	24 188,5
%	100,0	29,2	15,4	3,3	47,4	3,5	69,6

Source: EC Commission, Directorate-General for Agriculture.

[1] The expenditure items are taken from Member States' returns made under the advance payments system and are charged to a given financial year under Article 100 of th[e] Financial Regulation.
[2] Expenditure charged against the 1993 budget.
[3] Provisional data.
[4] Including the financial contribution from cereal and milk producers.
[5] Clearance of accounts + interest following reform of financial arrangements + free distribution of intervention products + income aid + anti-fraud measures + expenditure fro[m] appropriations carried over from 1992 (ECU 157,7 million in 1993 and ECU 441,8 million in 1994).

		1994 — Mio ECU (³)						
		Breakdown of economic nature of the measures						
				Interventions				
Other	Appropriations	Export refunds	Storage	Withdrawals from the market + similar operations	Price subsidies	Guidance premiums	Total	Other
9	10 = 11 + 16 + 17	11	12	13	14	15	16 = 12 + 13 + 13 + 15	17
–	12 652,3	1 513,2	178,1	–	10 961,0	–	11 139,1	–
–	2 061,5	1 377,4	551,2	–	133,0	–	684,1	–
4,0	1 819,3	52,8	36,0	–	1 722,7	–	1 758,7	7,8
–	378,4	–	–	–	378,4	–	378,4	–
–	863,5	–	–	–	863,5	–	863,5	–
–	–	–	–	–	–	–	–	–
8,3	1 556,9	216,7	–	369,0	861,7	88,6	1 319,3	20,9
–	1 220,0	80,4	261,8	339,2	141,3	397,2	1 139,5	–
1,2	1 057,4	49,9	18,5	–	965,3	20,4	1 004,2	3,3
–	287,1	18,9	–	–	127,2	–	127,2	141,0
–	–	–	–	–	–	–	–	–
–	(13)	–	–	–	–	–	–	–
–	4 248,8	1 926,8	226,5	–	1 639,8	424,7	2 291,0	31,0
–	–	–	–	–	–	–	–	–
–	3 466,5	1 708,4	– 209,1	–	1 570,3	–	1 361,2	396,9
–	1 279,8	–	1,7	–	1 278,1	–	1 279,8	–
–	416,3	259,1	21,9	–	–	–	21,9	135,3
–	239,6	239,6	–	–	–	–	–	–
45,9	117,3	–	–	–	–	–	–	117,3
–	631,4	631,4	–	–	–	–	–	–
–	35,5	–	–	17,5	15,6	–	33,1	2,4
59,4	32 331,6	8 074,6	1 086,6	725,7	20 657,9	930,9	23 401,0	855,9
–	p.m.	–	–	–	–	–	–	–
–	4,5	–	–	–	4,5	–	4,5	–
–	p.m.	–	–	–	–	–	–	–
–	86,0	86,0	–	–	–	–	–	–
–	339,7	–	–	–	–	–	–	339,7
59,4	32 762,0	8 160,6	1 086,6	725,7	20 662,4	930,9	23 405,5	1 195,6
35,8	32,0	–	–	–	–	–	–	32,0
221,7	490,1	–	–	–	–	–	–	490,1
83,7	173,1 (⁵)	–	–	–	–	–	–	173,1
400,4	33 457,2	8 160,6	1 086,6	725,7	20 662,4	930,9	23 405,5	1 890,8
1,2	100,0	24,4	3,3	2,2	61,7	2,8	70,0	5,6

3.4.5   Quantity and value of products in public storage

EUR 12

	Situation at 31.12.1991		Situation at 31.12.1992		Situation at 31.12.1993	
	Quantity (1 000 t)	Value (Mio ECU)	Quantity (1 000 t)	Value (Mio ECU)	Quantity (1 000 t)	Value (Mio ECU)
1	2	3	4	5	6	7
Common wheat	5 303,5	334,9	7 919,5	517,8	8 902,6	554,9
Non-breadmaking common wheat	164,2	10,9	167,2	10,5	470,3	28,3
Barley	4 632,7	273,0	5 764,0	358,1	7 137,6	403,6
Rye	3 583,2	219,9	2 536,1	159,9	2 357,8	130,3
Durum wheat	3 376,9	182,6	4 343,0	261,4	2 329,9	127,6
Maize	0,9	0,1	1 065,0	76,0	2 782,7	228,5
Sorghum	–	–	33,4	2,6	149,2	12,1
Rice	175,9	23,4	14,7	1,8	75,0	11,5
Triticale	–	–	p.m.	p.m.	p.m.	p.m.
Total cereals, rice included	17 237,3	1 044,8	21 842,9	1 388,1	24 205,1	1 496,8
Olive oil	18,3	23,6	56,7	57,2	243,2	217,1
Rape	13,4	2,4	–	–	–	–
Sunflower	0,4	0,1	–	–	–	–
Leaf tobacco	p.m.	p.m.	–	–	–	–
Processed tobacco	21,6	2,8	p.m.	p.m.	p.m.	p.m.
Baled tobacco	85,1	24,6	10,1	3,9	13,5	4,9
Total tobacco	106,7	27,4	10,1	3,9	13,5	4,9
Skimmed-milk powder	416,4	312,8	47,3	36,2	37,0	28,3
Butter	266,3	245,8	172,5	161,8	160,7	128,9
Grana Padano (cheese)	0,8	4,2	7,2	34,9	6,3	22,0
Total milk products	683,5	562,8	227,0	232,6	204,0	179,2
Beef carcasses	434,0	431,3	448,4	436,9	156,1	90,8
Boned beef	576,8	565,4	717,9	679,4	563,4	333,9
Total beef	1 010,8	996,7	1 166,3	1 116,3	719,5	424,7
Alcohol	2399,6 (²)	13,7	2359,3 (²)	15,5	3 031,5	21,7
General total	×	2 671,5	–	2 813,6	–	2 344,4

Source: EC Commission, Directorate-General for Agriculture.

(¹) The product values take account of financial depreciation.

(²) 1 000 hl.

(Mio ECU)

3.4.8   Implementation of budget by Member State (1993)

Type of financing	Total	Belgique/België	Danmark	BR Deutschland	Ellada	España	France	Ireland	Italia	Luxembourg	Nederland	Portugal	United Kingdom
1	2	3	4	5	6	7	8	9	10	11	12	13	14
I — Commitments													
Direct (1)	46,798						3,327		43,471				
— regional	46,798						3,327		43,471				
— general													
Indirect (2)	1 335,247	24,469	14,777	219,226	214,251	135,790	292,924	122,319	124,256	7,558	6,594	109,685	63,398
— regional	121,714				52,378	12,884	0,183	0,942	26,703			25,586	3,038
— general	1 213,533	24,469	14,777	219,226	161,873	122,906	292,741	121,377	97,553	7,558	6,594	84,099	60,360
Operational programmes (3)	1 613,874	12,747	5,209	128,021	168,993	261,703	300,168	38,884	450,374	1,456	12,932	202,176	31,211
— regional	1 119,029	7,680	2,729	104,904	116,165	159,777	247,547	22,767	303,713	1,080	5,376	138,368	8,923
— marketing/processing	494,845	5,067	2,480	23,117	52,828	101,926	52,621	16,117	146,661	0,373	7,556	63,808	22,288
Pilot projects, etc.													
(Art. 22/R. 797 - Art. 8/R. 4256)	30,004	2,074			3,167	5,523	8,415	1,834	5,288			2,085	0,643
Global grant	67,474	2,412		1,477	16,442	9,980	28,668	2,753	1,589				4,243
Community initiatives	67,474	2,412		1,477	16,442	9,980	28,668	2,753	1,589				4,243
Grand total	3 093,397	41,702	19,986	348,724	402,853	412,906	633,502	165,790	624,978	9,014	19,526	313,946	99,495
II — Payments													
Direct (1)	161,247	4,700	1,336	5,196	11,978	40,427	31,228	7,852	45,667		1,331	9,745	1,787
— regional	38,327	0,784		0,230			15,305		22,008				
— general	122,920	3,916	1,336	4,966	11,978	40,427	15,923	7,852	23,659		1,331	9,745	1,787
Indirect (2)	1 369,961	39,955	14,777	219,226	206,326	139,676	317,560	122,319	122,887	7,558	6,594	109,685	63,398
— regional	109,544				44,453	12,884	0,183	0,942	22,458			25,586	3,038
— general	1 260,417	39,955	14,777	219,226	161,873	126,792	317,377	121,377	100,429	7,558	6,594	84,099	60,360
Operational programmes (3)	1 199,984	5,908	6,108	112,917	157,718	225,075	238,416	43,933	210,104	0,345	7,174	156,456	35,830
— regional	889,708	2,293	2,176	94,751	124,076	166,531	202,106	28,222	132,915		2,422	114,540	19,519
— marketing/processing	310,276	3,615	3,932	18,166	33,642	58,544	36,310	15,711	77,189	0,188	4,752	41,916	16,311
Pilot projects, etc.													
(Art. 22/R. 797 - Art. 8/R. 4256)	15,888	1,033		0,043	2,016	2,209	3,481	0,734	4,479			1,483	0,410
Global grant	105,614	2,046	0,273	4,049	14,759	35,028	33,874	2,181	0,684	0,116	0,043	8,881	3,680
Community initiatives	105,614	2,046	0,273	4,049	14,759	35,028	33,874	2,181	0,684	0,116	0,043	8,881	3,680
Grand total	2 852,694	53,642	22,494	341,431	392,797	442,415	624,559	177,019	383,821	8,019	15,142	286,250	105,105

Source: EC Commission, Directorate-General for Agriculture.

(1) Direct measures: project-type measures (mainly for investments in the processing and marketing of agricultural products (Regulation (EEC) No 355/77) and, for some measures, under integrated Mediterranean programme (Regulation (EEC) No 2088/85)). This type of financing will be replaced by programmes.

(2) Indirect measures: Community part-financing of aid schemes introduced by the Member States within a Community legal framework (e.g. Regulation (EEC) No 2328/91).

(3) Operational programmes: decided upon within the framework of Regulation (EEC) No 2052/88, implemented by the Member States with a financial contribution from the Community.

3.4.9 Implementation of budget by 'objective' (1993)

(Mio ECU)

	Commitments					Payments					
1	Total	Objective 1 (¹)	Objective 5a (²)	Objective 5b (³)	Transitional (⁴)	Total	Objective 1 (¹)	Objective 5a (²)	Objective 5b (³)	Transitional (⁴)	Pre-1989 commitments not classifiable (⁵)
	2	3	4	5	6	7	8	9	10	11	12
Belgique/België	41,702		29,536	7,680	4,486	53,642		46,359	2,293	3,859	1,131
Danmark	19,986		17,257	2,729		22,494		19,758	2,449		0,287
BR Deutschland	348,724		242,343	106,381		341,431		239,121	98,800	0,043	3,467
Elláda	402,853	399,686			3,167	392,797	387,034			2,016	3,747
España	412,906	297,126	64,405	44,021	7,354	442,415	292,350	62,414	62,919	4,040	20,692
France	633,502	80,089	335,296	209,145	8,972	624,559	64,882	330,371	190,575	7,706	31,025
Ireland	165,790	163,956			1,834	177,019	173,786			0,734	2,499
Italia	624,978	316,518	153,496	122,829	32,135	383,821	170,687	116,809	66,853	21,134	8,338
Luxembourg	9,014		7,934	1,080		8,019		7,746	0,273		:
Nederland	19,526		14,150	5,376		15,142		12,264	2,465		0,413
Portugal	313,946	311,861			2,085	286,250	282,180			1,483	2,587
United Kingdom	99,495	29,982	59,467	9,403	0,643	105,105	40,790	55,298	8,369	0,410	0,238
Total	3 093,397 (⁶)	1 599,218	923,884	508,644	61,651 (⁶)	2 852,694	1 411,709	890,140	434,996	41,425	74,424

Source: EC Commission, Directorate-General for Agriculture.

(¹) The contribution of the EAGGF Guidance Section to all the agricultural measures implemented in the countries/regions given in Annex I to Regulation No 2052/88.
(²) The contribution of the EAGGF Guidance Section to measures applicable in all Member States (horizontal measures) implemented in countries/regions not covered by Objective 1.
(³) The contribution of the EAGGF Guidance Section to the measures implemented specifically in the regions designated by the Commission in Decision 89/426/EEC.
(⁴) The contribution of the EAGGF Guidance Section to measures applicable in some countries/regions (regional measures) covered neither by Objective 1 nor Objective 5b, and applied before 1 January 1989, as well as certain expenditure covered by Article 8 of Regulation No 4256/88 which cannot be charged to any of the three Objectives to which this Fund contributes.
(⁵) Payments made for commitments given before 1989 when the rules did not provide for a division into objectives.
(⁶) Including ECU 0,975 million 'Multi-State'.

3.5.1.1 Employment in agriculture: statistical sources and applications

There are several sources of Community statistics enabling employment in agriculture to be measured from various viewpoints, including employment statistics proper (sample survey of the labour force, annual employment estimates) and agricultural statistics (structural surveys of agricultural holdings). Methods and concepts vary from one source to another, and the purpose of this introduction is to help the user to choose, among the statistics given in the subsequent tables, those which will provide him with the information he seeks.

EMPLOYMENT IN AGRICULTURE AND IN THE OTHER SECTORS

One approach to the problem of employment in agriculture consists in considering it as part of overall employment and comparing it with employment in the other economic sectors. The relevant information comes from employment statistics; in these figures, the persons employed are assigned to that economic sector in which they mainly work, and the characteristics of employment are measured according to identical concepts from one sector to another.

Changes over time in numbers employed in the various sectors, and, in particular, in agriculture, are measured on the basis of annual employment estimates (Tables 3.5.1.2 and 3.5.1.3). For detailed information on the structure of employment in agriculture compared with that of other sectors (breakdown by sex, by occupational status, by working time, or by age), reference must be made to the sample survey of manpower, which provides a 'photograph' of employment in any given year (Table 3.5.1.4).

EMPLOYMENT IN AGRICULTURAL HOLDINGS

Only the statistics which have just been presented allow a proper comparison of employment in agriculture with employment in the other sectors. However, they do not cover all persons employed in agriculture: an important feature of farming is that so many farmers and farm workers work only part-time and often also have other jobs. In the employment statistics, such persons are not classified as working in agriculture.

A full measure of employment in agriculture is provided by the surveys on the structure of agricultural holdings; it should be noted that the information from this source enables employment in agriculture to be analysed as such but that, as it is established according to specific definitions, it cannot be compared with employment data for other sectors.

These surveys cover all persons employed on holdings, whether farming is their main activity or not; they also record working hours and any other remunerated work outside farming. They thus enable employment on agricultural holdings to be measured fully, and part-time and combined other employment to be analysed. By conversion of the numbers of persons employed into full-time equivalent workers ('annual work units' — AWU), the data on working hours give information on the actual volume of labour devoted to farming, the only valid measure of the labour contribution to agriculture, in view of the scale of part-time working (Tables 3.5.1.5 and 3.5.1.6).

3.5.1.2   'Persons employed' (¹) in 'agriculture, hunting, forestry and fishing' (1970-93)

	×1000				% TAV			
	1970	1980	1990	1993	$\frac{1980}{1970}$	$\frac{1990}{1970}$	$\frac{1990}{1980}$	$\frac{1993}{1992}$
1	2	3	4	5	6	7	8	9
EUR 12	16 230	11 830	8 599	7 198	– 3,1	– 3,1	– 3,1	– 5,8
Belgique/België	176	116	100	:	– 4,1	– 2,8	– 1,5	x
Danmark	303	199	150	140	– 4,1	– 3,5	– 2,8	– 2,3
BR Deutschland	2 262	1 403	990	849	– 4,7	– 4,0	– 3,4	– 5,0
Elláda	1 279	1 016	889	794	– 2,3	– 1,8	– 1,3	– 3,7
España	3 310	2 228	1 486	1 198	– 3,9	– 3,9	– 4,0	– 6,9
France	2 647	1 784	1 248	1 101	– 3,9	– 3,7	– 3,5	– 4,1
Ireland	283	209	169	144	– 3,0	– 2,5	– 2,1	– 5,2
Italia	3 878	2 899	1 863	1 504	– 2,9	– 3,6	– 4,3	– 6,9
Luxembourg	12	7	6	6	– 5,2	– 3,4	– 1,5	– 0,0
Nederland	289	244	289	:	– 1,7	0,0	1,7	x
Portugal	984	1 052	805	:	0,7	– 1,0	– 2,6	x
United Kingdom	806	669	603	547	– 1,8	– 1,4	– 1,0	– 3,2

Source: Eurostat, annual employment and labour force statistics.

(¹) 'Persons employed' includes all persons working for remuneration or self-employed, plus unpaid family workers. Persons employed in more than one economic sector are counted only in the sector in which they mainly work.

3.5.1.3   Employment in agriculture and in the other sectors

		EUR 12	Belgique/ Belgïe	Danmark	BR Deutsch- land	Ellada	España	France
1	2	3	4	5	6	7	8	9
Total civilian employ-ment (1 000 persons)	1970	121 158	3 555	2 345	26 169	3 133	12 219	20 623
	1980	125 036	3 607	2 492	26 528	3 355	11 549	21 644
	1990	132 847	3 675	2 630	27 988	3 719	12 578	22 098
	1991	132 992	3 686	2 607	28 537	3 632	12 608	22 142
	1992	132 219	3 753	2 584	28 708	3 685	12 359	22 008
	1993	129 440	×	2 574	28 254	3 720	11 826	21 781
Agriculture (% of total civilian employment)	1970	13,4	5,0	12,9	8,6	40,8	27,1	12,8
	1980	9,5	3,2	8,0	5,3	30,3	19,3	8,2
	1990	6,5	2,7	5,7	3,5	23,9	11,8	5,6
	1991	6,2	2,7	5,5	3,3	22,2	10,7	5,4
	1992	5,8	2,5	5,5	3,1	21,9	10,1	5,2
	1993	5,6	×	5,4	3,0	21,3	10,1	5,1
Industry (% of total civilian employment)	1970	41,6	43,2	36,5	49,3	25,0	35,5	39,3
	1980	37,8	35,2	29,4	43,7	30,2	36,1	36,0
	1990	32,4	28,7	26,6	39,8	27,7	33,4	30,0
	1991	31,9	28,5	26,4	39,2	27,5	33,1	29,5
	1992	31,2	27,5	25,8	38,3	27,1	32,4	28,7
	1993	30,5	×	25,3	37,1	24,2	30,7	27,7
Services (% of total civilian employment)	1970	44,9	52,2	50,9	42,0	34,2	37,4	47,5
	1980	52,7	61,7	62,7	51,0	39,5	44,6	55,7
	1990	61,1	68,5	67,6	56,7	48,3	54,8	64,4
	1991	62,0	68,8	68,1	57,5	50,3	56,3	65,1
	1992	63,0	70,0	68,7	58,5	51,0	57,5	66,0
	1993	64,0	×	69,3	59,9	54,5	59,2	67,2
Share of paid employment in agriculture (%)	1970	×	10,7	20,5	13,0	×	31,6	21,3
	1980	26,4	13,1	25,1	18,1	5,0	27,8	18,3
	1990	28,7	17,9	36,1	22,8	3,9	31,8	21,1
	1991	29,1	18,2	36,5	23,4	3,8	34,5	21,7
	1992	29,1	18,9	36,6	23,4	3,5	32,0	22,5
	1993	28,2	×	37,0	22,9	4,5	31,0	23,5

Source : Eurostat annual employment and labour force statistics and OECD annual labour force statistics.

Ireland	Italia	Luxem-bourg	Nederland	Portugal	United Kingdom	USA	Japan
10	11	12	13	14	15	16	17
1 047	19 312	139	4 679	3 373	24 583	78 678	50 940
1 143	20 413	156	4 970	3 973	25 211	99 303	55 360
1 123	21 215	186	6 268	4 479	26 888	117 914	62 490
1 122	21 410	194	6 444	4 602	26 007	116 877	63 690
1 127	21 270	199	6 576	4 512	25 438	117 598	64 360
1 134	20 152	203	×	4 424	25 044	×	×
27,0	20,1	8,7	6,2	29,2	3,3	4,5	17,4
18,3	14,2	4,8	4,9	26,5	2,7	3,6	10,4
15,0	8,8	3,3	4,6	18,0	2,2	2,8	7,2
13,8	8,5	3,1	4,5	17,6	2,3	2,9	6,7
13,6	8,2	3,0	4,6	11,6	2,3	2,9	6,4
12,7	7,5	3,0	×	11,7	2,2	×	×
29,8	39,3	44,2	38,9	32,8	45,1	34,4	35,7
32,5	37,7	38,1	31,4	36,5	38,0	30,5	35,3
28,6	32,7	30,8	26,3	34,6	29,0	26,2	34,1
28,8	32,3	29,7	25,5	33,7	27,8	25,3	34,4
28,2	32,2	29,3	25,0	33,2	26,9	24,6	34,6
27,5	33,0	28,8	×	33,0	26,2	×	×
43,1	40,5	46,7	54,9	37,7	51,8	61,1	46,9
49,2	48,0	57,0	63,6	36,8	59,4	65,9	54,2
56,3	58,5	65,9	69,1	47,4	68,7	70,9	58,7
57,4	59,2	67,1	69,9	48,7	69,8	71,8	58,9
58,2	59,6	67,7	70,3	55,2	70,8	72,5	59,0
59,8	59,6	68,2	×	55,4	71,6	×	×
13,1	31,9	13,3	×	×	×	×	5,3
12,9	37,5	16,2	22,5	22,8	59,7	×	7,8
13,6	42,2	24,2	35,6	18,9	55,2	×	9,3
14,8	40,8	24,6	36,5	16,7	55,5	×	10,1
×	42,8	25,0	36,7	18,4	48,3	×	11,2
×	40,2	26,7	×	16,7	50,5	×	×

3.5.1.4 Employment in agriculture and in the other sectors: structures compared (1992)

		Unit	EUR 12	Belgique/België	Danmark	BR Deutsch-land
1	2	3	4	5	6	7
Agriculture	numbers	1 000	7 804	109	136	1 044
	— men	%	65,2	66,2	76,1	58,6
	— women	%	34,8	33,8	23,9	41,4
Industry	numbers	1 000	43 189	1 164	715	11 719
	— men	%	76,3	80,1	73,4	74,8
	— women	%	23,7	19,9	26,6	25,2
Services	numbers	1 000	81 918	2 498	1 780	16 952
	— men	%	50,6	51,3	44,1	48,1
	— women	%	49,4	48,7	55,9	51,9
Agriculture	paid workers	%	28,6	10,4	36,4	30,9
	self-employed	%	71,2	89,6	63,6	69,1
Industry	paid workers	%	88,1	90,1	91,7	94,5
	self-employed	%	11,8	9,9	8,3	5,5
Services	paid workers	%	83,5	81,6	92,3	88,8
	self-employed	%	16,4	18,4	7,7	11,2
Agriculture	full-time	%	85,4	93,5	79,6	80,7
	part-time	%	14,5	6,5	19,9	19,3
Industry	full-time	%	93,8	96,4	89,0	92,5
	part-time	%	6,0	3,6	11,0	7,5
Services	full-time	%	80,9	83,3	72,7	78,2
	part-time	%	18,9	16,7	27,2	21,8
Agriculture	less than 25 years	%	9,3	8,9	14,3	9,1
	25 to 34	%	18,1	22,9	17,8	19,0
	35 to 44	%	19,5	18,4	15,6	19,6
	45 to 54	%	22,7	23,4	20,5	22,5
	55 to 64	%	23,5	24,2	20,0	22,5
	65 and over	%	6,9	:	11,8	7,3
Industry	less than 25 years	%	15,8	13,4	19,2	14,8
	25 to 34	%	28,1	33,9	27,0	27,4
	35 to 44	%	25,1	28,5	23,7	22,6
	45 to 54	%	21,4	19,0	19,2	23,9
	55 to 64	%	9,1	5,1	9,7	10,8
	65 and over	%	0,6	:	1,3	0,5
Services	less than 25 years	%	13,8	10,0	16,5	14,2
	25 to 34	%	28,1	34,2	23,7	27,0
	35 to 44	%	27,1	30,6	25,5	25,2
	45 to 54	%	20,4	18,0	23,3	22,8
	55 to 64	%	9,3	6,6	9,4	9,7
	65 and over	%	1,3	0,6	1,6	1,0

Source: Eurostat (Community survey of manpower).

Ellada	España	France	Ireland	Italia	Luxem-bourg	Nederland	Portugal	United Kingdom
8	9	10	11	12	13	14	15	16
804	1 257	1 301	157	1 657	5	247	517	569
58,1	72,4	64,2	91,0	63,4	71,3	76,4	50,5	77,2
41,9	27,6	35,8	9,0	36,6	28,7	23,6	49,5	22,8
933	4 075	6 497	322	6 962	47	1 571	1 468	7 715
76,4	83,5	75,2	76,8	75,9	87,0	83,3	67,0	76,1
23,6	16,5	24,8	23,2	24,1	13,0	16,7	33,0	23,9
1 942	7 126	14 187	667	12 396	107	4 503	2 523	17 237
62,8	56,8	47,6	52,6	59,0	52,2	53,0	50,6	45,6
37,2	43,2	52,4	47,4	41,0	47,8	47,0	49,4	54,4
3,5	31,6	20,5	14,5	42,2	22,7	39,6	18,8	42,3
96,5	67,3	79,5	85,5	57,8	77,4	60,4	81,2	57,5
69,1	83,2	90,1	88,4	81,9	94,4	94,9	84,4	85,6
30,9	16,3	9,9	11,6	18,1	5,6	5,1	15,6	14,0
65,0	75,2	88,5	84,5	68,6	90,6	90,4	79,9	88,3
35,0	23,7	11,5	15,5	31,4	9,4	9,6	20,1	11,5
91,6	93,6	84,0	94,5	84,7	87,8	71,7	82,0	78,6
8,4	6,4	16,0	5,5	15,3	9,4	28,3	18,0	19,3
96,8	98,0	95,8	96,4	96,5	98,0	85,1	96,3	90,4
3,2	2,0	4,2	3,6	3,5	1,9	14,9	3,7	8,6
96,0	92,0	83,8	87,4	94,0	90,7	61,0	92,8	69,0
4,0	8,0	16,2	12,6	6,0	9,1	39,0	7,2	30,0
7,8	10,5	6,3	10,3	8,6	11,3	18,6	8,1	14,3
11,3	15,8	20,2	16,5	20,4	20,8	23,0	11,5	22,6
16,2	18,5	24,7	19,7	20,1	20,8	19,1	14,2	18,3
24,1	22,6	23,4	19,5	24,5	22,7	19,9	20,3	19,0
31,0	28,9	21,9	19,7	20,7	:	16,2	28,0	15,9
9,6	3,6	3,5	14,3	5,7	:	3,2	17,9	9,9
13,5	18,0	11,7	22,3	17,5	12,6	17,1	24,2	16,3
24,8	26,1	29,8	30,8	28,9	30,7	30,9	26,6	26,9
27,8	25,1	31,0	23,6	24,4	29,9	25,7	23,9	23,8
22,7	19,8	20,7	15,9	21,1	20,6	20,3	16,0	21,1
10,4	10,6	6,5	6,8	7,2	5,9	5,5	8,1	10,8
0,8	0,3	0,2	:	0,8	:	0,4	1,1	1,1
10,5	14,6	10,6	18,9	10,3	16,7	18,1	12,7	17,6
28,6	29,5	30,2	30,1	28,5	33,6	30,4	27,0	25,5
29,4	25,6	31,5	24,8	28,8	26,7	27,5	27,4	24,2
19,7	17,0	19,8	16,6	21,3	16,1	17,7	20,4	20,2
10,0	12,2	7,2	7,8	9,6	6,2	5,6	10,3	10,5
1,7	1,1	0,6	1,7	1,6	0,7	0,8	2,3	2,0

3.5.1.5   Employment in agriculture: persons working on agricultural holdings (¹)

1	2	Unit	EUR 12	Belgique/ Belgïe	Danmark	Deutsch- land
		3	4	5	6	7
Total number of persons working on agricultural holdings (¹)	1980	× 1 000	:	186	234	1 983
	1985		:	158	158	1 740
	1987		17 708	147	148	1 624
	1989		16 798	141	123	1 519
Total number of AWU (equivalent full-time workers)	1980	× 1 000	:	124	172	1 051
	1985		:	107	122	918
	1987		9 135	101	112	851
	1989		7 945	94	56	787
Average AWU/persons working on agricultural holdings	1980	1	:	0,66	0,73	0,52
	1985		:	0,67	0,77	0,53
	1987		0,52	0,69	0,76	0,52
	1989		0,47	0,66	0,45	0,52
Breakdown by type of labour:						
● Numbers:						
– farm heads	1980	%	:	61,2	51,2	41,8
	1985		:	61,9	58,3	42,4
	1987		46,7	62,6	57,3	42,5
	1989		48,7	60,2	66,3	43,8
– spouses	1980	%	:	24,2	33,5	27,7
	1985		:	23,5	24,3	26,8
	1987		22,8	22,0	23,8	26,2
	1989		24,4	23,8	13,2	25,6
– other family members	1980	%	:	10,9	4,0	25,4
	1985		:	10,6	2,8	25,1
	1987		22,7	10,4	2,7	25,3
	1989		22,3	11,3	3,9	25,1
– regularly employed non-family members	1980	%	:	3,7	11,3	5,1
	1985		:	4,0	14,6	5,7
	1987		7,7	5,1	16,0	6,1
	1989		5,1	5,0	16,9	5,7
● AWU:						
– farm heads	1980	%	:	69,3	53,9	47,6
	1985		:	68,7	62,5	48,4
	1987		44,5	69,5	62,3	49,4
	1989		47,8	66,0	44,3	50,3
– spouses	1980	%	:	15,2	26,3	22,3
	1985		:	:	:	:
	1987		20,3	14,7	11,4	21,3
	1989		19,7	16,4	10,6	20,6
– other family members (²)	1980	%	:	10,8	4,6	20,8
	1985		:	26,6	13,9	40,7
	1987		36,0	25,3	13,7	40,4
	1989		36,4	27,0	18,7	20,6
– regularly employed non-family members	1980	%	:	4,7	15,2	8,0
	1985		:	3,9	19,0	9,5
	1987		12,3	4,2	19,3	8,8
	1989		8,7	6,1	35,2	9,3
– irregularly employed non-family members	1980	%	:	0,0	0,0	1,3
	1985		:	0,8	4,6	1,4
	1987		7,2	1,0	4,8	1,4
	1989		8,6	1,2	2,2	1,7
● Volume of labour in agriculture:						
– family members	1980	× 1 000	7 487,4 (³)	108,7	109,8	881,0
	1985	AWU	6 651,6 (³)	97,2	84,7	780,0
	1987		5 607,3 (⁴)	94,1	86,2	763,8
	1989		6 568,7	86,7	34,9	700,2
– non-family members	1980	× 1 000	1 639,6 (³)	6,9	27,8	106,0
	1985	AWU	1 349,4 (³)	7,7	27,1	110,0
	1987		1 360,8 (⁴)	5,2	27,3	86,9
	1989		1 376,1	6,8	20,8	86,6
– Total	1980	× 1 000	9 127,0 (³)	115,6	137,6	987,0
	1985	AWU	8 001,0 (³)	104,8	111,8	890,0
	1987		6 968,2 (⁴)	99,3	113,6	850,7
	1989		7 945,2	93,5	55,7	786,8

Source: Eurostat – Surveys of the structure of agricultural holdings + national data..

(¹) Without irregularly employed non-family members.
(²) Including spouses.
(³) Not including Portugal.
(⁴) Not including Italy.

Ellada	España	France	Ireland	Italia	Luxem-bourg	Nederland	Portugal	United Kingdom
8	9	10	11	12	13	14	15	16
1 841	:	2 659	469	5 301	12	302	:	724
:	:	2 246	428	5 134	10	295	:	713
2 082	3 436	2 034	400	5 155	10	293	1 666	714
2 022	2 839	2 027	313	5 279	9	289	1 561	659
797	:	1 848	310	2 158	9	242	:	583
:	:	1 565	276	2 126	7	234	:	543
849	1 627	1 459	255	2 134	7	234	983	524
749	1 143	1 390	250	1 926	6	226	847	474
0,43	:	0,66	0,64	0,36	0,73	0,77	:	0,75
:	:	0,69	0,64	0,41	0,70	0,79	:	0,76
0,41	0,49	0,72	0,64	0,41	0,69	0,80	0,59	0,73
0,37	0,40	0,69	0,80	0,36	0,67	0,78	0,54	0,72
54,1	:	45,5	45,7	52,1	39,4	48,1	:	32,8
:	:	44,7	51,2	54,3	42,5	45,0	:	32,6
45,8	46,6	45,3	52,4	53,3	39,7	43,9	37,2	30,8
45,7	56,1	50,2	54,5	50,5	42,3	43,2	38,4	36,9
30,1	:	27,5	20,9	22,4	32,0	24,8	:	13,6
:	:	25,5	18,5	21,1	24,8	22,2	:	15,5
33,8	16,2	24,5	18,0	21,0	24,7	21,5	26,8	16,3
33,6	17,8	24,9	23,0	24,1	23,4	21,5	28,1	18,3
15,4	:	13,0	27,6	22,1	25,1	16,5	:	18,3
:	:	14,7	22,4	22,7	27,5	18,1	:	18,4
20,2	27,5	13,5	20,4	23,4	28,6	18,1	28,6	20,0
20,5	22,4	16,8	18,3	24,2	28,3	18,6	28,3	17,3
0,4	:	8,0	5,8	3,4	3,5	10,6	:	35,3
:	:	15,1	7,9	1,9	5,2	14,7	:	33,5
0,2	9,7	16,8	9,2	2,3	7,1	16,6	7,4	32,9
0,2	4,6	8,9	4,3	1,5	6,2	17,6	5,5	30,0
56,8	:	47,6	53,0	46,7	42,9	51,4	:	34,1
43,4	:	47,1	56,3	49,3	46,3	49,7	:	31,9
43,5	42,8	46,2	58,1	:	45,7	48,5	38,2	32,3
48,2	50,1	51,3	57,8	46,9	48,3	47,0	40,3	36,7
28,4	:	22,0	15,5	18,7	26,7	18,6	:	11,4
:	:	:	:	:	:	:	:	:
31,3	16,1	19,5	12,4	20,1	23,7	15,6	27,3	12,1
30,5	13,4	19,9	21,2	17,6	18,9	14,7	28,0	12,8
14,0	:	16,4	23,6	19,7	25,9	15,4	:	16,1
43,4	:	30,8	31,0	36,7	46,6	31,9	:	26,3
42,4	36,0	28,7	29,9	36,7	45,2	31,3	49,6	27,4
46,9	29,1	33,1	36,5	37,0	44,2	31,8	46,2	28,1
0,8	:	9,6	7,9	4,1	4,4	11,4	:	38,4
0,4	:	17,6	11,0	4,0	7,0	15,8	:	36,6
0,4	8,4	18,9	10,6	3,7	8,3	16,5	9,5	35,3
0,5	8,8	10,8	4,4	3,4	8,9	18,8	7,4	33,2
0,0	:	4,4	0,0	10,8	0,1	3,2	:	0,0
12,8	:	4,5	1,7	10,0	0,1	2,6	:	6,8
13,7	12,7	6,2	1,4	9,9	0,8	3,7	2,7	5,0
4,5	16,5	5,7	1,8	13,6	0,8	4,1	7,5	5,8
858,0	1 229,4	1 552,0	274,9	1 950,5	8,6	203,7	:	310,8
798,0	879,0	1 368,0	240,3	1 904,8	6,4	189,4	:	303,8
727,3	1 281,1	1 110,0	223,5	:	3,0	186,6	821,0	310,6
711,1	852,7	1 161,7	234,2	1 598,3	5,7	173,6	720,9	289,0
98,0	350,4	282,0	35,4	463,5	0,6	50,6	:	218,4
133,0	271,4	235,0	34,7	293,7	0,6	53,2	:	183,0
119,9	343,5	371,9	30,5	:	3,7	47,3	114,5	210,1
37,6	290,3	228,4	15,5	327,4	0,6	51,5	126,0	184,7
956,0	1 579,8	1 834,0	310,3	2 414,0	9,2	254,3	:	529,2
931,0	1 150,4	1 603,0	275,0	2 198,5	7,0	242,7	:	486,8
847,3	1 624,7	1 481,9	254,0	:	6,7	233,9	935,6	520,7
748,7	1 143,4	1 390,1	249,7	1 925,6	6,3	225,0	846,9	473,7

3.5.1.6 Employment in agriculture: working hours and combined other employment of farmers (^{1})

1		Unit	EUR 12	Belgique/België	Danmark	BR Deutschland
	2	3	4	5	6	7
Total						
Numbers	1980	× 1 000	:	114	120	828
	1985		:	97	91	723
	1987		8 272	92	85	690
	1989		8 171	85	81	665
No other gainful employment	1980	%	:	67,5	80,3	56,8
	1985		:	68,1	68,9	57,5
	1987		69,8	67,4	67,2	57,0
	1989		68,0	65,8	66,4	56,5
With other main gainful employment	1980	%	:	29,5	13,2	37,3
	1985		:	29,2	9,3	37,6
	1987		23,0	29,6	10,3	38,3
	1989		26,1	32,0	11,8	34,7
With other secondary gainful employment	1980	%	:	3,1	6,5	5,9
	1985		:	2,6	21,8	4,9
	1987		7,1	3,0	22,5	4,7
	1989		5,8	2,2	21,8	8,8
Working hours = 100% (^{2})						
Numbers	1980	× 1 000	:	72	75	365
	1985		:	61	57	324
	1987		2 256 (^{3})	60	–	304
	1989		1 939	49	15	280
No other gainful employment	1980	%	:	94,6	95,2	95,2
	1985		:	98,2	82,8	95,5
	1987		91,6 (^{3})	96,9	–	95,7
	1989		94,3	100,0	85,0	95,6
With other main gainful employment	1980	%	:	2,0	0,0	0,0
	1985		:	0,0	1,4	0,0
	1987		0,3 (^{3})	0,0	–	0,0
	1989		0,0	0,0	1,2	0,1
With other secondary gainful employment	1980	%	:	3,4	4,9	4,9
	1985		:	1,8	15,6	4,5
	1987		8,1 (^{3})	3,1	–	4,3
	1989		5,3	0,0	13,8	3,9
Working hours from 50 to 100% (^{2})						
Numbers	1980	× 1 000	:	9	16	64
	1985		:	8	15	58
	1987		1 349 (^{3})	6	–	58
	1989		1 182	9	7	53
No other gainful employment	1980	%	:	54,1	65,4	37,2
	1985		:	59,1	59,9	36,3
	1987		77,0 (^{3})	50,3	–	37,4
	1989		76,3	46,5	68,8	35,5
With other main gainful employment	1980	%	:	39,0	15,0	26,6
	1985		:	25,5	14,1	40,5
	1987		8,5 (^{3})	36,9	–	42,0
	1989		9,7	36,8	9,5	39,6
With other secondary gainful employment	1980	%	:	7,0	19,5	36,2
	1985		:	15,4	26,1	23,2
	1987		14,5 (^{3})	12,8	–	20,5
	1989		14,1	16,8	21,8	25,0
Working hours of < 50% (^{2})						
Numbers	1980	× 1 000	:	33	29	400
	1985		:	28	19	341
	1987		4 582 (^{3})	26	–	328
	1989		4 448	25	23	319
No other gainful employment	1980	%	:	11,1	49,8	25,0
	1985		:	5,7	33,8	25,0
	1987		57,1 (^{3})	3,9	–	24,6
	1989		54,3	6,9	46,1	25,8
With other main gainful employment	1980	%	:	87,7	46,8	73,1
	1985		:	93,6	29,2	72,8
	1987		38,8 (^{3})	95,5	–	73,2
	1989		41,7	91,8	25,1	63,9
With other secondary gainful employment	1980	%	:	1,2	3,4	1,9
	1985		:	0,7	3,8	2,2
	1987		4,2 (^{3})	0,6	–	2,2
	1989		3,9	1,4	28,7	10,4

Source: Eurostat — Surveys of the structure of agricultural holdings.

(^{1}) Farmers who are at the same time farm heads. The farmer is the person for whom and on whose behalf the holding is farmed; the farm head is the person responsible for the current, day-to-day management of the holding. In EUR 10, 97% of agricultural holdings are farmed by farmers who are at the same time farm heads.
(^{2}) Farmers working their farms for respectively 100%, 50 to 100%, and less than 50% of the annual working hours of a full-time worker.
(^{3}) Not including Denmark.

Ellada	España	France	Ireland	Italia	Luxem-bourg	Nederland	Portugal	United Kingdom
8	9	10	11	12	13	14	15	16
997	:	1 210	214	2 760	5	145	:	237
951	:	998	213	2 754	4	132	:	219
953	1 601	921	210	2 750	4	129	619	220
924	1 594	1 017	171	2 665	4	125	599	243
70,4	:	79,8	73,7	70,6	78,6	79,1	:	78,9
65,6	:	67,7	66,5	73,8	82,4	79,9	:	78,8
66,6	70,4	68,2	63,5	76,0	81,3	76,4	61,7	76,1
67,6	65,8	72,7	73,8	69,7	81,6	100,0	63,6	70,4
20,6	:	15,8	18,8	26,0	14,7	14,7	:	14,8
27,5	:	13,0	23,4	22,5	15,0	14,6	:	12,3
26,9	23,2	11,8	26,0	20,5	14,4	15,5	31,8	13,9
26,4	28,5	15,6	20,5	27,9	14,2	0,0	32,3	17,3
9,0	:	4,4	7,5	3,4	6,7	6,2	:	6,3
6,8	:	19,3	10,0	3,7	5,4	5,4	:	9,0
6,5	6,4	19,9	10,5	3,4	4,4	8,1	6,4	10,0
6,1	5,7	11,8	5,7	2,5	4,5	0,0	4,1	12,3
165	:	666	109	322	3	108	:	150
103	:	565	95	371	3	98	:	138
89	418	532	91	350	2	96	180	133
115	249	506	115	293	2	81	123	112
95,4	:	98,5	94,2	97,9	93,5	93,3	:	95,3
97,1	:	76,8	91,0	97,7	94,8	93,9	:	95,4
97,1	97,1	76,2	92,4	97,8	94,6	93,1	97,4	94,5
97,5	97,7	90,2	88,9	94,6	95,0	100,0	99,1	92,1
0,0	:	0,1	0,9	0,0	0,7	1,9	:	1,7
0,0	:	0,1	1,2	0,0	0,2	2,4	:	0,6
0,0	0,0	0,0	0,9	0,0	0,0	2,8	0,6	0,8
0,0	0,0	0,0	5,8	0,0	0,0	0,0	0,0	0,0
4,6	:	1,4	5,0	2,1	5,9	4,8	:	3,5
2,9	:	23,1	7,8	2,3	5,0	3,7	:	4,0
2,9	2,9	23,8	6,7	2,2	5,4	4,1	2,0	4,7
2,5	2,3	9,7	5,3	5,4	5,0	0,0	0,1	7,9
265	:	184	51	440	1	21	:	31
236	:	150	56	467	1	18	:	28
204	208	132	51	492	1	18	149	29
183	199	171	27	306	1	21	181	25
76,2	:	71,0	72,0	87,3	42,5	42,5	:	55,8
82,9	:	60,4	66,9	88,6	62,0	41,9	:	50,2
85,8	77,5	60,7	63,9	89,8	72,7	32,2	68,8	47,2
84,9	84,8	62,0	43,7	83,0	79,0	100,0	80,5	44,7
2,7	:	9,7	14,2	4,4	42,9	52,4	:	24,9
5,6	:	7,7	15,6	2,7	27,5	44,5	:	8,8
4,4	6,9	6,9	15,9	2,8	22,9	45,1	14,6	9,9
5,4	4,1	8,2	4,7	8,3	17,1	0,0	9,6	6,4
21,1	:	19,3	13,8	8,3	1,4	16,8	:	19,3
11,4	:	31,9	17,5	8,7	10,5	13,6	:	41,0
9,8	15,6	32,3	20,2	7,3	4,4	22,7	16,6	43,0
9,6	11,1	29,8	8,8	8,7	3,9	0,0	9,9	48,9
566	:	360	54	1 997	1	17	:	56
612	:	283	61	1 916	1	14	:	53
659	975	256	68	1 907	1	15	291	57
624	790	323	24	1 975	1	18	265	63
60,3	:	49,7	34,1	62,5	38,6	34,2	:	47,6
53,7	:	53,4	28,1	65,6	41,7	34,6	:	50,1
56,5	57,4	55,5	24,5	68,5	45,6	22,1	36,1	48,1
57,0	49,7	50,9	34,3	63,9	38,1	100,0	35,7	41,9
35,0	:	47,8	59,3	35,0	55,6	60,1	:	45,8
40,6	:	41,7	65,3	31,8	55,5	60,3	:	44,5
37,6	36,7	38,9	67,2	28,9	53,2	61,7	59,9	46,6
37,4	43,7	43,6	61,2	35,0	58,7	0,0	62,5	52,4
4,6	:	2,5	6,6	2,5	5,8	5,7	:	6,6
5,7	:	4,8	6,6	2,6	2,8	5,1	:	5,4
5,9	5,9	5,6	8,2	2,6	1,2	16,2	4,0	5,3
5,7	6,6	5,5	4,4	1,1	3,2	0,0	1,9	5,8

3.5.2.2 Main crops in 1992

1	EUR 12		Belgique/België		Danmark		BR Deutschland		Elláda [1]		España [1]	
	Area	Share in UAA (%)	Area	Share in UAA (%)	Area	Share in UAA (%)	Area	Share in UAA (%)	Area	Share in UAA (%)	Area	Share in UAA (%)
	2	3	4	5	6	7	8	9	10	11	12	13
Cereals (total, excl. rice)	128 034 [4]	100,0	1376	100,0	2 976	100,0	11 868 [2]	100,0	5 741 [4]	100,0	26 389 [2]	100,0
of which:	33 061	27,2	311	22,6	1 451	54,1	4 336	37,7	1 392	25,0	7 318	29,3
common wheat	12 590	9,9	209	15,2	583	1,9	1 672	14,0	322	6,7	1 613	6,7
durum wheat	3 240	2,1	–	–	0	0,0	9	0,1	616	8,8	630	1,7
grain maize	3 812	3,2	10	0,7	–	–	275	1,9	211	4,2	393	1,8
barley	10 685	9,5	72	5,3	910	30,6	1 573	14,3	171	4,1	4 112	16,7
rye	747	3,5	2	0,1	88	3,0	308	3,5	17	0,4	180	0,7
Rice	368	0,3	–	–	–	–	–	–	14	0,4	86	0,4
Sugarbeet	1 856	1,4	101	7,4	65	2,2	381	3,4	49	0,6	163	0,6
Oilseeds (total)	4 923 [2]	3,7	6	0,4	180	6,1	656 [2]	5,1	16	5,2	1 557	4,4
of which: rape	1 855	1,4	6	0,4	180	6,1	555	4,8	–	–	8	0,0
sunflower	2 361 [2]	1,7	–	–	–	–	36	0,0	14	0,7	1 456	4,1
Olive trees	4 269 [4]	3,3	–	–	–	–	–	–	683 [4]	11,8	2 141	8,1
Cotton	375	0,3	–	–	–	–	–	–	300	4,4	75	0,3
Tobacco	201	0,2	0	0 0	–	–	3	0,0	75	1,6	24	0,1
Hops	29 [2]	0,0	0	0 0	–	–	21	0,2	0	0 0	1	0,0
Potatoes	1 456	1,1	65	4,7	54	1,8	252	1,8	43	0,9	257	1,0
Dry pulses	1 846 [4]	1,5	4	0,3	118	4,0	34	0,4	38 [4]	0,6	244	1,1
Fresh vegetables (total)	1 714 [3]	1,3	32	2,4	16	0,5	67	0,4	134	2,5	454	1,8
of which: tomatoes [1]	227	0,2	1	0,1	0	0,0	0	0,0	39	0,7	56	0,2
onions [1]	91	0,1	0	0,0	0	0,0	4	0,0	10	0,2	28	0,1
Fresh fruit (tot.) excl. citr. fr.	11 409 [6]	8,9	15	1,1	8	0,3	161	1,3	1 036 [6]	18,0	4 755	18,2
of which: apples	336 [3]	0,3	9	0,6	3	0,1	27	0,2	17 [3]	0,3	54	0,2
pears	136 [3]	0,1	3	0,2	0	0,0	2	0,2	5 [3]	0,1	37	0,1
peaches	229 [3]	0,2	0	0,0	–	–	0	0,0	42 [3]	0,7	66	0,3
apricots	64 [3]	0,0	–	–	–	–	0	0,0	6 [3]	0,1	26	0,1
melons	103	0,1	0	0,0	0	0,0	–	–	8	0,2	56	0,2
Citrus fruit (total)	520 [6]	0,4	–	–	–	–	–	–	53 [6]	0,9	268	1,0
of which: oranges and mandarins	333 [3]	0,2	–	–	–	–	–	–	43 [3]	0,7	151	0,5
lemons	103 [3]	0,1	–	–	–	–	–	–	13 [3]	0,2	46	0,2
Almonds	811 [3]	0,6	–	–	–	–	–	–	30 [3]	0,5	605	2,4
Vines	3 889	3,2	0	0,0	–	–	102	0,8	161	2,9	1 405	5,4
Flowers and ornamental plants	63 [5]	0,0	1	0,0	0	0,0	8	0,1	1	0,0	5	0,0
Green fodder	4 625 [7]	3,6	157	11,4	87	2,9	888	7,6	86 [7]	1,1	598	2,2

Source: Eurostat.

[1] Harvested area.
[2] 1991.
[3] 1990.
[4] 1989.
[5] 1988.
[6] 1987.
[7] 1986.

(1 000 ha)

	France		Ireland		Italia		Luxembourg		Nederland		Portugal (¹)		United Kingdom	
	Area	Share in UAA (%)	Area	Share in UAA (%)	Area	Share in UAA (%)	Area	Share in UAA (%)	Area	Share in UAA (%)	Area	Share in UAA (%)	Area	Share in UAA (%)
	14	15	16	17	18	19	20	21	22	23	24	25	26	27
	30 347	100,0	4 444 (²)	100,0	17 215 (⁴)	100,0	125	100,0	1 995	100,0	4 532	100,0	17 723	100,0
	9 321	30,7	300	6,8	4 009	25 3	30	23,6	184	9,2	763	16,8	3 489	19,7
	4 658	15,4	91	1,9	988	6,6	8	6,5	127	6,4	245	5,4	2 065	11,7
	424	1,4	–	–	1 530	10,5	–	–	–	–	30	0,7	1	0,0
	1 871	6,2	–	–	854	4,7	–	–	8	0,0	190	4,2	0	0,0
	1 800	5,9	184	4,4	450	2,7	14	10,9	34	1,7	67	1,5	1 297	7,3
	54	0,2	0	0,0	8	0,0	0	0,0	6	0,3	75	1,7	8	0,0
	24	0,1	–	–	216	1,2	–	–	–	–	27	0,6	–	–
	461	1,5	31	0,7	296	1,7	0	0,0	121	6,1	1	0,0	197	1,1
	1 694	5,6	6	0,1	329	2,8	2	1,2	4	0,2	78	1,7	576	3,3
	665	2,2	6	0,1	7	0,1	2	1,2	4	0,2	–	–	421	2,4
	979	3,2	–	–	91	0,7	–	–	–	–	78	1,7	–	–
	15	0,0	–	–	1 140	6,7	–	–	–	–	317	7,0	–	–
	–	–	–	–	0	0,0	–	–	–	–	–	–	–	–
	11	0,0	–	–	85	0,5	–	–	–	–	3	0,1	–	–
	1	0,0	3 (²)	0,1	–	–	–	–	–	–	–	–	4	0,0
	184	0,6	22	0,5	106	0,7	1	0,8	187	9,4	105	2,3	180	1,0
	716	2,4	3	0,1	134	0,9	1	0,6	10	0,5	230 (³)	5,8	208	1,2
	279	0,9	4	0,1	411 (³)	2,3	0	0,0	70	3,5	80 (²)	1,8	180	0,8
	12	0,0	0	0,0	118	0,8	0	0,0	2	0,1	20 (³)	0,4	1	0,0
	7	0,0	0	0,0	17	0,1	0	0,0	14	0,7	2	0,0	9	0,1
	1 202	4,0	2	0,0	3 197	19,0	1	1,2	26	1,3	748	16,5	47	0,3
	77	0,3	1	0,0	82	0,5	0	0,0	17	0,9	25	0,1	24	:
	16	0,1	0	0,0	52	0,3	0	0,0	5	0,3	14	0,3	4	0,0
	24	0,1	–	–	79	0,5	–	–	0	0,0	17	0,4	–	–
	19	0,1	–	–	16	0,1	–	–	–	–	1	0,0	–	–
	19	0,1	–	–	20	0,1	0	0,0	0	0,0	3	0,1	–	–
	3	0,0	–	–	182	1,1	–	–	–	–	26	0,6	–	–
	0	0,0	–	–	122	0,7	–	–	–	–	24	0,5	–	–
	0	0,0	–	–	39	0,2	–	–	–	–	2	0,0	–	–
	3	0,0	–	–	117	0,7	–	–	–	–	43	0,9	–	–
	948	3,1	–	–	1 005	6,2	1	1,0	0	0,0	265	5,8	0	0,0
	7	0,0	0	0,0	9 (²)	0,1	0	0,0	24	1,2	0	0,0	6	0,0
	1 591	5,1	564 (²)	10,0	836 (²)	4,9	12	9,9	218	10,9	–	–	51	0,3

3.5.2.3   Utilized agricultural area, woods and forests

			Arable land		Permanent meadow and pasture	
			1 000 ha	% of the UAA of the country	1 000 ha	% of the UAA of the country
1	2	3	4	5	6	7
EUR 12	1 000 ha	1992	66 561	52,5	48 618	38,3
	% TAV	1992/1985	− 0,2	−	− 0,4	−
	% TAV	1992/1991	− 1,2	−	0,0	−
Belgique/België	1 000 ha	1992	809	58,8	528	38,3
	% TAV	1992/1985	1,2	−	− 2,7	−
	% TAV	1992/1991	0,3	−	− 5,2	−
Danmark	1 000 ha	1992	2 756	92,6	208	7,0
	% TAV	1992/1985	0,8	−	− 0,9	−
	% TAV	1992/1991	− 0,5	−	− 1,9	−
BR Deutschland	1 000 ha	1992	7 318	61,3	4 294	36,9
	% TAV	1992/1985	0,2	−	− 0,9	−
	% TAV	1992/1991	0,2	−	− 0,7	−
Elláda	1 000 ha	1992	2 925	50,9	1 789	31,2
	% TAV	1992/1985	0,1	−	0,0	−
	% TAV	1992/1991	0,0	−	0,0	−
España	1 000 ha	1992	15 201	57,8	5 844	23,7
	% TAV	1992/1985	− 0,3	−	− 2,0	−
	% TAV	1992/1991	− 0,4	−	− 6,4	−
France	1 000 ha	1992	17 813	58,7	11 093	36,6
	% TAV	1992/1985	0,1	−	− 1,3	−
	% TAV	1992/1991	0,1	−	− 1,0	−
Ireland	1 000 ha	1992	754	17,0	3 687	83,0
	% TAV	1992/1985	− 5,3	−	− 3,1	−
	% TAV	1992/1991	− 22,8	−	− 21	−
Italia	1 000 ha	1992	8 917	51,8	4 508	28,4
	% TAV	1992/1985	− 0,7	−	− 1,5	−
	% TAV	1992/1991	− 1,0	−	− 7,4	−
Luxembourg	1 000 ha	1992	55	43,9	69	54,9
	% TAV	1992/1985	0,3	−	− 0,4	−
	% TAV	1992/1991	− 0,6	−	0,4	−
Nederland	1 000 ha	1992	920	46,1	1 030	51,6
	% TAV	1992/1985	1,0	−	− 1,3	−
	% TAV	1992/1991	1,2	−	− 1,3	−
Portugal	1 000 ha	1992	2 906	64,1	761	16,8
	% TAV	1992/1985	0,0	−	0,0	−
	% TAV	1992/1991	0,0	−	− 0,5	−
United Kingdom	1 000 ha	1992	6 544	36,9	11 124	62,8
	% TAV	1992/1985	− 1,0	−	− 0,6	−
	% TAV	1992/1991	− 0,1	−	− 0,5	−

Source : Eurostat.

Permanent crops		Total UAA		Woods and forests	
1 000 ha	% of the UAA of the country	1 000 ha	% of the UAA of EUR 12	1 000 ha	% of the UAA of the country
8	9	10	11	12	13
11 771	9,3	126 813	100,0	57 279	25,3
− 0,2	−	− 0,3	−	1,0	−
− 0,5	−	− 0,5	−	0,3	−
18	1,3	1 388	1,1	617	20,2
3,7	−	− 0,4	−	0,0	−
5,7	−	− 1,8	−	0,0	−
12	0,4	2 785	2,2	445	10,3
− 0,1	−	0,7	−	− 1,5	−
6,5	−	− 0,6	−	0,0	−
184	1,5	11 893	9,4	7 401	29,8
0,2	−	− 0,3	−	0,1	−
0,1	−	− 0,2	−	0,0	−
1 052	18,3	5 741	4,5	5 755	43,6
0,4	−	0,0	−	0,0	−
0,3	−	0,0	−	0,0	−
4 889	18,1	27 039	21,3	15 915	31,5
− 0,2	−	− 0,5	−	3,5	−
− 0,1	−	− 2,2	−	0,4	−
1 209	3,9	30 690	24,2	14 870	27,1
− 1,3	−	− 0,5	−	0,2	−
− 0,4	−	− 0,3	−	0,3	−
2	0,0	5 668	4,5	327	4,7
3,0	−	− 3,5	−	0,0	−
21,0	−	− 21,3	−	0,0	−
3 323	19,3	17 215	13,6	6 434	21,4
− 0,7	−	− 0,5	−	0,8	−
− 0,1	−	− 0,7	−	0,2	−
1	0,8	126	0,1	617	34,3
− 9,4	−	− 0,2	−	0,0	−
− 8,3	−	− 0,2	−	0,0	−
40	2,0	2 013	1,6	330	8,0
1,5	−	− 0,2	−	1,7	−
2,1	−	− 0,1	−	0,0	−
865	19,1	4 532	3,6	2 968	32,3
0,0	−	0,0	−	0,0	−
0,0	−	0,0	−	0,0	−
55	0,3	17 723	14,0	2 438	10,0
− 1,2	−	− 0,7	−	1,0	−
− 1,6	−	− 0,4	−	0,5	−

3.5.2.4   Area used for the principal agricultural products

1	2	3	Cereals including rice	Fresh vegetables	Roots and brassicas	
					Potatoes	Sugarbeet
	2	3	4	5	6	7
EUR 12 1 000 ha		1992	33 429	1 714 ([2])	1 436	1 865
% TAV		1992/1985	− 1	0,7	− 0,8	− 0
% TAV		1992/1991	− 2	0,7	3,9	0,8
Belgique/België 1 000 ha		1992	311	32	65	101
% TAV		1992/1985	− 1,5	0,9	4,1	− 2,2
% TAV		1992/1991	− 1,4	7,7	13,0	− 1,4
Danmark 1 000 ha		1992	1 609	16	54	65
% TAV		1992/1985	0	− 1,7	8,8	− 1,6
% TAV		1992/1991	2,5	0,1	23,0	0,7
BR Deutschland 1 000 ha		1992	4 336	67	252	381
% TAV		1992/1985	− 1,7	5,9	2,0	− 0,8
% TAV		1992/1991	− 1,5	5,5	12,1	− 1,6
Elláda ([1]) 1 000 ha		1992	1 407	134	43	49
% TAV		1992/1985	− 0,8	− 2	− 0,4	2,2
% TAV		1992/1991	− 5,9	1,7	− 4,8	8,9
España ([1]) 1 000 ha		1992	7 404	454	257	162
% TAV		1992/1985	− 0,4	− 0,6	− 3,6	− 1,4
% TAV		1992/1991	− 5,2	− 4,3	7,8	− 1,5
France 1 000 ha		1992	9 345	279	184	461
% TAV		1992/1985	− 0,5	1,6	− 1,9	− 0,9
% TAV		1992/1991	1,3	− 0,9	8,0	0,8
Ireland 1 000 ha		1992	300	4	22	31
% TAV		1992/1985	− 3,5	3,2	− 5,2	− 0,4
% TAV		1992/1991	− 0,4	− 3,1	− 2,0	− 6,0
Italia 1 000 ha		1992	4 225	411 ([3])	106	296
% TAV		1992/1985	− 1,7	0,1	− 2,0	3,5
% TAV		1992/1991	− 4	− 2,2	9,0	7,1
Luxembourg 1 000 ha		1992	30	0	1	0
% TAV		1992/1985	− 1,8	0	0,0	0
% TAV		1992/1991	− 4,5	− 7,5	− 3,4	16,0
Nederland 1 000 ha		1992	184	70	187	121
% TAV		1992/1985	0	0,2	1,5	− 1,1
% TAV		1992/1991	1,5	3,7	4,1	− 2,0
Portugal ([1]) 1 000 ha		1992	790	80	105	1
% TAV		1992/1985	− 2,7	− 2,3	− 3,2	0,0
% TAV		1992/1991	− 7,6	− 2,4 ([2])	− 3,4	0,0
United Kingdom 1 000 ha		1992	3 489	134	180	197
% TAV		1992/1985	− 2	− 1	− 0,8	− 0,6
% TAV		1992/1991	− 0,3	− 1,2	2,1	0,5

Source : Eurostat.

([1]) Harvested area.
([2]) 1991.
([3]) 1990.
([4]) 1989.
([5]) 1988.
([6]) 1986.
([7]) 1987.

Oilseeds	Green fodder	Dry pulses	Fruit trees	Vines
8	9	10	11	12
4 923 (²)	13 852 (⁷)	1 846 (⁴)	2 378 (³)	3 884
6,3	− 0,4	5,5	0,5	− 1,5
− 5,7	− 0,2	− 4,2	0,4	− 1,4
6	249	4	14	0
17,0	6,5	21,9	3,5	0,0
− 16,8	10,2	− 23,2	5,3	0,0
180	353	118	3	−
2,8	− 0,2	− 1,0	9,4	−
− 35,9	4,7	− 13,0	16,3	−
656 (²)	1 156	34	49	102,0
14,5	− 0,8	0,0	0,4	0,1
8,6	0,6	3,6	2,2	− 0,1
16	220 (⁷)	38 (⁴)	145	161
5,3	− 6,7	− 7,4	0,5	− 1,0
3,3	− 9,1	7,9	0,7	0,0
1 557	1 189	244	924	1 405
5,3	1,0	− 7,2	1,1	− 1,8
33,6	0,2	− 18,5	− 2,7	− 1,8
1 694	4 353	716	217	943,0
5,8	− 2,2	− 70,7	0,7	− 1,7
− 8,8	− 4,6	− 10,1	2,3	− 1,0
6	564 (³)	3	2	−
4,1	− 0,6	2,3	1,9	−
− 4,8	− 1,0	− 2,5	31,8	−
329	2 378 (³)	134	828	1 005
14,6	− 1,7	− 3,8	− 0,3	− 1,3
− 21,4	− 1,2	− 10,8	− 0,8	− 1,8
2	30	1	0	1,0
10,4	8,1	−	0,0	0,0
− 42,8	3,3	28,8	0,0	− 3,3
4	257	10	24	0,0
− 13,5	2,4	− 12,3	0,6	0,0
− 41,7	5,3	− 29,1	0,6	− 6,7
78	450	230 (²)	137	265,0
9,6	− 0,6	− 3,2	1,2	− 0,4
27,3	0,0	− 4,2	0,0	0,0
576	1 613	208	33	1
10,0	− 1,7	6,1	− 2,0	−
29,5	− 1,0	2,4	− 2,1	0,0

3.5.3.3   **Cattle numbers and number of holders (1991)**

(%)

	EUR 12	Belgique/België	Danmark	BR Deutschland	Elláda	España	France	Ireland	Italia	Luxembourg	Nederland	Portugal	United Kingdom
1	2	3	4	5	6	7	8	9	10	11	12	13	14
Average size of stocks	37,4	58,2	62,6	40,9	10,4	17,7	51,1	35,3	23,6	81,9	78,5	7,0	85,6
Total – Animals	100	100	100	100	100	100	100	100	100	100	100	100	100
– Holders	100	100	100	100	100	100	100	100	100	100	100	100	100
1- 2 – Animals	0,5	0,1	0,1	0,2	5,1	1,7	0,1	0,1	1,2	0,0	0,0	10,1	0,1
– Holders	12,3	4,1	4,8	4,4	34,0	18,2	3,4	1,9	17,1	2,0	2,0	44,3	2,5
3- 4 – Animals	1,0	0,3	0,3	0,5	6,4	2,8	0,3	0,5	2,6	0,1	0,1	12,2	0,1
– Holders	10,1	4,7	4,8	5,7	18,8	14,0	4,4	5,0	17,5	2,8	2,5	23,2	3,2
5- 9 – Animals	2,8	1,0	1,0	2,1	13,5	8,6	1,3	2,9	6,6	0,4	0,6	17,2	0,6
– Holders	15,1	8,7	8,7	12,3	21,6	21,8	9,8	14,1	22,4	4,4	6,5	18,8	7,1
10- 14 – Animals	3,3	1,4	1,3	2,8	9,5	10,5	1,8	5,2	7,2	0,6	0,9	8,6	0,9
– Holders	10,3	6,9	6,9	9,6	8,6	15,8	7,8	15,5	13,2	4,4	5,8	5,1	6,7
15- 19 – Animals	2,9	1,7	1,5	3,2	7,5	8,1	2,1	4,5	4,0	0,6	1,0	5,9	1,1
– Holders	6,4	6,0	5,6	7,7	4,7	8,4	6,3	9,3	5,6	3,1	4,8	2,5	5,5
20- 29 – Animals	6,1	4,2	3,6	7,4	10,9	11,2	5,1	10,3	7,5	2,1	2,5	8,7	2,7
– Holders	9,5	10,1	9,1	12,5	4,8	9,1	10,7	15,4	7,2	7,4	8,1	2,5	9,5
30- 39 – Animals	6,2	4,9	3,9	8,2	7,0	8,5	6,4	10,6	5,3	1,8	3,0	4,1	3,0
– Holders	6,7	8,3	7,2	9,8	2,2	4,2	9,5	11,0	3,6	4,6	6,8	1,0	7,5
40- 49 – Animals	6,2	5,6	4,2	8,8	5,5	6,1	7,2	7,7	5,9	2,8	2,1	3,6	3,2
– Holders	5,2	7,4	5,9	8,1	1,3	2,1	8,3	6,1	3,1	5,6	6,5	0,5	6,2
50- 59 – Animals	6,2	6,3	4,6	8,7	4,9	4,1	7,6	8,2	3,9	3,3	4,5	3,3	3,3
– Holders	4,3	6,8	5,3	6,6	1,0	1,4	7,1	5,4	1,7	5,4	7,2	0,5	5,2
60- 99 – Animals	22,0	26,3	24,1	27,6	14,6	9,2	28,4	21,5	13,5	20,4	23,1	6,5	15,2
– Holders	10,7	19,7	19,3	14,9	2,1	2,1	18,8	9,9	4,1	22,3	23,1	0,5	16,7
100-199 – Animals	26,5	32,8	40,9	23,7	11,0	10,9	30,6	19,9	18,0	51,6	37,9	8,3	31,8
– Holders	7,4	14,4	19,0	7,4	0,9	1,4	12,0	5,3	3,1	32,8	22,2	0,5	19,4
200-299 – Animals	8,4	8,1	10,2	4,6	1,9	7,8	6,3	5,0	8,5	12,1	9,3	3,5	18,8
– Holders	1,4	2,0	2,7	0,8	0,1	1,1	1,5	0,8	0,8	4,4	3,1	0,1	6,7
≥300 – Animals	8,0	7,2	4,3	2,2	2,3	10,5	2,7	3,7	15,7	4,3	13,5	7,9	19,2
– Holders	0,6	0,9	0,7	0,2	0,0	0,4	0,5	0,2	0,6	1,1	2,1	0,1	3,8

Source : Eurostat.

3.5.3.4 Changing structure of cattle farms, by Member State

	EUR 12	Belgique/ België	Danmark	B.R. Deutsch-land	Elláda	España	France	Ireland	Italia	Luxem-bourg	Neder-land	Portugal	United Kingdom
1	2	3	4	5	6	7	8	9	10	11	12	13	14
Holdings (× 1 000)													
1987	2 536	64	41	431	85	370	498	169	446	3	70	211	147
1989	2 418	60	37	391	72	432	471	167	372	3	66	202	145
1991	2 107	56	36	355	60	285	410	167	339	3	64	196	136
% TAV 1991/1987	− 4,5	− 3,2	− 3,4	− 4,7	− 8,2	− 6,3	− 4,8	− 0,4	− 6,6	− 3,9	− 2,2	− 1,8	− 1,9
% TAV 1991/1989	− 6,6	− 5,9	− 1,4	− 4,7	− 8,7	− 18,8	− 6,7	0,0	− 4,5	− 3,6	− 1,5	− 1,3	− 3,2
Animals (× 1 000)													
1987	80 248	3 079	2 351	15 291	741	5 076	21 052	5 580	8 794	208	4 895	1 332	11 849
1989	80 288	3 127	2 221	14 650	690	5 312	21 394	5 899	8 747	208	4 772	1 335	11 933
1991	78 833	3 264	2 222	14 526	631	5 046	20 970	5 899	8 005	205	5 062	1 381	11 623
% TAV 1991/1987	− 0,4	1,5	− 1,4	− 1,3	− 3,9	− 0,1	− 0,1	1,4	− 2,3	− 0,4	0,8	0,9	− 0,5
% TAV 1991/1989	− 0,9	2,2	0,0	− 0,4	− 4,3	− 2,5	− 1,0	0,0	− 4,3	− 0,8	3,0	1,7	− 1,3
Average number of animals per holding													
1987	31,6	48,2	57,7	35,5	8,7	13,7	42,3	32,9	19,7	71,1	69,6	6,3	80,7
1989	33,2	52,4	59,8	37,5	9,6	12,3	45,5	35,3	23,5	77,4	71,9	6,6	82,1
1991	37,4	58,2	62,6	40,9	10,4	17,7	51,1	35,3	23,6	81,9	78,5	7,0	85,6

Source : Eurostat.

3.5.3.5   **Changing structure of cattle farms, by herd size class**

EUR 12

	1-2	3-4	5-9	10-14	15-19	20-29	30-39	40-49	50-59	60-99	100-199	200-299	≥300	All classes
1	2	3	4	5	6	7	8	9	10	11	12	13	14	15
Holdings (× 1 000)														
1987	336	291	422	267	168	247	169	131	98	239	135	22	12	2 536
1989	294	267	417	266	154	228	158	121	97	233	146	24	12	2 418
1991	259	214	317	216	136	199	142	110	90	225	156	30	14	2 107
% TAV $\frac{1991}{1987}$	-6,3	-7,4	-6,9	-5,2	-5,2	-6,0	-4,3	-4,3	-2,0	-1,5	3,8	7,6	3,7	-4,5
% TAV $\frac{1991}{1989}$	-6,1	-10,5	-12,9	-9,8	-6,2	-6,5	-5,4	-4,6	-3,5	-1,8	3,5	10,9	6,9	-6,6
Animals (× 1 000)														
1987	554	1 026	2 876	3 183	2 851	5 972	5 769	5 821	5 294	18 147	17 943	5 208	5 604	80 248
1989	504	957	2 857	3 194	2 621	5 491	5 412	5 357	5 238	17 831	19 482	5 670	5 673	80 288
1991	427	758	2 179	2 599	2 292	4 772	4 862	4 873	4 869	17 334	20 902	6 610	6 275	78 833
% TAV $\frac{1991}{1987}$	-6,3	-7,3	-6,7	-4,9	-5,3	-6,0	-4,2	-4,3	-2,1	-1,1	3,9	6,1	2,9	-0,4
% TAV $\frac{1991}{1989}$	-8,0	-11,0	-12,7	-9,8	-6,5	-6,8	-5,2	-4,6	-3,6	-1,4	3,5	8,0	5,2	-0,9

Number of animals

Source : Eurostat.

(%)

3.5.3.6 Dairy cow numbers and number of holders (1991)

	EUR 12	Belgique/België	Danmark	BR Deutschland	Elláda	España	France	Ireland	Italia	Luxembourg	Nederland	Portugal	United Kingdom
1	2	3	4	5	6	7	8	9	10	11	12	13	14
Average size of herds	18,4	27,6	35,8	17,3	4,5	8,2	25,0	24,5	12,9	30,9	40,1	3,9	65,6
Total – Animals	100	100	100	100	100	100	100	100	100	100	100	100	100
– Holders	100	100	100	100	100	100	100	100	100	100	100	100	100
1- 2 – Animals	1,6	0,2	0,1	0,7	17,0	5,7	0,3	0,6	3,3	0,1	0,2	22,8	0,1
– Holders	20,5	4,4	3,2	7,7	53,6	31,4	5,5	10,5	28,1	2,0	5,5	63,9	3,5
3- 4 – Animals	2,0	0,4	0,2	1,6	15,2	6,7	0,4	0,5	5,3	0,1	0,2	17,0	0,1
– Holders	10,8	3,0	2,1	7,8	20,5	15,7	3,0	3,7	19,7	1,2	2,9	20,0	1,1
5- 9 – Animals	5,6	2,1	1,0	7,8	22,6	19,7	2,5	3,0	10,2	1,0	0,7	14,5	0,3
– Holders	15,4	8,4	4,9	19,5	16,2	24,3	9,0	10,0	20,1	4,6	4,9	9,0	2,8
10- 14 – Animals	7,4	5,1	2,5	11,4	10,7	18,2	5,9	7,5	9,1	1,8	1,7	11,9	0,6
– Holders	11,5	11,8	7,3	16,6	4,3	13,5	12,6	15,7	10,0	4,7	5,5	3,0	3,3
15- 19 – Animals	7,6	7,0	4,0	13,1	7,6	11,3	8,4	7,7	5,8	4,1	2,7	7,4	0,9
– Holders	8,3	11,6	8,4	13,5	2,1	5,9	12,1	11,4	4,5	7,9	6,5	2,0	3,4
20- 29 – Animals	16,6	18,4	12,8	25,6	8,2	13,5	23,5	17,6	10,6	16,8	8,7	8,4	3,5
– Holders	12,9	21,3	18,9	18,7	1,6	4,9	24,6	18,4	5,9	22,2	14,2	1,0	9,3
30- 39 – Animals	14,9	19,4	17,5	16,8	4,0	6,3	23,6	15,1	10,5	27,9	12,8	4,1	6,0
– Holders	8,1	16,0	18,4	8,7	0,6	1,6	17,6	11,2	4,1	26,5	15,0	0,4	11,4
40- 49 – Animals	11,2	16,3	16,9	10,3	4,4	3,7	15,2	12,9	7,5	25,4	15,6	3,6	7,6
– Holders	4,8	10,4	13,8	4,1	0,5	1,1	8,5	7,3	2,2	19,1	14,1	0,3	11,4
50- 59 – Animals	7,8	12,0	13,9	5,8	4,0	2,5	9,0	8,8	5,0	10,8	14,5	3,0	7,5
– Holders	2,7	6,2	9,2	1,9	0,4	0,5	4,0	4,0	1,2	6,6	10,7	0,2	9,2
60- 99 – Animals	14,7	15,9	23,1	6,1	3,3	5,1	9,8	18,5	15,2	10,9	31,4	3,6	29,6
– Holders	3,7	6,2	11,5	1,5	0,2	0,5	3,5	6,3	2,7	5,0	17,1	0,2	25,3
≥100 – Animals	10,6	3,2	8,0	1,0	3,1	7,5	1,4	7,7	17,6	1,0	11,4	3,8	43,9
– Holders	1,4	0,7	2,2	0,1	0,1	0,5	0,5	1,4	1,4	0,3	3,5	0,1	19,3

Source : Eurostat.

3.5.3.7 Changing structure of dairy farms, by Member State

1	EUR 12	Belgique/België	Danmark	BR Deutschland	Elláda	España	France	Ireland	Italia	Luxembourg	Neder-land	Portugal	United Kingdom
	2	3	4	5	6	7	8	9	10	11	12	13	14
Holdings (× 1 000)													
1987	1 600	38	27	337	61	251	291	69	311	2	58	108	48
1989	1 397	34	23	308	55	232	241	57	242	2	55	103	45
1991	1 202	29	21	275	47	185	199	57	197	2	48	100	42
% TAV 1991/1987	−6,9	−6,4	−6,1	−5,0	−6,1	−7,3	−9,1	−4,6	−10,1	−4,3	−4,6	−1,8	−3,2
% TAV 1991/1989	−7,2	−7,2	−4,4	−5,5	−7,6	−10,7	−9,1	0,0	−9,8	−4,1	−6,6	−1,4	−3,4
Animals (× 1 000)													
1987	25 116	922	811	5 390	232	1 783	5 841	1 444	3 024	64	2 166	388	3 052
1989	23 921	872	759	5 023	233	1 822	5 494	1 400	2 930	60	1 996	398	2 932
1991	22 086	806	742	4 769	214	1 516	4 968	1 400	2 536	52	1 911	394	2 779
% TAV 1991/1987	−3,2	−3,3	−2,2	−3,0	−2,0	−4,0	−4,0	−0,8	−4,3	−5,0	−3,1	0,4	−2,3
% TAV 1991/1989	−3,9	−3,9	−1,1	−2,6	−4,2	−8,8	−4,9	0,0	−7,0	−6,4	−2,2	−0,5	−2,6
Average number of animals per holding													
1987	15,7	24,2	30,4	16,0	3,8	7,1	20,0	20,9	9,7	31,8	37,6	3,6	63,2
1989	17,1	25,7	33,0	16,3	4,2	7,9	22,8	24,5	12,1	32,4	36,4	3,9	65,0
1991	18,4	27,6	35,8	17,3	4,5	8,2	25,0	24,5	12,9	30,9	40,1	3,9	65,6

Source: Eurostat.

3.5.3.8 Changing structure of dairy farms, by herd size class

		1-2	3-4	5-9	10-14	15-19	20-29	30-39	40-49	50-59	60-99	≥100	All classes
	1	2	3	4	5	6	7	8	9	10	11	12	13
Holdings (× 1 000) 1987		375	203	274	182	132	182	100	59	34	45	15	1 600
1989		275	169	246	166	120	168	101	59	34	44	15	1 397
1991		247	130	186	139	100	155	98	57	32	44	17	1 202
% TAV 1991/1987		-9,9	-10,5	-9,3	-6,6	-6,6	-3,9	-0,6	-0,8	-1,6	-3,3	1,7	-6,9
% TAV 1991/1989		-5,2	-12,3	-13,2	-8,6	-8,7	-3,9	-1,7	-1,6	-2,3	1,0	3,7	-7,2
Animals (× 1 000) 1987		548	712	1 858	2 149	2 214	4 315	3 386	2 582	1 823	3 265	2 264	25 116
1989		424	602	1 688	1 961	2 020	4 007	3 411	2 573	1 801	3 181	2 253	23 921
1991		359	449	1 246	1 627	1 677	3 673	3 280	2 475	1 722	3 244	2 336	22 086
% TAV 1991/1987		-10,0	-9,0	-9,5	-6,7	-6,7	-3,9	-0,8	-1,1	-1,4	-0,2	0,8	-3,2
% TAV 1991/1989		-8,0	-13,6	-14,1	-8,9	-8,9	-4,3	-1,9	-1,9	-2,2	1,0	1,8	-3,9

Source: Eurostat.

3.5.3.9 Pig numbers and number of holders (1991)

(%)

1	EUR 12	Belgique/ België	Danmark	BR Deutsch-land	Elláda	España	France	Ireland	Italia	Luxem-bourg	Neder-land	Portugal	United Kingdom
	2	3	4	5	6	7	8	9	10	11	12	13	14
Average size of stocks	66,1	393,5	345,2	76,4	20,1	40,1	81,5	397,8	23,5	75,7	472,3	15,4	453,4
Total – Animals	100	100	100	100	100	100	100	100	100	100	100	100	100
– Holders	100	100	100	100	100	100	100	100	100	100	100	100	100
1- 2 – Animals	1,3	0,0	0,0	0,5	4,8	2,8	1,2	0,1	4,6	0,5	0,0	5,5	0,0
– Holders	57,0	6,2	2,4	21,4	78,5	74,8	66,2	32,0	68,5	19,9	1,7	61,8	14,6
3- 9 – Animals	1,4	0,1	0,1	1,7	2,5	1,7	1,1	0,2	4,3	1,8	0,0	6,1	0,2
– Holders	18,3	6,9	5,2	26,7	10,6	10,7	14,7	20,0	23,8	30,4	2,6	22,9	14,0
10- 19 – Animals	1,1	0,2	0,3	1,9	2,7	1,4	0,4	0,4	1,8	2,0	0,1	6,2	0,3
– Holders	5,1	4,7	6,4	10,5	3,9	3,3	2,2	12,0	3,3	9,3	2,5	7,3	10,5
20- 49 – Animals	2,5	0,8	1,4	5,6	4,9	2,4	1,1	0,7	2,6	6,2	0,4	8,5	0,8
– Holders	5,3	9,5	15,0	13,4	3,5	3,0	2,8	8,0	2,0	13,3	6,1	4,1	11,1
50- 99 – Animals	3,8	2,0	3,0	8,5	5,1	3,9	1,8	0,6	2,1	9,9	1,7	6,8	1,2
– Holders	3,6	10,9	14,2	9,2	1,6	2,3	1,9	4,0	0,7	9,4	10,7	1,5	7,4
100-199 – Animals	7,0	5,4	6,2	13,9	4,2	6,0	4,9	1,4	2,3	15,3	5,5	12,1	2,5
– Holders	3,2	14,7	14,8	7,6	0,6	1,6	2,6	4,0	0,4	7,9	18,0	1,3	7,7
200-399 – Animals	12,9	11,9	13,2	22,5	6,9	11,4	10,8	2,0	5,5	17,3	12,3	8,3	6,0
– Holders	3,0	16,4	15,9	6,0	0,5	1,6	2,9	4,0	0,4	4,7	20,2	0,5	9,5
400-999 – Animals	29,5	34,1	32,4	37,7	16,0	23,9	37,9	11,1	13,1	37,0	34,6	8,5	17,8
– Holders	3,2	20,8	17,6	4,8	0,5	1,9	4,7	8,0	0,5	4,4	25,6	0,3	12,2
≥1 000 – Animals	40,4	45,5	43,4	7,8	52,8	46,4	40,8	83,4	63,8	10,1	45,4	38,0	71,3
– Holders	1,3	10,0	8,5	0,4	0,4	0,7	2,0	8,0	0,4	0,5	12,6	0,2	13,1

Source : Eurostat.

3.5.3.10   Changing structure of pig farms, by Member State

	EUR 12	Belgique/België	Danmark	BR Deutschland	Elláda	España	France	Ireland	Italia	Luxembourg	Nederland	Portugal	United Kingdom
1	2	3	4	5	6	7	8	9	10	11	12	13	14
Holdings (× 1 000)													
1987	1 873	27	38	392	56	362	187	5	487	1	35	263	21
1989	1 779	22	31	330	59	484	164	3	469	1	30	168	18
1991	1 535	17	28	288	48	429	148	3	362	1	28	166	17
% TAV 1991/1987	-4,9	-11,0	-6,9	-7,5	-3,5	4,4	-5,7	-15,0	-7,1	-12,2	-5,7	-10,8	-5,4
% TAV 1991/1989	-7,1	-12,2	-5,0	-6,6	-9,8	-5,8	-5,1	0,0	-12,1	-13,9	-4,0	-0,4	-2,8
Animals (× 1 000)													
1987	105 017	5 861	9 266	24 470	1 138	17 228	11 914	960	9 383	77	14 349	2 456	7 915
1989	102 536	6 474	9 190	22 556	1 160	16 850	12 275	995	9 254	71	13 729	2 599	7 383
1991	101 450	6 550	9 783	21 989	974	17 209	12 068	995	8 523	64	13 217	2 560	7 519
% TAV 1991/1987	-0,9	2,8	1,4	-2,6	-3,8	0,0	0,3	0,9	-2,4	-4,4	-2,0	1,0	-1,3
% TAV 1991/1989	-0,5	0,6	3,2	-1,3	-8,4	1,1	-0,8	0,0	-4,0	-4,6	-1,9	-0,8	9,2
Average number of animals per holding													
1987	56,1	221,1	245,9	62,4	20,4	47,6	63,7	200,0	19,3	53,8	405,9	9,3	382,8
1989	57,6	299,5	294,5	68,4	19,5	34,8	74,7	397,8	19,7	61,6	451,9	15,5	410,5
1991	66,1	393,5	345,2	76,4	20,1	40,1	81,5	397,8	23,5	75,7	472,3	15,4	463,4

Source : Eurostat.

3.5.3.11   Changing structure of pig farms, by herd size class

EUR 12

	1-2	3-9	10-19	20-49	50-99	100-199	200-399	400-999	≥1 000	All classes
	2	3	4	5	6	7	8	9	10	11
Holdings (× 1 000)										
1987	933	384	158	128	80	66	55	51	17	1 873
1989	995	339	120	96	62	54	48	47	18	1 779
1991	875	281	79	81	55	50	46	49	19	1 535
% TAV 1991/1987	−1,6	−7,6	−16,0	−10,9	−8,7	−7,0	−4,4	−0,9	3,0	−4,9
% TAV 1991/1989	−6,2	−9,0	−18,9	−8,0	−5,8	−4,0	−2,0	2,0	4,7	−7,1
Animals (× 1 000)										
1987	1 474	1 776	2 164	4 073	5 627	9 400	15 699	31 313	33 490	105 017
1989	1 512	1 638	1 669	3 034	4 406	7 612	13 582	29 487	39 597	102 536
1991	1 326	1 373	1 126	2 582	3 902	7 138	13 115	29 914	40 973	101 450
% TAV 1991/1987	−2,6	−6,2	−15,1	−10,8	−8,7	−6,6	−4,4	−1,1	5,2	−0,9
% TAV 1991/1989	−6,3	−8,4	−17,9	−7,7	−5,9	−3,2	−1,7	7,2	1,7	−0,5

Number of animals

Source : Eurostat.

3.5.4.1 Number and area of holdings ([1])

	Farm size class (ha UAA)	Holdings						
		× 1 000			% of total		% TAV	
		1980	1987	1989	1987	1989	$\frac{1989}{1980}$	$\frac{1989}{1987}$
1	2	3	4	5	6	7	8	9
EUR 12	1- 5	:	3 411,0	3 173,1	49,2	49,3	×	− 3,6
	5-10	:	1 163,0	1 038,7	16,8	16,1	×	− 5,5
	10-20	:	936,0	848,3	13,5	13,2	×	− 4,8
	20-50	:	946,0	894,8	13,7	13,9	×	− 2,7
	≥ 50	:	473,0	483,4	6,8	7,5	×	1,1
	Total	:	6 929,0	6 438,3	100,0	100,0	×	− 3,6
Belgique/België	1- 5	25,9	21,8	20,1	27,7	27,5	− 2,8	− 4,0
	5-10	18,1	14,3	12,7	18,1	17,4	− 3,9	− 5,8
	10-20	24,3	19,3	16,8	24,5	22,9	− 4,0	− 6,7
	20-50	19,1	18,8	18,7	23,9	25,5	− 0,2	− 0,3
	≥ 50	3,8	4,6	4,9	5,8	6,7	2,9	3,2
	Total	91,2	78,8	73,3	100,0	100,0	− 2,4	− 3,6
Danmark	1- 5	12,9	1,5	1,3	1,7	1,6	− 22,5	− 6,9
	5-10	20,5	14,0	12,1	16,3	15,1	− 5,7	− 7,0
	10-20	30,8	21,8	20,1	25,3	25,0	− 4,6	− 4,0
	20-50	40,4	33,9	31,3	39,4	38,9	− 2,8	− 3,9
	≥ 50	11,8	14,8	15,6	17,2	19,4	3,2	2,7
	Total	116,3	86,0	80,4	100,0	100,0	− 4,0	− 3,3
BR Deutschland	1- 5	275,8	196,9	183,1	29,4	29,0	− 4,4	− 3,6
	5-10	149,1	118,4	108,4	17,6	17,2	− 3,5	− 4,3
	10-20	181,3	148,5	134,6	22,1	21,3	− 3,3	− 4,8
	20-50	177,9	166,2	159,8	24,8	25,3	− 1,2	− 1,9
	≥ 50	31,3	40,7	45,6	6,1	7,2	4,3	5,8
	Total	797,4	670,7	631,5	100,0	100,0	− 2,6	− 3,0
Elláda	1- 5	541,3	488,0	478,1	69,4	70,0	− 1,4	− 1,1
	5-10	149,9	140,7	130,1	20,0	19,0	− 1,6	− 3,8
	10-20	46,6	53,5	53,8	7,6	7,9	1,6	0,3
	20-50	12,4	17,5	18,0	2,5	2,6	4,2	1,4
	≥ 50	1,6	3,8	3,2	0,5	0,5	8,0	− 8,2
	Total	751,8	703,5	683,2	100,0	100,0	− 1,1	− 1,4
España	1- 5	849,5 ([2])	821,1	750,4	53,3	54,7	− 1,4	− 4,4
	5-10	274,2 ([2])	292,0	244,7	19,0	17,8	− 1,3	− 8,5
	10-20	183,1 ([2])	189,5	165,7	12,3	12,1	− 1,1	− 6,5
	20-50	132,8 ([2])	144,6	124,9	9,4	9,1	− 0,7	− 7,1
	≥ 50	84,4 ([2])	92,7	87,0	6,0	6,3	0,3	− 3,1
	Total	1 524,0 ([2])	1 539,9	1 372,7	100,0	100,0	− 1,2	− 5,6
France	1- 5	234,0	166,0	187,9	18,2	20,3	− 2,4	6,4
	5-10	165,5	107,2	113,9	11,7	12,3	− 4,1	3,1
	10-20	240,0	174,7	163,3	19,1	17,6	− 4,2	− 3,3
	20-50	345,0	299,2	288,3	32,8	31,2	− 2,0	− 1,8
	≥ 50	151,0	164,7	172,1	18,1	18,6	1,5	2,2
	Total	1 135,0	911,8	925,5	100,0	100,0	− 2,2	0,7

Average size		UAA						
ha		1 000 ha			% of total		% TAV	
1987	1989	1980	1987	1989	1987	1989	$\dfrac{1989}{1980}$	$\dfrac{1989}{1987}$
10	11	12	13	14	15	16	17	18
×	×	:	8 080	7 490	7,1	6,5	×	− 3,7
×	×	:	8 116	7 280	7,1	6,3	×	− 5,3
×	×	:	13 237	12 056	11,6	10,5	×	− 4,6
×	×	:	29 505	28 132	25,8	24,5	×	− 2,4
×	×	:	55 624	59 702	48,6	52,1	×	3,6
16,5	17,8	:	114 562	114 660	100,0	100,0	×	0,0
×	×	66,7	57	54	4,2	4,0	− 2,3	− 2,7
×	×	132,6	103	91	7,6	6,8	− 4,1	− 6,0
×	×	349,7	281	244	20,6	18,1	− 3,9	− 6,8
×	×	566,0	570	569	41,8	42,3	0,1	− 0,1
×	×	292,0	352	395	25,8	29,4	3,4	5,9
17,3	18,3	1 407,0	1 363	1 345	100,0	100,0	− 0,5	− 0,7
×	×	37,1	5	4	0,2	0,1	− 21,9	− 10,6
×	×	150,0	102	90	3,6	3,2	− 5,5	− 6,1
×	×	447,5	317	291	11,3	10,5	− 4,7	− 4,2
×	×	1 248,7	1 082	569	38,7	20,5	− 8,4	− 27,5
×	×	1 202,7	1 292	1 825	46,2	65,7	4,7	18,9
32,5	34,6	3 086,0	2 798	2 779	100,0	100,0	− 1,2	− 0,3
×	×	659,5	507	469	4,3	4,0	− 3,7	− 3,8
×	×	1 086,0	864	790	7,3	6,7	− 3,5	− 4,4
×	×	2 635,2	2 163	1 961	18,3	16,7	− 3,2	− 4,8
×	×	5 342,9	5 117	4 960	43,3	42,1	− 0,8	− 1,5
×	×	2 448,8	3 175	3 588	26,8	30,5	4,3	6,3
17,6	18,6	12 172,4	11 826	11 768	100,0	100,0	− 0,4	− 0,2
×	×	1 342,0	1 196	1 166	32,1	31,8	− 1,5	− 1,3
×	×	1 004,2	949	877	25,5	23,9	− 1,5	− 3,9
×	×	609,0	712	723	19,1	19,7	1,9	0,8
×	×	340,8	493	518	13,2	14,1	4,8	2,5
×	×	133,4	372	387	10,0	10,5	12,6	2,0
5,3	5,4	3 429,3	3 722	3 671	100,0	100,0	0,8	− 0,7
×	×	2 007,0 [2]	1 947	1 769	7,9	7,2	×	− 4,7
×	×	1 894,7 [2]	2 010	1 696	8,1	6,9	×	− 8,1
×	×	2 522,9 [2]	2 607	2 282	10,6	9,3	× ´	− 6,4
×	×	4 070,2 [2]	4 441	3 845	18,0	15,7	×	− 7,0
×	×	12 881,4 [2]	13 676	14 939	55,4	60,9	×	4,5
16,0	17,9	23 376,2 [2]	24 681	24 531	100,0	100,0	×	− 0,3
×	×	620,0	432	479	1,5	1,7	− 2,8	5,3
×	×	1 215,0	785	818	2,8	2,9	− 4,3	:
×	×	3 550,0	2 562	2 436	9,1	8,5	− 4,1	− 2,5
×	×	10 960,0	9 632	9 347	34,4	32,8	− 1,8	− 1,5
×	×	12 500,0	14 613	15 456	52,1	54,2	2,4	2,8
30,7	30,8	28 845,0	28 024	28 536	100,0	100,0	− 0,1	0,9

3.5.4.1 *(cont.)*

	Farm size class (ha UAA)	Holdings						
		× 1 000			% of total		% TAV	
		1980	1987	1989	1987	1989	1989/1980	1989/1987
1	2	3	4	5	6	7	8	9
Ireland	1- 5	33,9	34,9	17,6	16,1	10,5	− 7,0	− 2,9
	5-10	35,4	32,9	24,1	15,2	14,3	− 4,2	− 14,4
	10-20	67,7	63,3	48,3	29,2	28,8	− 3,7	− 12,6
	20-50	66,6	66,3	58,4	30,5	34,8	− 1,4	− 6,1
	⩾50	19,7	19,5	19,6	9,0	11,6	− 0,1	0,3
	Total	223,3	216,9	168,0	100,0	100,0	− 3,1	− 12,0
Italia	1- 5	1 312,3	1 340,1	1 170,2	67,9	67,4	− 1,3	− 6,6
	5-10	322,3	333,0	284,3	16,9	16,4	− 1,4	− 7,6
	10-20	166,8	171,3	155,2	8,7	9,0	− 0,8	− 4,8
	20-50	86,9	91,6	87,6	4,6	5,0	0,1	− 2,2
	⩾50	38,0	38,0	38,4	1,9	2,2	0,1	0,5
	Total	1 926,3	1 974,0	1 735,7	100,0	100,0	− 1,2	− 6,2
Luxembourg	1- 5	0,9	0,7	0,7	18,9	19,1	− 2,8	0,0
	5-10	0,5	0,4	0,4	9,9	9,7	− 2,4	0,0
	10-20	0,7	0,5	0,4	12,4	11,6	− 6,0	− 10,6
	20-50	1,8	1,2	1,1	32,5	29,4	− 5,3	− 4,3
	⩾50	0,8	1,0	1,1	26,2	30,2	3,6	4,9
	Total	4,7	3,8	3,6	100,0	100,0	− 2,9	− 2,7
Nederland	1- 5	31,0	29,2	27,8	24,9	24,8	− 1,2	− 2,4
	5-10	26,1	21,6	21,5	18,4	19,1	− 2,1	− 0,2
	10-20	37,3	29,3	25,4	25,0	22,6	− 4,2	− 6,9
	20-50	30,8	32,0	31,6	27,3	28,1	0,3	− 0,6
	⩾50	3,8	5,2	6,0	4,4	5,4	5,2	7,4
	Total	129,0	117,3	112,3	100,0	100,0	− 1,5	− 2,2
Portugal	1- 5	272,4	278,4	309,2	72,5	74,4	1,4	5,4
	5-10	43,9	57,8	56,1	15,0	13,5	2,8	− 1,5
	10-20	18,3	27,5	27,4	7,2	6,6	4,6	− 0,2
	20-50	8,7	12,9	13,6	3,4	3,3	5,1	2,7
	⩾50	6,2	7,4	9,3	1,9	2,2	4,6	12,1
.	Total	349,5	384,0	415,6	100,0	100,0	1,9	4,0
United Kingdom	1- 5	29,4	32,8	26,7	13,5	11,3	− 1,1	− 9,8
	5-10	31,2	30,2	30,5	12,4	12,9	− 0,3	0,5
	10-20	39,8	37,1	37,4	15,3	15,8	− 0,7	0,4
	20-50	67,6	61,8	60,7	25,4	25,7	− 1,2	− 0,9
	⩾50	81,3	81,0	81,0	33,3	34,3	− 0,1	0,0
	Total	249,2	242,9	236,3	100,0	100,0	− 0,6	− 1,4

Source: Eurostat: harmonized national data + community surveys of the structure of agricultural holdings.

[1] Holdings of 1 ha UAA or more.
[2] 1982 survey. TAV 1987/1982.

Average size		UAA						
ha		1 000 ha			% of total		% TAV	
1987	1989	1980	1987	1989	1987	1989	1989/1980	1989/1987
10	11	12	13	14	15	16	17	18
×	×	98,0	99	57	2,0	1,3	− 5,8	− 24,1
×	×	264,9	248	181	5,0	4,1	− 4,1	− 14,6
×	×	977,7	916	701	18,6	15,8	− 3,6	− 12,5
×	×	2 037,6	2 027	1 840	41,2	41,4	− 1,1	− 4,7
×	×	1 670,2	1 626	1 663	33,1	37,4	− 0,0	1,1
22,7	26,4	5 048,4	4 916	4 442	100,0	100,0	− 1,4	− 4,9
×	×	3 022,5	3 045	2 678	20,1	17,9	− 1,3	− 6,2
×	×	2 229,4	2 277	1 967	15,0	13,1	− 1,4	− 7,1
×	×	2 278,9	2 339	2 130	15,4	14,2	− 0,7	− 4,6
×	×	2 594,7	2 715	2 637	17,9	17,6	0,2	− 1,4
×	×	5 279,6	4 765	5 588	31,5	37,3	0,6	8,3
7,7	8,6	15 405,1	15 141	15 000	100,0	100,0	− 0,3	− 0,5
×	×	2,4	2	2	1,6	1,6	− 2,0	0,0
×	×	3,8	3	3	2,4	2,4	− 2,6	0,0
×	×	10,1	7	6	5,6	4,7	− 5,6	− 7,4
×	×	61,8	44	38	34,9	29,9	− 5,3	− 7,1
×	×	51,7	70	78	55,6	61,4	4,7	5,6
33,2	35,3	129,8	126	127	100,0	100,0	− 0,2	0,4
×	×	82,1	76	75	3,8	3,7	− 1,0	− 0,7
×	×	191,7	157	154	7,8	7,7	− 2,4	− 1,0
×	×	536,6	425	365	21,1	18,2	− 4,2	− 7,3
×	×	902,6	963	963	47,7	47,9	0,7	0,0
×	×	300,2	396	452	19,6	22,5	4,7	6,8
17,2	17,9	2 013,2	2 017	2 011	100,0	100,0	− 0,0	− 0,1
×	×	581,6	626	666	19,6	16,6	1,5	3,1
×	×	300,7	397	387	12,4	9,7	2,8	− 1,3
×	×	247,6	372	375	11,6	9,4	4,7	0,4
×	×	261,5	383	407	12,0	10,2	5,0	3,1
×	×	1 723,6	1 424	2 171	44,5	54,2	2,6	23,5
8,3	9,6	3 115,0	3 202	4 006	100,0	100,0	2,8	11,9
×	×	82,9	88	74	0,5	0,4	− 1,3	− 8,3
×	×	230,0	221	226	1,3	1,4	− 0,2	1,1
×	×	581,4	536	539	3,2	3,3	− 0,8	0,3
×	×	2 228,9	2 038	2 001	12,2	12,1	− 1,2	− 0,9
×	×	13 999,2	13 863	13 659	82,8	82,8	− 0,3	− 0,7
68,9	69,8	17 123,2	16 746	16 499	100,0	100,0	− 0,4	− 0,7

3.5.6.1 Agricultural products sold through cooperatives (1992)

(%)

1	Belgique/ België (¹)	Danmark	BR Deutsch-land	Elláda (¹)	España	France (¹)	Ireland (¹)	Italia (¹)	Luxem-bourg (¹)	Neder-land	Portugal (¹)	United Kingdom
	2	3	4	5	6	7	8	9	10	11	12	13
Pigmeat	15	95	23 (¹)	3	5	80 (³)	55	10	35	25	–	20
Beef/veal	1	59	25 (¹)	2	6	30 (³)	9	13	25	16	–	5,1 (¹)
Poultrymeat	–	0	:	20	8	30	20	0	–	21	–	0,2 (¹)
Eggs	–	59	:	3	18	25	0	8	–	17	–	18
Milk	65	92	56	20 (²)	16	50	98	33	81	82	–	4
Sugarbeet	–	0	:	0	20	16 (⁴)	0	0	–	63	–	0,4 (¹)
Cereals	25 – 30	48	:	49	17	70	26	20 (⁸)	79	65	–	19
All fruit	60 – 65	90	20 – 40	51	32	45	14	40	10	78	–	31
All vegetables	70 – 75	90	55 – 65	12	15	35 (⁵)	8	13	–	70	–	19

Source: EC Commission, Directorate-General for Agriculture.

(¹) 1991.
(²) Cows', ewes' and goats' milk.
(³) Finished animals; young cattle not included 70%; store animals not included 40%.
(⁴) Processed into sugar.
(⁵) Excl. potatoes (seed potatoes, 65%; early potatoes and ware potatoes, 25%).
(⁶) 15 % maize not included in the percentage.
(⁷) 43 % citrus fruits not included in the percentage.
(⁸) 28% maize.

3.5.6.2 Products sold under contracts concluded in advance (1992)

(%)

1	Belgique/ België (⁵)	Danmark	BR Deutsch-land (¹)(⁵)	Elláda (⁵)	España (²)	France (⁵)	Ireland (⁵)	Italia (²)	Luxem-bourg (⁵)	Neder-land	Portugal (⁵)	United Kingdom
	2	3	4	5	6	7	8	9	10	11	12	13
Pigmeat	55	0	:	5	:	30	0	:	15	35	:	70
Calves	90	0	:	3	:	35	0	:	–	85	:	1 (⁵)
Poultrymeat	90	–	:	15	:	50	90	:	–	90	:	95
Eggs	70	–	:	10	:	20	30	:	–	50	:	70
Milk	–	–	99	30 (³)	:	1 (⁴)	10	:	–	90	:	98
Sugarbeet	100	100	100	100	100	100	100	100	–	100	:	100
Potatoes	20 – 25	65	:	2,5	:	10	10	:	–	50	0	30
Peas	98	100	90	85	:	90	100	:	–	85	95	60
Canned tomatoes	–	–	:	100	:	–	0	100	–	–	100	– (⁵)

Source: EC Commission, Directorate-General for Agriculture.

(¹) Including producers' group.
(²) 1990.
(³) Cows', ewes' and goats' milk.
(⁴) Milk production is not subject to contracts. Only the prices are set by contract (for nearly all farmers).
(⁵) 1991.

3.5.6.5   **Amount of assistance provided for in the Community support frameworks adopted pursuant to Regulations (EEC) Nos 866/90 and 867/90 for the period 1991-93 (¹) (breakdown by Member State and by sector)**

(1 000 ECU — at constant 1991 prices)

	EUR 12	Belgique/België	Danmark	BR Deutschland	Elláda	España	France	Ireland	Italia	Luxembourg	Nederland	Portugal	United Kingdom
1	2	3	4	5	6	7	8	9	10	11	12	13	14
Forestry products	29 222	521	336	571	4 666	2 388	7 384	–	10 610	–	–	2 746	–
Meat	274 474	3 491	9 680	59 802	12 540	53 332	30 208	41 150	16 590	–	7 210	19 214	21 258
Knackers' products	24 294	–	–	24 294	–	–	–	–	–	–	–	–	–
Milk and milk products	218 928	9 168	4 546	110 350	27 663	11 329	10 665	6 486	9 235	–	–	17 800	11 686
Eggs and poultry	43 245	670	749	6 686	5 308	4 300	8 914	3 189	–	–	3 707	1 344	8 378
Sundry animals	12 653	–	–	–	450	6 836	–	2 204	1 095	–	–	1 588	480
Cereals	86 427	320	–	42 508	15 470	8 005	–	1 182	10 010	–	–	6 655	2 277
Sugar	6 275	–	–	–	–	–	6 275	–	–	–	–	–	–
Oilseeds	35 827	–	–	–	9 248	14 800	–	–	8 650	–	–	2 802	328
Protein plants	6 085	–	–	–	–	–	–	–	5 290	–	–	–	795
Wine and spirits	76 725	–	–	1 101	4 571	14 976	12 454	–	11 840	994	–	30 788	–
Fruit and vegetables	209 835	3 029	206	50 698	15 739	39 145	25 185	2 050	40 510	–	4 142	22 504	6 627
Flowers and plants	14 237	–	518	4 842	–	194	1 309	–	6 520	–	–	237	616
Seeds	8 086	–	316	547	–	1 045	2 509	–	3 670	–	–	–	–
Potatoes	59 141	–	–	38 049	–	2 365	3 877	4 125	–	210	2 841	1 303	6 372
Tobacco	0	–	–	–	–	–	–	–	–	–	–	–	–
Sundry vegetables	10 165	–	–	–	–	–	1 674	–	5 480	–	1 131	1 439	440
Multi-purpose markets and distribution	9 902	–	–	–	–	–	3 467	–	5 910	–	525	–	–
Animal feedingstuffs	15 169	–	–	–	7 965	–	–	–	6 850	–	–	–	354
Miscellaneous marketing and processing	0	–	–	–	–	–	–	–	–	–	–	–	–
Total	**1 140 690**	**17 199**	**16 351**	**339 448**	**103 620**	**158 715**	**113 921**	**60 386**	**142 260**	**1 204**	**19 556**	**108 420**	**59 611**

Source : EC Commission, Directorate-General for Agriculture.

(¹) Provisional data.

3.5.6.6   Specific measures to assist mountain and hill farming and farming in certain less-favoured areas — Article 19 of Regulation (EEC) No 2328/91

	Compensatory allowances granted in respect of less-favoured areas								
	Number of holdings			Amounts of allowances paid in 1992			Amounts of allowances per LU		
	1990	1991	1992	Total (ECU)	Average allowance per holding (ECU)		Number of LU 1992 (1 000)	ECU/LU	
					1991	1992		1991	1992
1	2	3	4	5	6	7	8	9	10
Belgique/België	7 853	7 702	7 450	8 143 757	1 091	1 093	116	70	70
Danmark	–	–	–	–	–	–	–	–	–
BR Deutschland	245 679	238 331	234 292	437 356 694	1 314	1 867	3 229	82	85
Elláda	214 151	214 151	197 789	94 720 400	390	479	1 242	50	56
España	228 039	210 940	209 068	80 713 437	365	386	1 528	:	32
France	161 559	154 295	149 098	231 263 922	1 531	1 551	4 903	55	47
Ireland	100 010	80 772	111 274	127 488 634	920	1 146	2 112	:	60
Italia	92 000	68 980	34 595	12 040 084	571	348	192	59	41
Luxembourg	2 507	2 820	2 779	9 199 707	3 272	3 310	58	89	90
Nederland	1 557	1 750	1 866	975 838	493	523	14	72	70
Portugal	111 842	112 695	120 370	42 861 618	338	356	609	51	61
United Kingdom	55 935	55 164	55 257	177 193 029	3 255	3 207	2 834	65	63
Total	1 221 000	1 147 600	1 123 838	1 221 957 120	923	1 087	16 837		

Source : EC Commission, Directorate-General for Agriculture.

3.5.6.7 **Breakdown by type of region of aid granted by the EAGGF Guidance Section to operational programmes approved in 1993 under Regulations (EEC) Nos 866/90 and 867/90**
(at 1993 prices)

(1 000 ECU)

	EUR 12	Belgique/België	Danmark	BR Deutschland	Elláda	España	France	Ireland	Italia	Luxembourg	Nederland	Portugal	United Kingdom
1	2	3	4	5	6	7	8	9	10	11	12	13	14
Regions — Objective 1	264 646	–	–	71 317	27 461	83 022	3 483	15 098	25 224	–	–	34 717	4 323
Regions — Objective 5b	15 942	1 856	–	2 262	–	–	9 812	–	–	–	143	–	1 869
Other regions	170 620	3 182	2 480	16 959	–	18 564	39 326	–	70 347	385	7 413	–	11 964
Total	451 208	5 038	2 480	90 538	27 461	101 586	52 621	15 098	95 571	385	7 556	34 717	18 156

Source: EC Commission, Directorate-General for Agriculture.

3.5.6.8   Breakdown by region (NUTS 2) of aid granted by the EAGGF Guidance Section to operational programmes approved in 1993 under Regulations (EEC) Nos 866/90 and 867/90

(1 000 ECU)

Member State	Region	
Belgique/België	Several regions	–
	Flandre-Orientale/Oost-Vlaanderen	909
	Flandre-Occidentale/West-Vlaanderen	1 325
	Anvers/Antwerpen	–
	Limbourg/Limburg	522
	Brabant	163
	Hainaut/Henegouwen	442
	Namur/Namen	8
	Liège/Luik	1456
	Luxembourg/Luxembourg	214
	Total	5 038
Danmark	Several regions	–
	Storkøbenhavn	–
	Øst for Storebælt ekskl. Storkøbenhavn	2 480
	Vest for Storebælt	–
	Total	2 480
BR Deutschland (NUTS 1)	Several regions	–
	Schleswig-Holstein	5 323
	Hamburg	9
	Niedersachsen	926
	Bremen	–
	Nordrhein-Westfalen	1 926
	Hessen	574
	Rheinland-Pfalz	1 153
	Baden-Württemberg	4 218
	Bayern	4 318
	Saarland	775
	Berlin (West)	–
	Total	19 227

Member State	Region	
BR Deutschland (Neue Länder) (NUTS 1)	Berlin (Ost)	–
	Mecklenburg-Vorpommern	1 478
	Sachsen-Anhalt	38 308
	Brandenburg	–
	Sachsen	9 125
	Thüringen	22 406
	Total	71 317
Elláda	Several regions	27 461
	Total	27 461
España	Several regions	–
	Galicia	–
	Principado de Asturias	–
	Cantabria	–
	País Vasco	–
	Navarra	–
	Castilla-León	–
	La Rioja	–
	Madrid	–
	Cataluña	–
	Aragón	–
	Extremadura	–
	Castilla-La Mancha	–
	Comunidad Valenciana	–
	Murcia	–
	Andalucía	–
	Baleares	–
	Canarias	–
	Total	101 586

Member State	Region	
France	Several regions	–
	Île-de-France	49 138
	Champagne-Ardenne	–
	Picardie	–
	Haute-Normandie	–
	Centre	–
	Basse-Normandie	–
	Bourgogne	–
	Nord-Pas-de-Calais	–
	Lorraine	–
	Alsace	–
	Franche-Comté	–
	Pays de la Loire	–
	Bretagne	–
	Poitou-Charentes	–
	Aquitaine	–
	Midi-Pyrénées	–
	Limousin	–
	Rhône-Alpes	–
	Auvergne	–
	Languedoc-Roussillon	–
	Provence-Alpes-Côte d'Azur	–
	Corse	–
	DOM	3 483
	Total	52 621

3.5.6.8 (cont.)

(1 000 ECU)

Member State	Region	
Ireland (NUTS 3)	Several regions	–
	Donegal	84
	North-West	2 120
	North-East	785
	West	4 358
	Midlands	2 112
	East	1 050
	Midwest	154
	South-East	2 024
	South-West	2 412
	Total	15 098
Italia	Several regions	19 577
	Piemonte	4 885
	Valle d'Aosta	1 318
	Liguria	1 681
	Lombardia	7 205
	Trentino-Alto Adige	4 674
	Veneto	6 694
	Friuli-Venezia Giulia	5 080
	Emilia-Romagna	7 818
	Toscana	3 495
	Umbria	1 008
	Marche	2 755
	Lazio	4 155
	Campania	1 617
	Abruzzi	8 741
	Molise	2 129
	Puglia	1 534
	Basilicata	4 086
	Calabria	1 390
	Sicilia	3 310
	Sardegna	2 416
	Total	95 571

Member State	Region	
Luxembourg	Several regions	385
	Total	385
Nederland	Several regions	–
	Groningen	541
	Friesland	293
	Drenthe	268
	Overijssel	596
	Gelderland	1459
	Utrecht	–
	Noord-Holland	566
	Zuid-Holland	269
	Zeeland	453
	Noord-Brabant	1938
	Limburg	561
	Z.IJ.-polders	612
	Total	7 556
Portugal	Several regions	–
	Entre Douro e Minho	2 437
	Trás-os-Montes	4 719
	Beira Litoral	3 675
	Beira Interior	4 021
	Ribatejo e Oeste	12 394
	Alentejo	2 626
	Algarve	85
	Açores	2 025
	Madeira	2 735
	Total	34 717

Member State	Region	
United Kingdom (NUTS 1)	Several regions	–
	North	847
	Yorkshire-Humberside	1 250
	East Midlands	496
	East Anglia	1 749
	South-East	–
	South-West	2 539
	West Midlands	1 695
	North-West	443
	Wales	1 043
	Scotland	3 769
	Northern Ireland	4 324
	Total	18 155

Source: EC Commission, Directorate-General for Agriculture.

3.5.6.9 **Breakdown by product group of aid granted by the EAGGF Guidance Section to operational programmes**

1	EUR 12	Belgique/België	Danmark	BR Deutschland	New *Länder*	Elláda
1	2	3	4	5	6	7
Forestry products	18 430	577	37	–	:	5 174
Meat	149 791	2 347	1 524	25 040	:	6 471
Milk and milk products	54 852	2 084	614	14 359	:	752
Eggs and poultry	16 529	–	305	4 968	:	507
Sundry animals	4 210	–	–	–	:	213
Cereals	24 830	30	–	17 655	:	1 430
Sugar	692	–	–	–	:	–
Oilseeds	14 530	–	–	–	:	2 343
Protein plants	95	–	–	–	:	–
Wine and spirits	36 648	–	–	1 868	:	1 073
Fruit and vegetables	83 162	–	–	7 519	:	2 032
Flowers and plants	6 597	–	–	3 186	:	–
Seeds	2 946	–	–	257	:	–
Potatoes	20 693	–	–	14 729	:	–
Tobacco	601	–	–	–	:	–
Sundry vegetables	4 406	–	–	–	:	–
Multi-purpose markets and distribution	1 553	–	–	–	:	–
Animal feedingstuffs	9 628	–	–	–	:	7 466
Other	0	–	–	–	:	–
Miscellaneous marketing and processing	58	–	–	–	:	–
Total (¹)	450 251	5 038	2 480	89 580		:	27 461

Source : EC Commission, Directorate-General for Agriculture.

(¹) Total including the new *Länder*, for which a breakdown by sector is not possible.

under Regulations (EEC) Nos 866/90 and 867/90 in 1993 ([1])

(1 000 ECU)

España	France	Ireland	Italia	Luxembourg	Nederland	Portugal	United Kingdom
8	9	10	11	12	13	14	15
1 595	2 327	–	5 945	–	–	2 774	–
35 745	17 155	12 194	34 030	–	2 414	4 611	8 260
12 286	5 431	482	6 561	–	–	8 873	3 409
2 670	4 851	785	–	–	1 460	–	984
2 605	–	435	408	–	–	180	369
2 075	–	–	2 383	–	–	938	319
–	692	–	–	–	–	–	–
8 525	–	–	1 947	–	–	1 215	500
–	–	–	–	–	–	–	95
8 603	7 894	–	8 451	385	–	8 373	–
24 568	10 162	–	27 453	–	1 657	7 276	2 496
–	949	602	1 697	–	–	102	60
1 033	1 001	–	655	–	–	–	–
1 881	713	–	–	–	1 708	–	1 663
–	–	601	–	–	–	–	–
–	1 446	–	2 326	–	259	375	–
–	–	–	1 553	–	–	–	–
–	–	–	2 162	–	–	–	–
–	–	–	–	–	–	–	–
–	–	–	–	–	58	–	–
101 586	52 621	15 098	95 571	385	7 556	34 717	18 156

3.5.6.11   **Investment aid for agricultural holdings (1992)**
(Application of Council Regulation (EEC) No 2328/91)

| | Number of plans approved | Volume of eligible investment involved (1 000 ECU) | Total aid proposed under the Community scheme (1 000 ECU) | Average per plan (ECU) | | Total number of PIPs (physical improvement plans) on all main occupation holdings (¹) % |
| | | | | Eligible investments | Planned aid | |
1	2	3	4	5=3/2	6=4/2	7
Belgique/België	88	9 220	2 582	104 768	29 342	0,1
Danmark	2 073	135 256	38 641	65 247	18 640	3,6
BR Deutschland	5 084	277 620	78 426	54 607	15 426	1,5
Elláda	6 168	206 250	—	33 439	—	2,2
España	4 949	118 950	39 404	24 035	7 962	0,9
France	5 663	443 283	108 921	78 277	19 234	0,8
Ireland	3 291	78 917	30 429	23 980	9 246	2,9
Italia	3 367	172 880	60 816	51 345	18 062	0,4
Luxembourg	82	8 568	3 756	104 490	45 800	2,7
Nederland	954	74 090	8 650	77 662	9 067	0,9
Portugal	3 172	123 413	40 109	38 907	12 645	1,0
United Kingdom	986	31 919	13 265	32 373	13 453	0,6
Total	35 877	1 680 366	424 999	46 837	11 846	1,0

Source : EC Commission, Directorate-General for Agriculture.
(¹) Calculated on the basis of the 1989/90 Community survey of the structure of agricultural holdings.

3.5.6.12 Special aid for young farmers (1992)
(Application of Council Regulation (EEC) No 2328/92)

1	Number of beneficiaires or aids approved in 1992		Eligible amount of the start-up premium (1 000 ECU)	Volume of eligible investments provided for in the plans (1 000 ECU)	Total investment aid eligible under the Community regulation (1 000 ECU)	Of which additional investment aid (1 000 ECU)	Average per beneficiary (1 000 ECU)	
	Start-up premium (Article 10)	Investments aid (Article 11)					Start-up premium	Investment aid
1	2	3	4	5	6	7	8	9
Belgique/België	1 134	323	34 877	5 724	4 307	187	31	13
Danmark	2 073	387	3 499	37 991	13 619	2 637	2	35
BR Deutschland	4 058	390	15 594	67 403	59 103	4 052	4	152
Elláda	494	344	::	::	::	::	::	::
España	::	::	::	::	::	::	::	::
France	10 330	1 856	–	::	::	::	::	::
Ireland	387	410	987	13 357	13 357	5 252	::	::
Italia	2 191	722	11 914	42 129	16 111	13 617	5	22
Luxembourg	57	44	756	4 990	2 757	551	13	63
Nederland	80	–	–	5 827	713	143	–	::
Portugal	1 247	1 187	12 735	57 330	23 290	4 658	10	20
United Kingdom	(¹)	33	(¹)	1 126	443	111	(¹)	13
Total	22 051	5 696	80 362	235 877	133 700	31 208	4	23

Source : EC Commission, Directorate-General for Agriculture.

(¹) Scheme not implemented.

3.5.6.13 Community aid scheme for early retirement from agriculture (Regulation (EEC) No 2079/92)

1	Programmes approved (²)	Beneficiaries: farmers (³)	Beneficiaries: workers (³)	Area released (³) (ha)	Total cost forecast 1993-97 (Mio ECU)	EAGGF commitment 1993-97 (Mio ECU)
	2	3	4	5	6	7
Belgique/België	1	5 000	(⁴)	80 000	92,0	46,0
Danmark	1	4 500	(⁴)	150 000	72,0	36,0
BR Deutschland	1	2 500	25	:	38,4	19,2
Elláda	1	50 000	(⁴)	250 000	186,5	139,9
España	3	26 750	4 800	450 000	444,3	303,2
France	2	46 600	(⁴)	1 673 000	887,0	443,5
Ireland	1	7 000	500	210 000	125,2	93,8
Italia	1	26 500	1 000	276 500	291,9	176,6
Luxembourg (¹)	–	–	–	–	–	–
Nederland (¹)	–	–	–	–	–	–
Portugal	3	6 073	950	33 000	56,6	42,4
United Kingdom (¹)	–	–	–	–	–	–
Total	14	173 850	7 275	3 122 500	2 193,8	1 300,6

Source : EC Commission, Directorate-General for Agriculture.

(¹) Luxembourg, the Netherlands and the United Kingdom do not apply the scheme.
(²) As at 31.12.1994.
(³) Forecasts for the period 1993-97.
(⁴) Not applicable.

3.5.6.14 Forestry measures on farms (1991)
(Application of Council Regulation (EEC) No 2328/91)

Title VIII of Regulation (EEC) No 2328/91

	Existing woodland (1 000 ha)(¹)	Afforestation				Improvement			Total		
		New plantings (ha)		Eligible amount of aid (1 000 ECU)	Average premium per ha (ECU)	Improvement of woodland (ha)		Amount of aid (1 000 ECU)	Area (ha)		Eligible amount of aid (1 000 ECU)
		1990	1991			1990	1991		1990	1991	
1	2	3	4	5	6	7	8	9	10	11	12
Belgique/België	617	—	—	—	—	—	—	—	—	—	—
Danmark	494	—	—	—	—	29 710	41 966	761	29 710	41 966	761
BR Deutschland	10 512	565	1 240	366	5 155	5 603	12 928	659	6 168	14 148	1 025
Elláda	5 755	1 748	4 336	588	215	—	—	—	1 748	4 236	588
España	12 511	1 441	1 032	503	1 237	5 135	2 132	320	6 576	3 164	823
France	14 688	6 279	7 112	1 604	1 228	91	128	13	6 370	7 240	1 617
Ireland	327	57	7	21	1 682	—	—	—	57	7	21
Italia	6 410	1 207	109	710	617	—	962	120	1 207	1 071	830
Luxembourg	88	—	—	—	—	—	—	—	—	—	—
Nederland	330	—	—	—	—	—	—	—	—	—	—
Portugal	2 986	2	17	5	452	—	—	—	2	17	5
United Kingdom	2 297	9 420	11 621	5 582	1 029	—	—	—	9 420	11 621	5 582

Source : EC Commission, Directorate-General for Agriculture.
(¹) Woodland excluding other woodland (scrubland, heathland, etc.). *Source :* Eurostat.

3.5.6.16   **Areas set aside under the different set-aside schemes for arable land** (1993/94 marketing year)

Member States	Area set aside (1 000 ha)			
	Five-year set-aside (¹)	Annual set-aside		Total
		Total (²)	of which industrial set-aside (³)	
Belgique/België	1	19	3	20
Danmark	7	208	19	215
BR Deutschland	411	1 050	68	1 461
Elláda	1	15	0	16
España	85	875	6	960
France	214	1 578	73	1 792
Ireland	2	26	0	28
Italia	786	195	43	981
Luxembourg	0	2	0	2
Nederland	15	8	1	24
Portugal	0	61	0	61
United Kingdom	133	568	51	701
Total	1 655	4 605	264	6 260

Source: EC Commission, Directorate-General for Agriculture.

(¹) Regulation (EEC) No 2328/91.
(²) Regulation (EEC) No 1765/92.
(³) Regulations (EEC) Nos 1765/92 and 334/93; provisional data.

3.6.1   World exports and EC external trade in all products, agricultural products (¹) and other products

(Billion USD)

1	1986	1987	1988	1989	1990	1991 ∞	1992 ∞	1993 ∞
	2	3	4	5	6	7	8	9
World exports (²)								
– All products	1 712,1	1 966,0	2 211,7	2 390,6	2 550,7	2 587,2	2 737,4	:
of which: agricultural products	229,2	256,9	294,9	312,4	303,2	300,0	315,0	:
other products	1 482,9	1 709,1	1 916,8	2 078,2	2 247,5	2 287,2	2 422,4	:
External EC trade (²)								
Exports:								
– all products	339,1	391,7	429,0	455,0	528,9	524,7	567,7	565,1
of which: agricultural products	28,9	32,8	36,0	39,7	44,8	44,6	50,3	49,0
Imports:								
– all products	331,3	392,6	458,2	492,2	587,7	612,2	634,3	569,1
of which: agricultural products	52,8	58,7	64,6	63,5	71,2	70,5	73,8	63,9
World exports of agricultural products as percentage of total world exports	13,4	13,1	13,3	13,1	11,9	11,6	11,5	×
EC exports of agricultural products as percentage of total EC exports	8,5	8,4	8,4	8,7	8,5	8,5	8,9	8,7
EC imports of agricultural products as percentage of total EC imports	15,9	15,0	14,1	12,9	12,1	11,5	11,6	11,2
Index changes (1985 = 100)								
World exports:								
– all products	108,4	124,4	140,0	151,3	161,4	163,7	173,3	×
– agricultural products	105,8	118,6	136,1	144,2	139,9	138,4	145,4	×
– other products	108,8	125,4	140,6	152,4	164,9	167,8	177,7	×
External EC trade								
Exports:								
– all products	117,3	135,5	148,4	157,4	183,0	181,6	196,4	195,5
– agricultural products	110,7	125,7	137,9	152,0	171,6	170,9	192,7	187,7
Imports:								
– all products	106,8	126,6	147,8	158,7	189,5	197,4	204,5	183,5
– agricultural products	113,1	125,7	138,3	136,0	152,5	151,0	158,0	136,8

Sources: GATT statistics and Eurostat.

NB: When comparing statistical series for trade expressed in value terms, it is important to remember that, because of exchange rate movements, the use of one currency unit rather than another may alter the apparent trend. For example, between 1985 and 1986, the ratio of the USD to the ECU changed by 22,4 %, and, between 1986 and 1987, by a further 14,8 %.

(¹) SITC 0, 1, 21, 22, 232 (231 from 1988), 24, 261 to 265 + 268, 29, 4.

(²) Excl. intra-Community trade.

3.6.2 EC trade by product

EUR 12
(Mio ECU)

SITC codes	Products	Imports			Exports			Balances		
		1991 ∞	1992 ∞	1993 ∞	1991 ∞	1992 ∞	1993 ∞	1991 ∞	1992 ∞	1993 ∞
1	2	3	4	5	6	7	8	9	10	11
0	Food products	35 033	35 135	33 344	22 897	24 885	27 043	−12 136	−10 250	−6 301
04	of which: − cereals	1 274	1 165	1 214	4 099	5 325	5 194	2 824	4 160	3 979
05	− fruit and vegetables	11 482	11 706	10 174	3 547	3 410	3 805	−7 935	−8 296	−6 369
011	− beef and veal	801	847	768	1 114	1 136	1 049	314	289	281
1	Beverages and tobacco	3 097	3 212	3 430	8 200	8 763	9 347	5 103	5 551	5 917
21	Skins and furs	754	766	750	438	479	629	−316	−287	−121
22	Oilseeds	3 235	3 331	3 494	52	112	183	−3 183	−3 219	−3 312
231	Natural rubber	636	630	627	13	18	19	−623	−613	−608
24	Timber and cork	7 070	6 992	6 724	773	642	639	−6 297	−6 350	−6 085
261-265 + 268	Natural textile fibres	3 467	3 078	2 414	540	488	552	−2 928	−2 590	−1 862
29	Agricultural raw materials	1 884	1 998	2 053	1 666	1 785	1 872	−218	−214	−181
4	Oils and fats	1 684	1 724	1 757	1 332	1 438	1 366	−352	−287	−391
592.11 } 592.12 }	Starches, gluten	5	4	5	70	151	152	65	147	148
	Total	56 866	56 871	54 599	35 983	38 759	41 803	−20 883	−18 112	−12 796

Source: Eurostat and EC Commission, Directorate-General for Agriculture.

3.6.3 Exports of agricultural and food products by the EC and some other countries

(Mio USD)

SITC codes	Products	EUR 12		United States of America		Canada		Australia		New Zealand	
		1991 ∞	1992 ∞	1991	1992	1991	1992	1991	1992	1991	1992
1	2	3	4	5	6	7	8	9	10	11	12
0 to 9	All products	524 720	567 733	400 984	424 871	126 762	134 617	40 444	41 325	9 325	9 440
0, 1, 21, 22, 231, 24, 261-265, 268, 29, 4, 592.1	Agricultural and food products	44 589	50 313	52 129	56 204	16 799	19 061	10 412	10 897	5 848	6 023
0	Food products and live animals	28 373	32 303	29 596	32 864	9 618	10 660	6 502	6 950	4 345	4 510
	of which:										
00	Live animals	501	499	688	608	789	1 064	139	164	83	103
01	Meat	4 205	4 350	3 630	4 205	817	949	2 576	2 687	1 545	1 608
02	Milk and eggs	4 605	5 175	454	709	177	186	605	700	1 353	1 327
03	Fish	1 614	1 606	3 063	3 377	2 142	2 059	569	665	558	646
04	Cereals	5 079	6 912	10 919	12 179	4 238	4 711	1 651	1 594	27	30
05	Fruit and vegetables	4 395	4 427	5 342	5 718	586	631	515	602	650	664
06	Sugar and honey	2 291	2 421	492	475	174	205	52	67	26	32
07	Coffee, cocoa, tea, spices	1 872	2 188	395	494	194	240	68	75	30	20
08	Animal feed	1 346	1 733	3 279	3 621	352	411	228	297	39	33
09	Other food products	2 466	2 991	1 336	1 476	148	205	97	99	35	49
1	Beverages and tobacco	10 162	11 375	6 762	7 064	857	1 065	194	221	35	45
112	of which: Alcoholic beverages	7 682	8 486	597	713	528	562	165	200	32	42
21	Hides	543	621	1 383	1 355	258	230	227	240	210	195
22	Oilseeds	64	145	4 313	4 799	644	604	45	43	1	1
231	Natural rubber	16	23	36	31	2	1	2	2	:	:
24	Timber and cork	958	834	5 114	5 298	4 967	5 940	23	24	400	448
261 265 268	Natural textile fibres	669	633	2 630	2 135	18	18	3 263	3 197	607	546
29	Agricultural raw materials	2 065	2 316	1 101	1 157	198	236	130	139	187	213
4	Oils and fats	1 332	1 438	1 144	1 447	214	280	26	81	62	63

Sources: Eurostat and EC Commission, Directorate-General for Agriculture.
Other countries: Comtrade.

3.6.4   Imports of agricultural and food products by the EC and some other countries

(Mio USD)

SITC codes	Products	EUR 12 1991 ∞	EUR 12 1992 ∞	United States of America 1991	United States of America 1992	Canada 1991	Canada 1992	Australia 1991	Australia 1992	New Zealand 1991	New Zealand 1992
1	2	3	4	5	6	7	8	9	10	11	12
0 to 9	All products	612 211	634 336	508 944	553 496	118 088	122 584	40 132	42 949	8 497	9 205
0, 1, 21, 22, 231, 24, 261-265, 268, 29, 4, 592.1	Agricultural and food products	70 467	73 824	36 145	38 928	8 721	8 998	2 493	2 591	732	744
0	Food products and live animals	43 411	45 609	23 905	24 659	6 662	6 853	1 534	1 567	515	516
	of which:										
00	Live animals	801	799	1 197	1 458	117	117	48	43	21	16
01	Meat	3 462	3 877	3 129	2 931	814	789	30	29	15	18
02	Milk and eggs	874	853	486	540	152	162	102	111	9	8
03	Fish	8 943	9 308	5 950	5 975	663	665	353	353	34	33
04	Cereals	1 579	1 512	1 084	1 345	455	580	83	96	77	78
05	Fruit and vegetables	14 228	15 196	6 239	6 585	2 655	2 637	336	348	135	142
06	Sugar and honey	1 698	1 921	1 351	1 405	395	389	44	52	74	63
07	Coffee, cocoa, tea, spices	5 931	5 627	3 612	3 476	636	654	220	229	59	55
08	Animal feed	5 388	5 890	376	409	403	439	65	57	21	26
09	Other food products	507	625	481	537	372	420	253	249	69	76
1	Beverages and tobacco	3 838	6 170	5 128	5 713	615	646	302	304	99	95
112	of which: Alcoholic beverages	900	1 048	3 590	4 056	509	528	209	216	76	72
21	Hides	934	994	170	190	67	75	7	2	8	14
22	Oilseeds	4 008	4 324	150	146	145	122	45	37	17	17
231	Natural rubber	788	818	743	859	71	81	32	36	4	5
24	Timber and cork	8 761	9 077	3 342	4 291	648	682	340	379	20	19
261 265 268	Natural textile fibres	4 297	3 995	259	265	106	89	38	35	5	6
29	Agricultural raw materials	2 335	2 594	1 438	1 543	280	301	83	99	25	26
4	Oils and fats	1 684	1 724	919	1 148	117	136	108	127	33	42

Sources : Eurostat and EC Commission, Directorate-General for Agriculture.
Other countries : Comtrade.

EUR 12

3.6.5 '1991' world production and trade in the principal agricultural products — The EC share of the world market

| | World production 1 000 t | World trade (1) 1 000 t | (3/2) × 100 Proportion of production traded | % of world trade | | (6-5) Net EC share of world trade (2) |
| | | | | Imported by EC | Exported by EC | |
1	2	3	4	5	6	7
Total cereals (except rice) (3)	1 406 959	194 880	13,9	2,6	15,9	13,3
of which: total wheat	567 562	101 996	18,0	1,4	20,7	19,3
Feed grain (except rice) (3)	839 397	92 884	11,1	3,7	10,6	6,9
of which: maize	498 850	63 460	12,7	4,6	1,0	-3,6
Oil seeds (by weight produced)	260 689	35 263	13,5	43,1	0,5	-42,6
of which: soya	108 548	27 066	24,9	49,3	0,1	-49,2
Wine	27 624	1 502	5,4	18,2	75,2	57,0
Sugar	125 810	28 664	22,8	6,6	18,1	11,5
Total milk	468 559	790	0,2	1,6	32,5	30,9
Butter	7 330	765	10,4	8,9	31,6	22,7
Cheese	14 458	897	6,2	12,4	52,1	39,7
Milk powder (skimmed and whole)	6 092	1 916	31,5	0,5	46,6	46,1
Total meat (except offal)	179 594 (4)	7 545 (5)	4,2	9,9	22,1	12,2
of which: - beef and veal	54 128 (4)	3 239 (5)	6,0	6,5	27,2	20,7
- pigmeat	71 072 (4)	1 218 (5)	1,7	2,6	24,2	21,6
- poultrymeat	41 425 (4)	2 078 (5)	5,0	5,7	22,2	16,5
Eggs	38 071	335	0,9	8,1	32,5	24,4

Sources: FAO (World production and world trade); Eurostat and EC Commission, Directorate-General for Agriculture (EC share in world trade).

(1) Exports (excluding intra-EC trade) and excluding processed products.
(2) Net balance EC trade/world trade.
(3) Cereals as grain : processed products excluded.
(4) Including salted meat.
(5) Excluding salted meat for trade.

3.6.6   EUR 12 trade in agricultural and food products, (¹) according to principal customer countries

(Mio ECU)

No	Main client countries (based on 1993)	Exports			Corresponding imports			Trade balance		
		1991 ∞	1992 ∞	1993 ∞	1991 ∞	1992 ∞	1993 ∞	1991 ∞	1992 ∞	1993 ∞
1	2	3	4	5	6	7	8	9	10	11
1	United States	4 483	4 646	4 819	7 177	7 711	7 367	− 2 693	− 3 064	− 2 548
2	Russia	0	1 222	2 961	0	455	808	0	767	2 153
3	Switzerland	2 785	2 721	2 717	993	1 029	1 043	1 792	1 692	1 674
4	Japan	2 501	2 523	2 702	221	182	157	2 280	2 341	2 545
5	Austria	1 676	1 728	1 788	1 272	1 298	1 197	404	430	591
6	Sweden	1 633	1 718	1 663	1 643	1 750	1 738	− 10	− 32	− 75
7	Saudi Arabia	1 235	1 303	1 133	16	10	29	1 218	1 293	1 104
8	Poland	1 027	959	1 125	1 216	1 150	1 016	− 190	− 191	109
9	Hong Kong	854	984	1 124	49	52	69	805	932	1 055
10	Canary Islands	1 018	1 082	1 096	1 025	1 024	1 260	− 6	58	− 164
11	Algeria	909	866	922	22	28	27	887	838	895
12	Canada	842	834	876	1 579	1 496	1 352	− 737	− 661	− 476
13	Norway	621	675	722	1 401	1 604	1 525	− 780	− 929	− 803
14	Egypt	432	444	673	148	131	133	284	312	540
15	Taiwan	351	426	560	81	95	91	269	331	470
16	Libya	413	364	500	3	4	9	410	360	491
17	Finland	512	518	497	907	927	1 001	− 395	− 409	− 505
18	Turkey	311	322	470	1 253	1 151	1 194	− 942	− 830	− 724
19	United Arab Emirates	438	420	470	7	4	6	430	416	464
20	Mexico	324	496	458	248	218	204	77	278	254
21	Czech Republic	0	15	454	0	0	322	0	15	133
22	Australia	393	407	408	1 290	1 476	1 140	− 897	− 1 070	− 733
23	Singapore	328	366	380	153	158	163	175	208	217
24	Israel	312	303	371	755	735	593	− 443	− 431	− 222
25	Hungary	187	240	354	994	913	780	− 807	− 673	− 425
	Total of 25 countries (A)	23 585	25 582	29 245	22 454	23 600	23 225	1 130	1 982	6 021
	Total of third countries (B)	35 983	38 759	41 803	56 866	56 871	54 599	− 20 883	− 18 112	− 12 796
	% A/B	65,5	66,0	70,0						

Source : Eurostat and EC Commission, Directorate-General for Agriculture.

(¹) In SITC 0, 1, 21, 22, 231, 24, 261-265 + 268, 29, 4, 592.1.

3.6.7 EUR 12 trade in agricultural and food products, (¹) according to principal supplier countries

(Mio ECU)

No	Main supplier countries (based on 1993)	Imports			Corresponding exports			Trade balance		
		1991 ∞	1992 ∞	1993 ∞	1991 ∞	1992 ∞	1993 ∞	1991 ∞	1992 ∞	1993 ∞
1	2	3	4	5	6	7	8	9	10	11
1	United States	7 177	7 711	7 367	4 483	4 646	4 819	-2 694	-3 064	-2 548
2	Brazil	3 896	4 158	4 146	333	199	285	-3 563	-3 959	-3 861
3	Argentina	3 172	2 729	2 496	82	122	166	-3 091	-2 607	-2 330
4	Sweden	1 643	1 750	1 738	1 633	1 718	1 663	-10	-32	-75
5	Thailand	1 593	2 038	1 609	272	296	348	-1 321	-1 742	-1 260
6	Norway	1 401	1 604	1 525	621	675	722	-780	-929	-804
7	China	1 494	1 440	1 422	298	196	155	-1 196	-1 244	-1 267
8	Canada	1 579	1 496	1 352	842	835	876	-736	-661	-476
9	Malaysia	1 236	1 264	1 276	166	157	187	-1 070	-1 107	-1 088
10	Ivory Coast	1 419	1 341	1 270	177	175	190	-1 241	-1 166	-1 080
11	Indonesia	1 126	1 195	1 269	79	102	115	-1 047	-1 092	-1 154
12	Canary Islands	1 025	1 024	1 260	1 019	1 082	1 096	-6	58	-164
13	New Zealand	1 211	1 193	1 209	52	52	63	-1 160	-1 141	-1 147
14	Austria	1 272	1 298	1 197	1 676	1 728	1 788	404	430	591
15	Turkey	1 253	1 151	1 194	311	322	470	-942	-830	-724
16	Australia	1 290	1 476	1 141	393	407	408	-897	-1 070	-733
17	Switzerland	993	1 029	1 043	2 785	2 721	2 717	1 792	1 692	1 674
18	Poland	1 216	1 150	1 016	1 027	959	1 125	-190	-191	109
19	Finland	907	927	1 001	512	518	497	-395	-409	-505
20	Colombia	1 109	1 040	996	27	37	50	-1 082	-1 002	-946
21	Russia	0	455	808	0	1 222	2 961	0	767	2 153
22	South Africa	1 095	993	789	216	220	212	-878	-773	-577
23	Hungary	994	913	780	187	240	354	-807	-673	-426
24	Morocco	919	809	770	267	330	301	-651	-479	-469
25	India	678	654	750	33	66	59	-644	-588	-690
	Total of 25 countries (A)	39 697	40 835	39 425	17 490	19 023	21 630	-22 207	-21 812	-17 795
	Total of third countries (B)	56 866	56 871	54 599	35 983	38 759	41 803	-20 883	-18 112	-12 796
	% A/B	69,8	71,8	72,2						

Source: Eurostat and EC Commission, Directorate-General for Agriculture.

(¹) In SITC 0, 1, 21, 22, 231, 24, 261-265 + 268, 29, 4, 592.1.

3.6.8   Community imports, by product

	1 000 t			% TAV	
	1990/91 ∞	1991/92 ∞	1992/93 ∞	1991/92 / 1990/91	1992/93 / 1991/92
1	2	3	4	5	6
Total cereals (¹):	5 147	5 543	4 048	7,7	− 27,0
— Common wheat	1 485	1 343	1 053	− 9,6	− 21,6
— Durum wheat	283	304	313	7,4	3,0
— Rye	21	24	28	14,3	16,7
— Barley	210	119	130	− 43,3	9,2
— Oats	28	31	40	10,7	29,0
— Maize	2 661	3 284	1 914	23,4	− 41,7
— Other (including sorghum)	458	438	566	− 4,4	29,2
Husked rice	480	412	348	− 14,2	− 15,5
Sugar (²)	2 066	1 919	1 970	− 7,1	2,7
Wine (1 000 hl) (³)	3 065	3 143	3 255	2,5	3,6
Fresh fruit	5 833	:	:	×	×
Fresh vegetables	2 802	:	:	×	×
Rapeseed	281	142	375	− 49,5	164,1
Sunflower seed	309	533	786	72,5	47,5

	1991 ∞	1992 ∞	1993 ∞	1992 / 1991	1993 / 1992
Olive oil	139,8	92,2	95,3	− 34,0	3,4
Soya:					
— seed	12 529,4	14 263,2	12 833,2	13,8	− 10,0
— oil	12,2	6,5	2,9	− 46,7	− 55,4
— cake	10 458,5	10 763,7	11 176,6	2,9	3,8
Lucerne meal	82,1	111,9	13,3	36,3	− 88,1
Fibres:					
— flax	13,8	12,6	15,5	− 8,7	23,0
— hemp	1,0	0,2	0,3	− 80,0	50,0
Raw tobacco	527,7	526,8	417,4	− 0,2	− 20,8
Apples (fresh)	1 007,1	865,7	645,6	− 14,0	− 25,4
Pears (fresh)	262,4	297,6	247,6	13,4	− 16,8
Peaches	14,1	16,8	8,9	19,1	− 47,0
Oranges	922,9	885,5	799,3	− 4,1	− 9,7
Lemons	119,8	128,6	95,7	7,3	− 25,6
Tomatoes	350,5	356,7	390,9	1,8	9,6
Potatoes	535,9	502,0	395,3	− 6,3	− 21,3
Live plants (⁴)	585,5	628,9	674,0	7,4	7,2
Hops:					
— cones and powders	18,0	16,9	15,7	− 6,1	− 7,1
— saps and extracts	0,9	1,0	0,9	11,1	− 10,0
Butter and butteroil	68,2	48,3	65,2	− 29,2	35,0
Cheese	109,4	109,6	108,6	0,2	− 0,9
Skimmed-milk powder	5,1	3,2	19,0	− 37,3	493,8
Whole-milk powder	0,5	0,5	1,2	0,0	140,0
Condensed milk	2,2	1,1	0,3	− 50,0	− 72,7
Casein	58,2	53,6	58,7	− 7,9	9,5
Beef/veal (⁵)	303,8	312,2	274,9	2,8	− 11,9
Pigmeat (⁵)	35,3	35,6	12,3	0,8	− 65,4
Poultrymeat (⁵)	114,9	119,3	117,3	3,8	− 1,7
Sheepmeat (⁵)	224,6	229,4	209,7	2,1	− 8,6
Eggs (⁶)	23,6	19,2	18,6	− 18,6	− 3,1

Source: Eurostat and EC Commission, Directorate-General for Agriculture.

(¹) Incl. derived products, except rice.
(²) Incl. the sugar contained in processed products.
(³) Incl. vermouths and aromatized wines, except in the case of France.
(⁴) In million ECU; including horticultural products.
(⁵) Live animals and meat expressed as fresh carcass weight (incl. preserves).
(⁶) In terms of shell weight (from 1977, albumin and its derivatives included).

3.6.9 Community exports, by product

EUR 12

1	1 000 t			% TAV	
	1990/91 ∞	1991/92 ∞	1992/93 ∞	1991/92 / 1990/91	1992/93 / 1991/92
1	2	3	4	5	6
Total cereals (¹):	30 143	34 787	37 363	15,4	7,4
— Common wheat	18 281	19 526	21 496	6,8	10,1
— Durum wheat	2 023	4 006	3 421	98,0	− 14,6
— Rye	276	715	1 828	159,1	155,7
— Barley	9 442	9 509	8 809	0,7	− 7,4
— Oats	58	59	68	1,7	15,3
— Maize	63	933	1 735	1 381,0	86,0
— Other (including sorghum)	0	39	6	×	:
Husked rice	0	0	30	×	:
Sugar (²)	5 681	4 792	5 664	− 15,6	18,2
Wine (1 000 hl) (³)	8 570	8 345	9 776	− 2,6	17,1
Fresh fruit	1 416	:	:	×	:
Fresh vegetables	4 663	:	:	×	:
Rapeseed	1	3	602	200,0	19 966,7
Sunflower seed	3	4	54	33,3	1 250,0

	1991 ∞	1992 ∞	1993 ∞	1992 / 1991	1993 / 1992
Olive oil	192,9	233,3	212,1	20,9	− 9,1
Soya :					
— seed	22,7	16,2	18,3	− 28,6	13,0
— oil	598,4	634,1	513,0	6,0	− 19,1
— cake	793,7	1 092,9	1 035,1	37,7	− 5,3
Lucerne meal	59,4	62,3	106,7	4,9	71,3
Fibres :					
— flax	37,5	38,3	68,2	2,1	78,1
— hemp	0,2	0,2	0,2	0,0	0,0
Raw tobacco	210,6	200,3	209,7	− 4,9	4,7
Apples (fresh)	146,5	170,4	366,6	16,3	115,1
Pears (fresh)	52,2	60,0	61,0	14,9	1,7
Peaches	96,8	121,6	103,9	25,6	− 14,6
Oranges	541,5	616,3	822,1	13,8	33,4
Lemons	194,2	141,9	257,2	− 26,9	81,3
Tomatoes	131,6	147,4	210,4	12,0	42,7
Potatoes	1 090,4	697,0	901,9	− 36,1	29,4
Live plants (⁴)	990,7	1 043,9	1 044,2	5,4	0,0
Hops :					
— cones and powders	14,3	14,8	13,8	3,5	− 6,8
— saps and extracts	1,7	1,6	6,6	− 5,9	312,5
Butter and butteroil	301,7	223,6	184,5	− 25,9	− 17,5
Cheese	483,6	465,9	523,7	− 3,7	12,4
Skimmed-milk powder	252,8	391,3	283,7	54,8	− 27,5
Whole-milk powder	618,4	580,5	588,3	− 6,1	1,3
Condensed milk	316,0	343,0	351,2	8,5	2,4
Casein	59,7	71,2	59,8	19,3	− 16,0
Beef/veal (⁵)	1 107,1	1 057,3	974,0	− 4,5	− 7,9
Pigmeat (⁵)	366,6	244,5	362,2	− 33,3	48,1
Poultrymeat (⁵)	468,7	510,5	650,6	8,9	27,4
Sheepmeat (⁵)	19,7	5,8	6,3	− 70,6	8,6
Eggs (⁶)	104,9	106,3	110,1	1,3	3,6

Source : Eurostat and EC Commission, Directorate-General for Agriculture.

(¹) Incl. derived products, except rice.
(²) Incl. the sugar contained in processed products.
(³) Incl. vermouths and aromatized wines, except in the case of France.
(⁴) In million ECU; including horticultural products.
(⁵) Live animals and meat expressed as fresh carcass weight (incl. preserves).
(⁶) In terms of shell weight (from 1977, albumin and its derivatives included).

EUR 12

3.6.10 EC imports of agricultural products (1) from various groups of countries

1	Mio ECU			% TAV		% of total EUR 12		
	1991 ∞	1992 ∞	1993 ∞	1992/1991	1993/1992	1991	1992	1993
	2	3	4	5	6	7	8	9
1. World total (2)	155 062	159 068	147 462	2,6	−7,3	x	x	x
2. Total EUR 12, intra-EC	98 002	102 005	92 863	4,1	−9,0	x	x	x
3. Total EUR 12, extra-EC	56 867	56 872	54 599	0,0	−4,0	100,0	100,0	100,0
4. Industrialized countries (class I)	21 874	22 479	21 079	2,8	−6,2	38,5	39,5	38,6
of which: USA	7 177	7 711	7 367	7,4	−4,5	12,6	13,6	13,5
Canada	1 579	1 496	1 352	−5,3	−9,6	2,8	2,6	2,5
Japan	221	182	157	−17,6	−13,7	0,4	0,3	0,3
5. Developing countries (class II)	29 157	28 612	27 673	−1,9	−3,3	51,3	50,3	50,7
of which: Argentina	3 172	2 729	2 496	−14,0	−8,5	5,6	4,8	4,6
Brazil	3 895	4 158	4 146	6,8	−0,3	6,8	7,3	7,6
Morocco	919	809	770	−12,0	−4,8	1,6	1,4	1,4
6. Central and East European countries (class IV)	4 087	4 057	3 923	−0,7	−3,3	7,2	7,1	7,2
of which: Poland	1 216	1 150	1 016	−5,4	−11,7	2,1	2,0	1,9
Hungary	994	913	780	−8,1	−14,6	1,7	1,6	1,4
Romania	86	90	95	4,7	5,6	0,2	0,2	0,2
7. EFTA (3)	7 002	7 368	7 193	5,2	−2,4	12,3	13,0	13,2
8. Industrialized commonwealth (4)	5 174	5 158	4 491	−0,3	−12,9	9,1	9,1	8,2
9. Mediterranean basin (5)	3 727	3 344	3 234	−10,3	−3,3	6,6	5,9	5,9
10. Latin America	11 957	11 400	10 984	−4,7	−3,6	21,0	20,0	20,1
11. ACP (Lomé Convention)	6 715	6 526	6 338	−2,8	−2,9	11,8	11,5	11,6

Source: Eurostat and EC Commission. Directorate-General for Agriculture.

(1) SITC 0, 1, 21, 22, 231, 24, 261-265 + 268, 29, 4, 592.11 + 12.
(2) Not including confidential, ships' stores, etc.
(3) Iceland, Norway, Sweden, Finland, Switzerland, Austria.
(4) Canada, Australia, New Zealand; plus the Union of South Africa.
(5) Tunisia, Morocco, Algeria, Malta, Cyprus, Israel, Egypt, Syria, Jordan, Turkey, Lebanon, Libya.

3.6.11 EC exports of agricultural products (¹) to various groups of countries

EUR 12

1	Mio ECU			% TAV		% of total EUR 12		
	1991 ∞	1992 ∞	1993 ∞	$\frac{1992}{1991}$	$\frac{1993}{1992}$	1991	1992	1993
	2	3	4	5	6	7	8	9
1. World total (²)	132 143	138 799	138 500	5,0	−0,2	×	×	×
2. Total EUR 12, intra-EC	95 297	99 375	96 697	4,3	−2,7	×	×	×
3. Total EUR 12, extra-EC	35 985	38 776	41 803	7,8	7,8	100,0	100,0	100,0
4. Industrialized countries (class I)	16 948	17 283	18 048	2,0	4,4	47,1	44,6	43,2
of which: USA	4 483	4 646	4 819	3,6	3,7	12,5	12,0	11,5
Canada	842	834	876	−1,0	5,0	2,3	2,2	2,1
Japan	2 501	2 523	2 702	0,9	7,1	7,0	6,5	6,5
5. Developing countries (class II)	14 512	15 101	16 016	4,1	6,1	40,3	38,9	38,3
of which: Argentina	82	122	166	48,8	36,1	0,2	0,3	0,4
Brazil	333	199	285	−40,2	43,2	0,9	0,5	0,7
Morocco	267	330	301	23,6	−8,8	0,7	0,9	0,7
6. Central and East European countries (class IV)	4 076	5 861	6 600	43,8	12,6	11,3	15,1	15,8
of which: Poland	1 027	958	1 125	−6,7	17,4	2,9	2,5	2,7
Hungary	187	240	354	28,3	47,5	0,5	0,6	0,8
Romania	263	344	331	30,8	−3,8	0,7	0,9	0,8
7. EFTA (³)	7 294	7 424	7 451	1,8	0,4	20,3	19,1	17,8
8. Industrialized commonwealth (⁴)	1 503	1 512	1 559	0,6	3,1	4,2	3,9	3,7
9. Mediterranean basin (⁵)	3 689	3 666	4 359	−0,6	18,9	10,3	9,5	10,4
10. Latin America	1 396	1 527	1 713	9,4	12,2	3,9	3,9	4,1
11. ACP (Lomé Convention)	2 371	2 515	2 364	6,1	−6,0	6,6	6,5	5,7

Source: Eurostat and EC Commission, Directorate-General for Agriculture.

(¹) SITC 0, 1, 21, 22, 231, 24, 261-265 + 268, 29, 4, 592.11 + 12.
(²) Not including confidential, ships' stores, etc.
(³) Iceland, Norway, Sweden, Finland, Switzerland, Austria.
(⁴) Canada, Australia, New Zealand; plus the Union of South Africa.
(⁵) Tunisia, Morocco, Algeria, Malta, Cyprus, Israel, Egypt, Syria, Jordan, Turkey, Lebanon, Libya.

3.6.12   EC trade with ACP countries and Member States' overseas territories

EUR 12

(Mio ECU)

		Imports			Exports			Trade balance		
1	2	1991 ∞	1992 ∞	1993 ∞	1991 ∞	1992 ∞	1993 ∞	1991 ∞	1992 ∞	1993 ∞
		3	4	5	6	7	8	9	10	11
0-9	All products	19 835	18 661	15 368	17 563	19 254	18 754	-2 272	593	3 386
	Agricultural products (total) (¹)	7 005	6 809	6 574	2 643	2 807	2 664	-4 362	-4 002	-3 910
00	Live animals	7	6	13	8	7	7	1	1	-6
01	Meat	72	127	143	249	255	262	177	128	119
02	Milk and eggs	2	1	3	431	440	503	429	439	500
03	Fish	923	883	861	150	158	163	-773	-725	-698
04	Cereals	40	45	84	495	549	454	455	504	370
05	Fruit and vegetables	770	794	789	147	151	169	-623	-643	-620
06	Sugar and honey	737	770	746	221	198	169	-516	-572	-577
07	Coffee, cocoa, tea, spices	2 123	1 935	1 879	33	35	32	-2 090	-1 900	-1 847
08	Animal feed	67	57	58	28	29	27	-39	-28	-31
09	Food products	2	2	2	238	259	245	236	257	243
11	Beverages	116	135	159	349	412	362	233	277	203
12	Tobacco	369	370	238	108	119	112	-261	-251	-126
21	Hides	54	41	37	1	1	1	-53	-40	-36
22	Oilseeds	44	42	47	1	1	1	-43	-41	-46
231	Natural rubber	120	129	120	1	0	1	-119	-129	-119
24	Timber and cork	870	847	804	10	8	8	-860	-839	-796
261-265 + 268	Natural textile fibres	324	250	222	5	3	2	-319	-247	-220
29	Agricultural raw materials	156	169	187	22	22	23	-134	-147	-164
4	Oils and fats	206	203	182	146	157	122	-60	-46	-60
592.11 592.12	Starches, inuline Gluten	0	0	0	1	0	0	1	0	0

Source : Eurostat and EC Commission, Directorate-General for Agriculture.

(¹) 0, 1, 21, 22, 231, 24, 261-265 + 268, 29, 4, 592.11-12.

3.6.13   EC trade with Mediterranean countries (¹)

EUR 12

(Mio ECU)

		Imports			Exports			Trade balance		
		1991 ∞	1992 ∞	1993 ∞	1991 ∞	1992 ∞	1993 ∞	1991 ∞	1992 ∞	1993 ∞
1	2	3	4	5	6	7	8	9	10	11
0-9	All products	35 877	35 159	33 176	37 950	38 827	45 600	2 073	3 668	12 424
	Agricultural products (total) (²)	3 727	3 344	3 234	3 689	3 665	4 359	-38	321	1 125
00	Live animals	5	4	6	125	129	207	120	125	201
01	Meat	35	35	35	241	260	326	206	225	291
02	Milk and eggs	10	8	10	503	573	587	493	565	577
03	Fish	403	390	372	50	50	66	-353	-340	-306
04	Cereals	15	17	25	839	832	1 200	824	815	1 175
05	Fruit and vegetables	2 188	1 975	1 852	205	170	157	-1 983	-1 805	-1 695
06	Sugar and honey	28	37	35	515	340	405	487	303	370
07	Coffee, cocoa, tea, spices	48	51	51	77	80	92	29	29	41
08	Animal feed	18	15	17	198	224	237	180	209	220
09	Food products	32	31	42	188	219	248	156	188	206
11	Beverages	40	40	38	143	141	168	103	101	130
12	Tobacco	79	104	90	124	167	134	45	63	44
21	Hides	17	22	29	36	44	65	19	22	36
22	Oilseeds	28	44	41	4	12	19	-24	-32	-22
231	Natural rubber	1	1	2	3	2	2	2	1	0
24	Timber and cork	10	16	15	89	57	100	79	41	85
261-265 + 268	Natural textile fibres	272	152	163	50	56	70	-222	-96	-93
29	Agricultural raw materials	271	267	272	82	93	92	-189	-174	-180
4	Oils and fats	226	134	138	214	213	182	-12	79	44
592.11	Starches, inuline }	0	0	0	2	3	3	2	3	3
592.12	Gluten									

Source: Eurostat and EC Commission, Directorate-General for Agriculture.

(¹) Malta, Turkey, Morocco, Algeria, Tunisia, Libya, Egypt, Cyprus, Lebanon, Syria, Israel and Jordan.

(²) 0, 1, 21, 22, 231, 24, 261-265 + 268, 29, 4, 592.11-12.

3.6.14 EC trade in agricultural and food products (¹)

	Imports					Exports				
	Mio ECU			% TAV		Mio ECU			% TAV	
	1991 ∞	1992 ∞	1993 ∞	1992/1991	1993/1992	1991 ∞	1992 ∞	1993 ∞	1992/1991	1993/1992
1	2	3	4	5	6	7	8	9	10	11
Intra-Community										
EUR 12	98 001	102 005	92 863	4,1	−9,0	95 297	99 375	96 697	4,3	−2,7
BLEU/UEBL	9 174	9 519	9 119	3,8	−4,2	9 145	9 524	10 652	4,1	11,8
Danmark	1 852	1 988	1 983	7,3	−0,3	5 507	5 585	5 490	1,4	−1,7
BR Deutschland	24 082	25 334	20 784	5,2	−18,0	12 868	13 429	12 064	4,4	−10,2
Elláda	1 966	2 207	2 281	12,3	3,4	1 586	1 850	1 429	16,6	−22,8
España	4 656	4 984	5 332	7,0	7,0	5 729	5 707	6 155	−0,4	7,9
France	14 920	15 331	15 198	2,8	−0,9	20 773	21 651	21 508	4,2	−0,7
Ireland	1 704	1 794	1 624	5,3	−9,5	3 670	4 432	4 397	20,8	−0,8
Italia	15 933	15 428	13 890	−3,2	−10,0	6 759	6 686	6 832	−1,1	2,2
Nederland	9 434	10 377	8 817	10,0	−15,0	21 117	21 953	20 550	4,0	−6,4
Portugal	1 776	1 963	1 873	10,5	−4,6	850	847	731	−4,0	−13,7
United Kingdom	12 499	13 077	11 962	4,6	−8,5	7 292	7 711	6 889	5,7	−10,7
With non-EEC countries										
EUR 12	56 866	56 871	54 599	0,2	−4,0	35 983	38 759	41 803	7,7	7,9
BLEU/UEBL	2 942	2 825	2 949	−4,0	4,4	1 524	1 642	1 705	7,7	3,8
Danmark	2 088	2 188	2 212	4,8	1,1	3 174	3 230	3 431	1,8	6,2
BR Deutschland	12 741	12 853	12 246	0,9	−4,7	6 230	6 481	7 061	4,0	8,9
Elláda	852	735	658	−13,7	−10,5	709	817	954	15,2	16,8
España	5 327	5 516	5 098	3,5	−7,6	2 594	2 775	2 965	7,0	6,8
France	7 371	7 032	6 654	−4,6	−5,4	7 564	8 164	8 643	7,9	5,9
Ireland	407	412	428	1,2	3,9	976	1 150	1 278	17,8	11,1
Italia	8 568	7 862	7 323	−8,2	−6,9	3 246	3 673	3 760	13,2	2,4
Nederland	6 231	7 269	6 786	16,7	−6,6	4 996	5 599	6 485	12,1	15,8
Portugal	1 544	1 483	1 414	−4,0	−4,7	384	407	328	6,0	−19,4
United Kingdom	8 796	8 696	8 830	−1,1	1,5	4 585	4 821	5 192	5,1	7,7

Source: Eurostat and EC Commission, Directorate-General for Agriculture.
(¹) SITC 0, 1, 21, 22, 231, 24, 261-265 + 268, 29, 4, 592.1.

3.6.15   Intra-Community trade, by product, incoming merchandise

EUR 12

1	1 000 t			% TAV	
	1990/91 ∞	1991/92 ∞	1992/93 ∞	1991/92 / 1990/91	1992/93 / 1991/92
1	2	3	4	5	6
Total cereals (¹):	32 369	34 367	32 055	6,2	− 7
— Common wheat	13 271	13 997	13 895	5,5	− 1
— Durum wheat	2 609	3 183	2 633	22,0	− 17
— Rye	134	135	127	0,7	− 6
— Barley	6 281	5 943	5 098	− 5,4	− 14
— Oats	263	307	260	16,7	− 15
— Maize	9 486	10 444	9 700	10,1	− 7
— Other (including sorghum)	315	353	335	12,1	− 5
Husked rice	183	251	249	37,2	− 1
Sugar (²)	2 566	2 908	2 532	13,3	− 13
Wine (1 000 hl) (³)	26 697	27 728	25 642	3,9	− 8
Fresh fruit	5 247	:	:	×	×
Fresh vegetables	8 331	:	:	×	×
Rapeseed	1 864	1 885	:	1,1	×
Sunflower seed	981	1 226	:	25,0	×

	1991 ∞	1992 ∞	1993 ∞	1992 / 1991	1993 / 1992
Olive oil	480,8	339,4	368,5	− 29,4	8,6
Soya:					
— seed	497,5	516,2	589,1	3,8	14,1
— oil	582,1	558,0	422,0	− 4,1	− 24,4
— cake	2 952,6	2 992,1	2 852,4	1,3	− 4,7
Lucerne meal	493,1	468,7	682,6	− 4,9	45,6
Fibres:					
— flax	150,9	134,4	171,4	− 10,9	27,5
— hemp	2,9	3,9	3,1	34,5	− 20,5
Raw tobacco	142,0	136,0	124,8	− 4,2	− 8,2
Apples (fresh)	1 450,2	1 380,6	1 330,3	− 4,8	− 3,6
Pears (fresh)	379,2	396,6	356,1	4,6	− 10,2
Peaches	537,8	590,5	446,3	9,8	− 24,4
Oranges	1 297,5	1 346,1	1 160,6	3,7	− 13,8
Lemons	303,8	315,9	304,5	4,0	− 3,6
Tomatoes	955,1	1 031,8	838,2	8,0	− 18,8
Potatoes	4 384,4	4 087,2	3 776,6	− 6,8	− 7,6
Live plants (⁴)	3 672,6	3 709,1	2 639,4	1,0	− 28,8
Hops:					
— cones and powders	7,2	8,0	7,5	11,1	− 6,3
— saps and extracts	1,4	1,3	4,6	− 7,1	253,8
Butter and butteroil	543,0	612,3	581,4	12,8	− 5,0
Cheese	1 234,4	1 350,5	1 236,1	9,4	− 8,5
Skimmed-milk powder	579,2	695,2	595,1	20,0	− 14,4
Whole-milk powder	167,8	186,7	200,0	11,3	7,1
Condensed milk	393,9	484,7	319,6	23,1	− 34,1
Casein	72,8	68,7	65,5	− 5,6	− 4,7
Beef and veal (⁵)	1 888,6	1 931,4	1 727,2	2,3	− 10,6
Pigmeat (⁵)	2 111,2	2 428,4	2 254,5	15,0	− 7,2
Poultrymeat (⁵)	825,0	948,2	871,2	14,9	− 8,1
Sheepmeat (⁵)	224,9	259,8	245,8	15,5	− 5,4
Eggs (⁶)	575,5	561,0	412,3	− 2,5	− 26,5

Source: Eurostat and EC Commission, Directorate-General for Agriculture.

(¹) Incl. derived products, except rice.
(²) Incl. the sugar contained in processed products.
(³) Incl. vermouths and aromatized wines, except in the case of France.
(⁴) Million ECU; including horticultural products.
(⁵) Live animals and meat expressed as fresh carcass weight (incl. preserves).
(⁶) In terms of shell weight (from 1977, albumin and its derivatives included).

3.6.16 Intra-Community trade, by product, outgoing merchandise

EUR 12

1	1 000 t			% TAV	
	1990/91 ∞	1991/92 ∞	1992/93 ∞	1991/92 / 1990/91	1992/93 / 1991/92
	2	3	4	5	6
Total cereals (¹):	32 269	35 669	32 526	10,5	− 8,8
— Common wheat	14 115	14 298	14 549	1,3	1,8
— Durum wheat	2 155	3 721	2 629	72,7	− 29,3
— Rye	133	151	162	13,5	7,3
— Barley	7 206	7 063	5 783	− 2,0	− 18,1
— Oats	275	309	281	12,4	− 9,1
— Maize	8 071	9 751	8 781	20,8	− 9,9
— Other (including sorghum)	305	371	333	21,6	− 10,2
Husked rice	147	232	266	57,8	14,7
Sugar (²)	2 620	2 985	2 759	13,9	− 7,6
Wine (1 000 hl) (³)	26 702	27 342	26 237	2,4	− 4,0
Fresh fruit	:	:	:	×	×
Fresh vegetables	:	:	:	×	×
Rapeseed	:	:	:	×	×
Sunflower seed	:	:	:	×	×

	1991 ∞	1992 ∞	1993 ∞	1992 / 1991	1993 / 1992
Olive oil	454,9	372,3	336,2	− 18,2	− 9,7
Soya:					
— seed	435,0	377,4	238,7	− 13,2	− 36,8
— oil	589,3	592,5	415,8	0,5	− 29,8
— cake	2 975,5	3 017,1	3 057,2	1,4	1,3
Lucerne meal	434,6	427,5	727,0	− 1,6	70,1
Fibres:					
— flax	156,8	139,4	135,3	− 11,1	− 2,9
— hemp	0,8	0,5	0,8	− 37,5	60,0
Raw tobacco	141,3	118,4	113,0	− 16,2	− 4,6
Apples (fresh)	1 358,9	1 269,8	1 465,9	− 6,6	15,4
Pears (fresh)	322,1	377,0	438,4	17,0	16,3
Peaches	545,3	598,3	537,5	9,7	− 10,2
Oranges	1 299,8	1 373,1	1 375,1	5,6	0,1
Lemons	300,0	311,1	351,1	3,7	12,9
Tomatoes	956,1	1 042,1	1 034,9	9,0	− 0,7
Potatoes	4 444,4	4 050,3	3 819,3	− 8,9	− 5,7
Live plants (⁴)	3 585,9	3 690,7	3 149,2	2,9	− 14,7
Hops:					
— cones and powders	8,5	8,3	5,7	− 2,4	− 31,3
— saps and extracts	1,0	1,1	1,1	10,0	0,0
Butter and butteroil	593,9	598,4	598,9	0,8	0,1
Cheese	1 236,8	1 354,5	1 412,6	9,5	4,3
Skimmed-milk powder	585,0	718,4	615,2	22,8	− 14,4
Whole-milk powder	185,4	207,4	286,0	11,9	37,9
Condensed milk	397,7	477,0	421,3	19,9	− 11,7
Casein	62,0	60,4	57,6	− 2,6	− 4,6
Beef and veal (⁵)	1 934,2	1 941,9	1 861,5	0,4	− 4,1
Pigmeat (⁵)	2 135,6	2 438,5	2 358,4	14,2	− 3,3
Poultrymeat (⁵)	834,7	958,3	987,2	14,8	3,0
Sheepmeat (⁵)	225,6	257,6	241,7	14,2	− 6,2
Eggs (⁶)	582,9	561,8	504,3	− 3,6	− 10,2

Source: Eurostat and EC Commission, Directorate-General for Agriculture.

(¹) Incl. derived products, except rice.
(²) Incl. the sugar contained in processed products.
(³) Incl. vermouths and aromatized wines, except in the case of France.
(⁴) Million ECU; including horticultural products.
(⁵) Live animals and meat expressed as fresh carcass weight (incl. preserves).
(⁶) In terms of shell weight (from 1977, albumin and its derivatives included).

3.6.17   EC imports of agricultural and food products from EFTA countries

SITC codes	Products	Iceland		Norway	
		1992 ∞	1993 ∞	1992 ∞	1993 ∞
1	2	3	4	5	6
0-9	All products	888,5	820,1	17 227,4	17 013,4
	Agricultural products (total) (¹)	742,4	688,4	1 603,9	1 525,1
00	Live animals	1,7	1,7	0,3	0,3
01	Meat	0,6	0,2	1,4	0,9
02	Milk and eggs	0,1	0,0	10,6	8,2
03	Fish	663,0	602,5	1 286,8	1 225,5
04	Cereals	0,0	0,0	5,3	5,5
05	Fruit and vegetables	0,2	0,0	3,9	3,7
06	Sugar and honey	0,0	0,0	1,9	1,3
07	Coffee, cocoa, tea, spices	0,0	0,0	6,2	8,6
08	Animal feed	55,4	56,1	53,0	41,6
09	Food products	0,0	0,0	2,8	3,0
11	Beverages	1,2	1,3	3,3	3,2
12	Tobacco	0,0	0,0	0,2	0,1
21	Hides	3,9	2,8	40,7	36,3
22	Oilseeds	0,0	0,0	0,0	0,0
231	Natural rubber	0,0	0,0	0,0	0,0
24	Timber and cork	0,5	0,0	147,4	149,0
261-265 + 268	Natural textile fibres	1,1	1,1	2,5	2,5
29	Agricultural raw materials	1,6	1,5	11,5	7,9
4	Oils and fats	13,0	21,1	26,0	27,6
592.11	Starches, inuline				
592.12	Gluten	0,0	0,0	0,0	0,0

Sources: Eurostat and EC Commission, Directorate-General for Agriculture.

(¹) 0, 1, 21, 22, 231, 24, 261 to 265 + 268, 29, 4, 592.1.

(Mio ECU)

	Sweden		Finland		Switzerland		Austria	
	1992 ∞	1993 ∞	1992 ∞	1993 ∞	1992 ∞	1993 ∞	1992 ∞	1993 ∞
	7	8	9	10	11	12	13	14
	24 625,6	23 210,9	10 871,4	10 154,3	34 951,4	36 118,0	23 081,3	22 289,9
	1 749,5	1 738,1	926,7	1 001,5	1 028,5	1 042,9	1 298,1	1 196,9
	3,4	5,4	0,5	0,5	15,2	16,2	64,2	76,4
	8,5	8,4	0,9	0,4	11,1	11,1	163,3	145,2
	8,1	7,1	42,0	42,4	296,2	295,0	61,7	69,5
	63,9	47,5	1,0	0,8	2,7	1,5	1,1	1,0
	56,9	54,5	10,5	21,4	58,1	73,1	64,2	66,0
	31,6	26,9	6,1	5,1	30,4	22,9	45,3	38,8
	14,2	12,1	28,0	22,2	43,8	43,8	23,2	19,2
	65,6	85,8	13,4	15,4	94,2	98,3	33,3	54,9
	4,2	3,8	0,6	1,0	51,8	55,9	14,7	20,5
	28,7	30,9	1,8	1,9	139,3	159,0	32,7	28,1
	11,4	9,8	4,1	3,0	27,6	38,4	67,5	66,3
	0,6	0,6	0,0	0,3	7,7	4,2	0,6	1,4
	47,7	35,3	50,8	47,3	55,8	42,2	26,3	24,7
	3,0	3,5	0,0	0,1	0,3	0,4	5,6	9,4
	0,2	0,2	0,0	0,0	0,1	0,0	0,1	0,1
	1 353,6	1 355,9	756,0	829,2	115,5	107,9	660,0	538,8
	0,4	0,9	0,2	0,1	26,1	20,3	3,5	5,5
	27,9	26,7	3,6	3,3	40,0	39,9	18,7	18,1
	19,3	22,8	6,8	7,0	12,6	12,5	11,5	13,0
	0,0	0,0	0,0	0,0	0,1	0,2	0,0	0,0

3.6.18   EC exports of agricultural and food products to EFTA countries

SITC codes	Products	Iceland		Norway	
		1992 ∞	1993 ∞	1992 ∞	1993 ∞
1	2	3	4	5	6
0-9	All products	663,2	620,3	9 860,9	9 977,9
	Agricultural products (total) (¹)	64,8	64,1	674,9	721,9
00	Live animals	0,2	0,1	2,0	2,1
01	Meat	0,2	0,3	6,4	9,0
02	Milk and eggs	0,4	0,3	15,6	15,5
03	Fish	1,0	0,9	31,8	28,3
04	Cereals	11,6	9,5	62,7	67,0
05	Fruit and vegetables	9,7	9,9	110,0	119,2
06	Sugar and honey	5,3	5,5	74,9	76,4
07	Coffee, cocoa, tea, spices	7,9	7,9	42,7	48,7
08	Animal feed	2,1	2,0	62,2	58,6
09	Food products	7,4	9,1	69,3	83,3
11	Beverages	10,4	9,8	63,8	66,7
12	Tobacco	2,0	2,0	13,7	12,9
21	Hides	0,2	0,1	14,4	13,3
22	Oilseeds	0,1	0,2	2,8	5,4
231	Natural rubber	0,1	0,0	0,2	0,2
24	Timber and cork	2,4	2,4	15,4	18,5
261-265 + 268	Natural textile fibres	0,4	0,7	6,6	6,2
29	Agricultural raw materials	2,3	2,1	62,9	61,7
4	Oils and fats	1,1	1,3	16,3	27,7
592.11	Starches, inuline				
592.12	Gluten	0,0	0,1	0,9	1,1

Sources: Eurostat and EC Commission, Directorate-General for Agriculture.

(¹) 0, 1, 21, 22, 231, 24, 261 to 265 + 268, 29, 4, 592.1.

(Mio ECU)

Sweden		Finland		Switzerland		Austria	
1992 ∞	1993 ∞	1992 ∞	1993 ∞	1992 ∞	1993 ∞	1992 ∞	1993 ∞
7	8	9	10	11	12	13	14
21 426,1	20 578,7	7 519,8	6 815,1	38 627,2	38 919,6	29 680,7	29 430,8
1 717,6	1 663,5	517,9	496,9	2 720,8	2 716,8	1 727,8	1 788,3
7,7	6,5	3,1	1,9	25,0	26,3	6,1	6,5
91,1	61,4	6,6	8,1	126,2	130,8	24,0	26,1
72,6	72,3	14,9	17,3	194,3	205,9	81,4	94,0
97,7	82,2	6,7	4,9	145,7	141,5	67,8	71,0
101,3	98,5	33,9	27,6	144,2	163,3	133,5	156,7
361,7	385,6	114,6	121,4	486,1	483,8	324,7	334,1
50,5	59,3	18,5	20,6	82,9	82,5	39,3	49,4
100,9	105,2	36,9	41,7	117,3	121,7	123,5	139,9
109,7	117,2	31,3	29,9	126,1	133,7	148,0	136,7
117,5	130,2	35,0	33,3	133,6	151,7	134,9	149,7
216,0	224,7	69,5	65,0	462,3	440,8	89,6	87,2
12,8	12,9	4,5	4,6	100,8	106,3	15,4	16,0
26,0	17,8	39,3	40,4	61,8	31,6	27,3	26,3
8,3	4,1	2,0	1,0	3,6	3,9	6,6	9,0
0,4	0,6	0,5	0,3	0,4	0,6	1,9	1,7
82,5	52,1	29,2	13,2	107,2	106,5	220,4	189,5
4,6	3,9	4,0	4,6	87,9	58,1	31,2	29,3
211,8	182,0	59,4	50,7	280,1	293,3	208,5	218,1
36,9	39,8	6,4	8,0	32,7	31,6	42,1	46,0
7,1	7,0	1,0	2,6	1,1	3,1	0,4	1,0

3.7.1 **Share of consumer expenditure on food, beverages and tobacco in the final consumption of households**

1	% of total expenditure on final consumption by households (¹) in 1992					Foodstuffs, beverages and tobacco % TAV	Foodstuffs, beverages and tobacco % TAV (⁴)
	Foodstuffs, beverages and tobacco	Foodstuffs	Non-alcoholic beverages	Alcoholic beverages	Tobacco	1992 / 1985	1992 / 1985
1	2	3	4	5	6	7	8
EUR 12 (³)	19,8	14,9	0,6	2,7	1,7	4,5	1,2
Belgique/België	18,0	14,7	0,6	1,3	1,5	2,5	1,2
Danmark	21,2	14,8	0,7	2,9	2,8	2,5	0,6
BR Deutschland	15,4 (³)	11,4	0,5 (³)	2,5	1,5	3,4 (³)	1,2 (³)
Elláda	36,7	28,8	1,1	3,0	3,8	18,4	1,4
España (²)	21,8	18,5	0,5	1,4	1,4	7,7	1,2
France	18,6	15,0	0,5	1,9	1,2	4,3	1,5
Ireland	34,8	18,4	1,8	10,4	4,1	5,2	2,0
Italia	19,9	17,1	0,4	1,0	1,4	6,0	0,6
Luxembourg (⁵)	18,6	10,9	0,6	1,3	5,7	3,3	0,3
Nederland	14,9	11,3	0,6	1,5	1,5	2,5	2,2
Portugal (³)	32,4	25,5	0,3	4,4	2,2	20,0	1,3
United Kingdom	21,6	11,3	0,9	6,6	2,7	6,0	0,6

Source: Eurostat — SEC.
(¹) Within the economic territory, and based on current prices.
(²) 1990, 1990/1983.
(³) 1989, 1989/1982.
(⁴) On the basis of development at constant 1985 prices.
(⁵) 1991, 1991/1984.

3.7.2 Human consumption of certain agricultural products

(Kg/head)

		EUR 12	BLEU/UEBL	Danmark	BR Deutschland	Elláda	España	France	Ireland	Italia	Nederland	Portugal	United Kingdom
1	2	3	4	5	6	7	8	9	10	11	12	13	14
Cereals (1)													
— Total cereals	»1985/86«	83	73	71	73	107	74	79	102	114	60	86	75
(without rice)	1992/93	81 (11)	74	68	73 (11)	103	71 (12)	72	94	121	58	87	75
— Wheat (1)	»1985/86«	72	69	46	51	106	72	70	87	107	54	70	63
	1992/93	70 (11)	71	51	54 (11)	101	68 (12)	67	75	114	52	71	62
— Rye (1)	»1985/86«	3	1	18	13	0	1	0	0	0	3	6	0
	1992/93	4 (11)	1	13	13 (11)	1	2 (12)	0	1	0	3	5	0
— Grain/maize (1)	»1985/86«	6	2	2	7	1	1 (12)	4	13	8	2	9	11
	1992/93	6 (11)	1	1	5 (11)	1	1 (12)	4	17	7	2	10	10
— Total milled rice (2)	»1985/86«	4	3	1	2	5	6	4	2	5	2	14	3
	1992/93	4 (11)	3	1	2 (11)	6	6	4	2	5	5	15	4
Potatoes	»1985/86«	80	97	66	73	83	106	75	140	37	85	94	107
	1992/93	79 (11)	101	57	75 (11)	81	105	74	150	41	86	155	94 (12)
Sugar (3)	»1985/86«	33	36	41	35	28	25	35	39	27	37	26	37
	1992/93	34 (12)	40	41	35 (11)	20 (12)	27 (12)	35	34	28	40 (12)	28	39
Vegetables													
— Total vegetables	»1985/86«	116	85	163	72	194	150	118	86	174	91	115	85
(incl. preserved veg.),	1992/93	116 (8)	98	80 (8)	81 (11)	208	192 (12)	124 (9)	96 (12)	178	99 (12)	113	65 (9)
of which: Cauliflowers (4)	»1985/86«	5	5	3	3	3	5	5	4	5	6	2	6
	1992/93	5 (8)	9	3 (8)	3 (10)	4	5 (8)	5 (9)	3 (12)	5	5 (12)	2 (12)	8 (12)
Tomatoes (4)	»1985/86«	27	21	14	14	92	32	21	11	41	16	29	14
	1992/93	26 (8)	27	17 (8)	24 (10)	71	44	23 (9)	14 (12)	50	19	14	15 (9)
Fruit (5)													
— Total fresh fruit	»1985/86«	60	50	38	79	76	67	69	30	69	64	37	38
(including preserved fruit and fruit juice)	1992/93	61 (8)	63	49 (8)	61 (11)	90	66 (12)	58 (9)	37 (12)	93	38	74	38 (9)
of which: Apples (4)	»1985/86«	18	20	19	22	22	21	16	18	20	33	9	12
	1992/93	19 (8)	29	24 (8)	18 (10)	14	20 (8)	15 (9)	18 (12)	26	17 (12)	25	14 (9)
Pears (4)	»1985/86«	7	6	3	4	9	11	6	2	14	5	6	2
	1992/93	6 (8)	7	4 (8)	2 (10)	3	8 (8)	5 (9)	2 (12)	14	3 (12)	8	3 (9)
Peaches (4)	»1985/86«	7	4	3	5	9	10	7	1	16	3	8	3
	1992/93	7 (8)	4	4 (8)	5 (10)	9	11 (8)	7 (9)	2 (12)	16	2 (12)	8	3 (9)
Citrus fruit													
Total citrus fruit	»1985/86«	28	21	11	28	48	25	20	15	39	82	13	14
	1992/93	32 (8)	25	15 (8)	36 (11)	57	46 (12)	24 (9)	11 (12)	49	43 (12)	23	21 (9)
of which: Oranges (4)	»1985/86«	16	16	6	8	27	17	11	12	23	73	9	10
	1992/93	19 (8)	22	10 (8)	9 (10)	34	26 (8)	13 (9)	20 (12)	29	104 (12)	16	11 (9)
Wine (6)	»1985/86«	44	39	19	25	31	49	81	3	71	14	72	9
	1992/93	37	21	23	26 (11)	28	42	65	4	63	13	55	12

Milk products													
Fresh products (without cream)	»1985«	:	83	154	88	63	:	96	195	79	134	:	131
	1992	98 (¹⁰)	85	145	93	62	114	97	189	64	137	100	131
— Cheese	»1985«	:	12	12	14	22	:	20	4	15	13	:	6
	1992	14 (¹⁰)	14	16	17	21	7	23	7	19	14	7	7
— Butter (fats)	»1985«	:	7	6	6	1	:	8	8	2	3	:	4
	1992	5	7	6	7	1	0,4	8	3	2	3	1	3
— Margarine (fat)	»1985«	5	11	12	7	2	1	3	5	1	11	5	7
	1992	5 (⁹)	9	11	7	2 (⁹)	2	3 (⁹)	4 (⁹)	1	10 (⁹)	6	7 (¹⁰)
Eggs	»1985«	14	14	15	17	12	16	15	13	11	12	6	13
	1992	13 (¹⁰)	14	15	14	11	15	15	11	12	11 (¹⁰)	8	11
Meat (⁷)													
Total meat (without offal),	»1985«	83	94	85	95	72	78	96	80	79	75	51	70
	1992	87 (¹⁰)	95	101	95 (¹⁰)	79	97	100	86	86	88	74	71
of which: Total beef/veal	»1985«	23	26	15	23	22	11	32	23	27	18	11	22
	1992	20 (¹⁰)	16	21	19	22	13	24	17	21	19	16	17
Beef	»1985«	20	23	14	21	18	9	25	23	23	16	10	22
	1992	19	16	21	19	22	13	24	17	21	19	16	20
Veal	»1985«	3	3	1	2	4	2	7	0	4	2	1	0
	1992	2 (¹⁰)	5	0,2	1	1	0,3	6	0,3	4	2	1	0
Pigmeat	»1985«	37	47	58	60	21	37	35	34	28	42	21	24
	1992	41	50	65	56	23	50	37	38	33	44	30	23
Poultrymeat	»1985«	16	15	11	10	16	21	18	17	18	14	14	16
	1992	19	18	14	13	18	23	22	23	20	21	20	21
Sheepmeat and goatmeat	»1985«	4	2	1	1	14	5	4	7	2	1	3	7
	1992	4	2	1	1	15	7	6	8	2	1	4	7
Oils and fats													
Total fats and oils	»1985«	26	30	31	21	32	28	21	20	28	36	21	30
	1992	27 (⁹)	32	44 (¹⁰)	23	33 (⁹)	33	22 (⁹)	23 (⁹)	31 (¹⁰)	38 (⁹)	28 (⁹)	30 (¹⁰)
of which: vegetable	»1985«	14	5	17	6	24	22	13	12	23	7	14	11
	1992	15 (⁹)	8	26 (¹⁰)	10	26 (⁹)	25	13 (⁹)	16 (⁹)	24	7 (⁹)	21	11 (¹⁰)
of land animals	»1985«	6	8	2	6	3	3	4	1	4	13	2	9
	1992	6 (⁹)	10	13	5	4 (⁹)	3	5 (⁹)	1 (⁹)	5	15 (⁹)	2 (⁹)	9 (¹⁰)

Source : Eurostat.

(¹) Flour equivalent.
(²) Expressed in product weight.
(³) White-sugar equivalent.
(⁴) Human consumption based on marketed produce and including processed products.
(⁵) Not including citrus fruits.
(⁶) Litres/head.
(⁷) Including cutting-room fat.
(⁸) 1987/88.
(⁹) 1988/89.
(¹⁰) 1989/90.
(¹¹) 1990/91.
(¹²) 1991/92.

3.7.3   Self-sufficiency in certain agricultural products

(%)

		EUR 12	BLEU/UEBL	Danmark	BR Deutschland	Elláda	España	France	Ireland	Italia	Nederland	Portugal	United Kingdom
1	2	3	4	5	6	7	8	9	10	11	12	13	14
Cereals													
— Total cereals (excl. rice)	» 1985/86 «	110	54	117	94	104	83	201	90	80	28	33	120
	1992/93	120 (8)	54	113	114 (8)	114	99 (9)	249	98	92	29	34	125
— Total wheat	» 1985/86 «	124	69	117	101	123	94	233	61	81	55	38	107
	1992/93	133 (8)	77	152	121 (8)	132	98 (9)	273	83	84	46	21	131
— Rye	» 1985/86 «	111	76	201	108	101	101	103	0	81	34	98	79
	1992/93	122 (7)	48	118	123 (7)	108	100 (9)	101	0	85	43	99	168
— Barley	» 1985/86 «	119	75	116	102	53	104	191	124	58	23	49	155
	1992/93	123 (7)	65	88	105 (7)	82	115 (9)	270	124	79	21	25	147
— Grain/maize	» 1985/86 «	77	5	0	45	96	44	176	0	87	0	24	0
	1992/93	94 (7)	8	0	59 (7)	107	74 (9)	233	0	112	5	38	0
— Total milled rice	» 1985/86 «	75	0	0	0	125	97	11	0	215	0	72	0
	1992/93	75 (7)	0	0	0	110	124	28	0	246	0	66	0
Potatoes	» 1985/86 «	101	108	98	99	108	100	101	87	96	146	94	92
	1992/93	100 (7)	142	98	92 (7)	99	100	105	83	91	135	90	92 (9)
Sugar	» 1985/86 «	123	227	220	130	97	109	203	145	81	153	2	56
	1992/93	128 (7)	222	189	141 (7)	88 (9)	79 (9)	235	150	116	167 (9)	1	65
Fresh vegetables	» 1985/86 «	107	116	70	37	157	131	91	81	125	204	144	63
	1992/93	106 (5)	140	55 (5)	38 (7)	158	117 (9)	89 (6)	83	123	254 (9)	123	88 (6)
Fresh fruit (excl. citrus fruit)	» 1985/86 «	87	61	38	53	125	116	89	15	128	57	95	22
	1992/93	85 (5)	81	20 (5)	22 (7)	122	94 (9)	86 (7)	14 (9)	117	63 (9)	78	19 (6)
Citrus fruit	» 1985/86 «	75	0	0	0	163	299	3	0	113	0	100	0
	1992/93	70 (5)	0	0	0	145	231 (9)	2 (6)	0	106	0	88	0
Wine	» 1985/86 «	104	68 (3)	0	57	119	118	108	0	121	0	113	0
	1992/93	115 (7)	119	0	88 (8)	120	117	113	0	127	0	117	0
Milk products													
— Fats	» 1985 «	—	107	218	119	82	—	121	286 (5)	69	291	—	88
	1992	—	112	217	103 (6)	85 (4)	99 (8)	107	175 (5)	79 (6)	152 (5)	100	86 (7)
— Proteins	» 1985 «	—	98	428	132	27	96 (4)	137	202	61	1 147	101 (4)	86
	1992	—	119	437	134 (7)	83 (4)	99 (8)	107	147	65 (7)	1 221	103	85 (8)
— Fresh milk products (excl. cream)	» 1985 «	102 (4)	124	105	104	98	99 (4)	111	100	97	93	100 (4)	100
	1992	102 (4)	140	103	113 (7)	95	97	102	101	95	86	99	98
— Whole-milk powder	» 1985 «	316 (4)	239	2 517	140	33 (4)	106 (4)	721	2 767	13	573	92 (4)	234
	1992	272 (7)	211	10 300	136 (7)	0	99	500	3 100	9	2 667	120	150

	Year												
— Skimmed-milk powder	»1985«	123*(4)	172	106	271(7)	0	78(4)	135	1 024	0	42	117(4)	161
	1992	153(7)	123	81	462(7)	0	62	113	12 600	0	26	133	95
— Concentrated milk	»1985«	—	66	767	145	—	83(4)	232	—	58	375	—	112
	1992		183	0	129	93	100	96	0	9	302	0	134
— Cheese	»1985«	153(7)	36	436	96	88	91(4)	114	467	77	250	93(4)	71
	1992	106*(4)	40	365	96(7)	80	89	117	404	85	266	99	74
— Butter	»1985«	107(7)	120	183	121	56	155(4)	118	484	57	444	104(4)	74
	1992	110*(4)	101	194	96(7)	29	180	92	1 233	76	510	131	54
— Margarine	»1985«	121(7)	134	120	101	95	98	74	94	84	146	106	90
	1992	102	179	123	102(7)	92(6)	93	72(6)	121(6)	67	132(6)	107	95
Eggs	»1985«	102(6)	116	100	73	97	100	99	78	91	327	99	96
	1992	102	124	101	76(8)	99	97	99	89(8)	95	318(8)	102	95
Meat (1) — Total (2)	»1985«	101(8)	122	324	91	70	97	100	270	74	240	97	81
	1992	102(4)	152	329	90(7)	66	97	107	299	75	231	88	85
— Total beef/veal	»1985«	102(7)	130	324	118	35	95	119	655	62	200	87	87
	1992	107(4)	179	204	120(7)	28	100	122	977	69	179	70	85
— Beef	»1985«	108(7)	130	334	121	40	88	121	657	60	143	87	87
	1992	106(4)	196	205	123(7)	25	102	127	992	65	138	70	83
— Veal	»1985«	107(7)	130	100	80	17	98	110	33 399	76	735	86	147
	1992	113(4)	125	100	72(7)	97	11	100	100	88	567	75	917
— Pigmeat	»1985«	113(4)	145	370	87	71	97	81	116	71	270	97	71
	1992	102(4)	178	414	86(7)	65	98	91	126	65	278	88	75
— Poultrymeat	»1985«	104(7)	88	206	61	98	98	130	92	98	217	100	96
	1992	105(7)	103	229	58(7)	94	94	149	110	98	185	105	90
— Sheepmeat and goatmeat	»1985«	105(7)	22	33	45	87	100	70	190	58	261	100	76
	1992	80(4)	14	40	43(7)	84	89	49	325	56	156	69	104
Oils and fats — Total	»1985«	63	31	93	49	129	98	63	62	52	32	36	28
	1992	70(6)	30	99(8)	65(7)	117(6)	80(7)	82(6)	59(6)	42(7)	33(6)	32(6)	34(7)
— Vegetable	»1985«	56	3	20	20	144	105	52	0	47	1	25	20
	1992	65(6)	2	0	40(7)	127(6)	86(7)	83(6)	0	44	0.4(6)	27	31
— Cutting-room fat	»1985«	84	70	157	116	65	80	99	186	80	59	73	53
	1992	86(6)	74	120(7)	120(7)	57(6)	66	92(6)	240(6)	67	71(6)	69(6)	59(8)

Source : Eurostat.

(1) Excl. offal.
(2) Incl. cutting-room fat.
(3) Only Luxembourg.
(4) »1987« or 1987.
(5) 1988 or 1987/88.
(6) 1989 or 1988/89.
(7) 1990 or 1989/90.
(8) 1991 or 1990/91.
(9) 1992 or 1991/92.

4.1.1.1 Area, yield and production of common and durum wheat

1	Area (1 000 ha)					Yield (100 kg/ha)					Production (1 000 t)				
	1985	1992 ∞	1993 ∞	% TAV 1992/1985	% TAV 1993/1992	1985	1992 ∞	1993 ∞	% TAV 1992/1985	% TAV 1993/1992	1985	1992 ∞	1993 ∞	% TAV 1992/1985	% TAV 1993/1992
	2	3	4	5	6	7	8	9	10	11	12	13	14	15	16
Common wheat															
EUR 12	12 803	13 502	12 339	0,8	−8,6	51,0	56,1	60,1	1,4	7,1	65 338	75 739	74 176	2,1	−2,1
Belgique/België	188	209	203	1,5	−2,9	63,1	66,1	71,9	0,7	8,8	1 187	1 382	1 463	2,2	5,9
Danmark	340	586	621	8,1	6,0	58,0	61,1	70,0	0,7	14,6	1 972	3 583	4 349	8,9	21,4
BR Deutschland	1 609	2 583	2 385	7,0	−7,7	60,8	59,9	65,9	−0,2	10,0	9 779	15 472	15 720	6,8	1,6
Elláda	457	332	329	−4,5	−0,9	21,4	27,1	27,2	3,4	0,4	980	899	895	−1,2	−0,4
España	1 911	1 613	1 412	−2,4	−12,5	25,9	19,1	30,2	−4,3	58,1	4 958	3 078	4 260	−6,6	38,4
France	4 632	4 655	4 306	0,1	−7,5	60,6	65,8	66,0	1,2	0,3	28 091	30 613	28 427	1,2	−7,1
Ireland	78	91	77	2,2	−15,4	63,5	78,4	78,0	3,1	−0,5	495	713	597	5,4	−16,3
Italia	1 295	988	889	−3,8	−10,0	35,6	46,7	46,1	4,0	−1,3	4 610	4 610	4 096	0,0	−11,1
Luxembourg	7	8	8	1,9	0,0	40,0	57,5	58,4	5,3	1,6	28	46	49	7,3	6,5
Nederland	128	127	118	−0,1	−7,1	66,5	80,1	87,7	2,7	9,5	851	1 017	1 035	2,6	1,8
Portugal	262	245	232	−1,0	−5,3	13,9	9,8	17,2	−4,9	75,5	365	240	400	−5,8	66,7
United Kingdom	1 896	2 065	1 758	1,2	−14,9	63,4	68,2	73,3	1,0	7,5	12 022	14 086	12 884	2,3	−8,5
Durum wheat															
EUR 12	2 509	3 248	2 867	3,8	−11,7	23,4	27,9	24,5	2,5	−12,2	5 862	9 053	7 035	6,4	−22,3
BR Deutschland	15	16	10	0,9	−37,5	58,0	43,1	47,9	−4,2	11,1	87	69	46	−3,3	−33,3
Elláda	426	616	583	5,4	−5,4	19,4	23,5	21,4	2,8	−8,9	827	1 445	1 248	8,3	−13,6
España	133	630	624	24,9	−1,0	27,9	20,3	11,9	−4,4	−41,4	371	1 279	742	19,3	−42,0
France	165	425	221	14,5	−48,0	44,4	44,6	40,5	0,1	−9,2	732	1 895	897	14,6	−52,7
Italia	1 741	1 530	1 410	−1,8	−7,8	21,8	28,3	28,9	3,8	2,1	3 789	4 329	4 075	1,9	−5,9
Portugal	23	30	18	3,9	−40,0	13,9	10,0	12,1	−4,6	21,0	32	30	22	−0,9	−26,7
United Kingdom	6	1	1	−22,6	0,0	40,0	50,0	50,0	3,2	0,0	24	6	6	−18,0	0,0

4.1.1.2 Area, yield and production of rye and barley

	Area					Yield					Production				
	1 000 ha			% TAV		100 kg/ha			% TAV		1 000 t			% TAV	
	1985	1992 ∞	1993 ∞	1992/1985	1993/1992	1985	1992 ∞	1993 ∞	1992/1985	1993/1992	1985	1992 ∞	1993 ∞	1992/1985	1993/1992
1	2	3	4	5	6	7	8	9	10	11	12	13	14	15	16
Rye and meslin															
EUR 12	1 014	1 080	1 084	0,9	0,4	31,3	32,0	37,6	0,3	17,5	3 178	3 451	4 078	1,2	18,2
Belgique/België	5	2	2	-12,3	0,0	46,0	43,9	43,3	-0,7	-1,4	23	9	10	-12,5	11,1
Danmark	127	88	77	-5,1	-12,5	44,5	35,0	44,3	-3,4	26,6	565	308	339	-8,3	10,1
BR Deutschland	426	625	671	5,6	7,4	42,7	39,5	45,1	-1,1	14,2	1 821	2 473	3 031	4,5	22,6
Elláda	12	17	19	5,1	11,8	19,2	24,0	22,2	3,2	-7,5	23	42	42	9,0	0,0
España	211	194	173	-1,2	-10,8	12,9	12,1	17,7	-0,9	46,3	273	234	306	-2,2	30,8
France	87	55	48	-6,3	-12,7	34,1	37,8	39,8	1,5	5,3	297	208	189	-5,0	-9,1
Italia	9	8	8	-1,7	0,0	24,4	28,8	28,7	2,4	-0,3	22	23	23	0,6	0,0
Luxembourg	1	0	0	x	x	30,0	33,3	33,3	1,5	0,0	3	2	2	-5,6	0,0
Nederland	5	6	7	2,6	16,7	38,0	56,7	55,7	5,9	-1,8	19	34	41	8,7	20,6
Portugal	123	75	73	-6,8	-2,7	7,9	10,7	8,9	4,4	-16,8	97	80	65	-2,7	-18,8
United Kingdom	8	8	6	0,0	-25,0	43,8	46,4	52,1	0,8	12,3	35	37	30	0,8	-18,9
Barley															
EUR 12	12 852	11 520	10 149	-1,6	-11,9	40,1	37,5	42,2	-1,0	12,5	51 473	43 250	42 835	-2,5	-1,0
Belgique/België	118	72	66	-6,8	-8,3	58,1	63,4	64,2	1,3	1,3	685	460	425	-5,5	-7,6
Danmark	1 104	910	721	-2,7	-20,8	47,6	32,7	47,3	-5,2	44,6	5 251	2 974	3 407	-7,8	14,6
BR Deutschland	1 949	2 408	2 201	3,1	-8,6	49,7	50,7	50,0	0,3	-1,4	9 691	12 196	11 006	3,3	-9,8
Elláda	312	171	167	-8,2	-2,3	18,7	25,5	24,8	4,5	-2,7	583	436	415	-4,1	-4,8
España	4 246	4 112	3 485	-0,5	-15,2	25,2	14,9	27,3	-7,2	83,2	10 698	6 105	9 520	-7,7	55,9
France	2 256	1 800	1 623	-3,2	-9,8	50,7	58,2	55,4	2,0	-4,8	11 442	10 476	8 995	-1,3	-14,1
Ireland	298	184	177	-6,7	-3,8	50,1	63,3	55,1	3,4	-13,0	1 494	1 167	975	-3,5	-16,5
Italia	461	450	425	-0,3	-5,6	34,0	38,7	38,4	1,9	-0,8	1 566	1 742	1 634	1,5	-6,2
Luxembourg	17	14	14	-2,7	0,0	35,9	50,3	49,8	4,9	-1,0	61	70	68	2,0	-2,9
Nederland	39	34	40	-1,9	17,6	50,5	60,0	63,0	2,5	5,0	197	204	252	0,5	23,5
Portugal	86	67	67	-3,5	0,0	7,6	8,1	14,8	0,9	82,7	65	54	99	-2,6	83,3
United Kingdom	1 966	1 297	1 164	-5,8	-10,3	49,5	56,8	51,9	2,0	-8,6	9 740	7 366	6 038	-3,9	-18,0

4.1.1.3 Area, yield and production of oats and mixed cereals and maize

	Area					Yield					Production				
	1000 ha			% TAV		100 kg/ha			% TAV		1000 t			% TAV	
	1985	1992∞	1993∞	1992/1985	1993/1992	1985	1992∞	1993∞	1992/1985	1993/1992	1985	1992∞	1993∞	1992/1985	1993/1992
1	2	3	4	5	6	7	8	9	10	11	12	13	14	15	16
Oats and mixed cereals															
EUR 12	2 360	1 412	1 405	−7,1	−0,5	33,2	28,7	33,1	−2,1	15,3	7 825	4 050	4 654	−9,0	14,9
Belgique/België	24	10	13	−11,8	30,0	45,0	37,9	50,2	−2,4	32,5	108	37	64	−14,2	73,0
Danmark	41	28	32	−5,3	14,3	41,0	32,3	43,6	−3,3	35,0	168	89	139	−8,7	56,2
BR Deutschland	692	411	407	−7,2	−1,0	47,4	36,9	47,7	−3,5	29,3	3 278	1 518	1 941	−10,4	27,9
Elláda	43	43	43	0,0	0,0	14,9	17,0	17,2	1,9	1,2	64	73	75	1,9	2,7
España	459	314	328	−5,3	4,5	14,8	10,0	12,3	−5,4	23,0	680	313	405	−10,5	29,4
France	547	228	222	−11,8	−2,6	40,3	41,6	41,9	0,5	0,7	2 203	948	929	−11,3	−2,0
Ireland	23	20	20	−2,0	0,0	46,1	67,7	56,0	5,6	−17,3	106	136	113	3,6	−16,9
Italia	178	146	144	−2,8	−1,4	19,9	22,8	25,5	2,0	11,8	355	333	367	−0,9	10,2
Luxembourg	10	5	4	−9,4	−20,0	39,0	43,1	46,0	1,4	6,7	39	20	20	−9,1	0,0
Nederland	12	4	5	−14,5	25,0	49,2	52,0	59,5	0,8	14,4	59	19	30	−14,9	57,9
Portugal	190	98	92	−9,0	−6,1	6,3	4,6	8,3	−4,4	80,4	119	45	76	−13,0	68,9
United Kingdom	141	106	95	−4,0	−10,4	45,8	48,8	52,0	0,9	6,6	646	519	494	−3,1	−4,8
Maize															
EUR 12	3 984	3 830	3 790	−0,6	−1,0	64,9	78,2	79,9	2,7	2,2	25 847	29 951	30 279	2,1	1,1
Belgique/België	7	10	18	5,2	80,0	71,4	63,8	90,6	−1,6	42,0	50	65	168	3,8	158,5
BR Deutschland	181	296	331	7,3	11,8	66,5	72,4	80,2	1,2	10,8	1 204	2 139	2 656	8,6	24,2
Elláda	222	211	212	−0,7	0,5	85,9	97,0	99,0	1,8	2,1	1 908	2 048	2 099	1,0	2,5
España	526	393	274	−4,1	−30,3	64,9	70,2	61,9	1,1	−11,8	3 414	2 757	1 699	−3,0	−38,4
France	1 891	1 869	1 851	−0,2	−1,0	65,8	76,6	80,9	2,2	5,6	12 448	14 886	14 966	2,6	0,5
Italia	911	854	927	−0,9	8,5	68,8	86,6	86,6	3,3	0,0	6 271	7 394	8 029	2,4	8,6
Nederland	0	8	10	x	25,0	x	81,8	90,9	x	11,1	2	63	95	63,7	50,8
Portugal	246	190	167	−3,6	−12,1	22,4	31,6	34,1	5,0	7,9	550	600	568	1,3	−5,3

4.1.1.4 Area, yield and production of other cereals and total cereals (excl. rice)

	Area					Yield					Production				
	1000 ha			% TAV		100 kg/ha			% TAV		1000 t			% TAV	
1	1985	1992 ∞	1993 ∞	1992/1985	1993/1992	1985	1992 ∞	1993 ∞	1992/1985	1993/1992	1985	1992 ∞	1993 ∞	1992/1985	1993/1992
	2	3	4	5	6	7	8	9	10	11	12	13	14	15	16
Other cereals (¹)															
EUR 12	191	643	640	18,9	− 0,5	41,8	44,8	46,7	1,0	4,2	799	2 877	2 989	20,1	3,9
Belgique/België	3	7	9	12,9	28,6	40,0	63,2	54,8	6,8	− 13,3	12	46	49	21,2	6,5
BR Deutschland	12	175	219	46,6	25,1	45,0	50,9	52,5	1,8	3,1	54	891	1 147	49,3	28,7
Elláda	0	2	1	x	− 50,0	x	22,0	20,0	x	− 9,1	0	4	2	x	− 50,0
España	31	62	40	10,4	− 35,5	37,4	28,9	16,4	− 3,6	− 43,3	116	178	65	6,3	− 63,5
France	123	286	254	12,8	− 11,2	43,7	49,7	51,4	1,9	3,4	537	1 424	1 304	14,9	− 8,4
Italia	21	33	42	6,7	27,3	35,2	56,9	57,1	7,1	0,4	74	189	237	14,3	25,4
Luxembourg	0	2	3	x	50,0	x	55,3	52,3	x	− 5,4	1	13	14	44,3	7,7
Nederland	0	5	6	x	20,0	x	85,3	96,5	x	13,1	1	41	55	70,0	34,1
Portugal	0	58	56	x	− 3,4	x	7,2	15,1	x	109,7	0	42	85	x	102,4
United Kingdom	0	11	7	x	− 36,4	x	43,9	46,9	x	6,8	4	49	31	43,0	− 36,7
Total cereals (excl. rice)															
EUR 12	35 713	35 236	32 275	− 0,2	− 8,4	44,9	47,8	51,5	0,9	7,7	160 322	168 372	166 045	0,7	− 1,4
Belgique/België	345	311	312	− 1,5	0,3	59,9	64,3	69,8	1,0	8,6	2 065	1 998	2 178	− 0,5	9,0
Danmark	1 612	1 609	1 451	− 0,0	− 9,8	49,4	43,2	56,8	− 1,9	31,5	7 956	6 954	8 236	− 1,9	18,4
BR Deutschland	4 884	6 514	6 224	4,2	− 4,5	53,1	53,4	57,1	0,1	6,9	25 914	34 758	35 547	4,3	2,3
Elláda	1 472	1 392	1 354	− 0,8	− 2,7	29,8	35,5	35,3	2,5	− 0,6	4 385	4 946	4 776	1,7	− 3,4
España	7 517	7 318	6 336	− 0,4	− 13,4	27,3	19,1	26,8	− 5,0	40,3	20 510	13 945	16 996	− 5,4	21,9
France	9 701	9 318	8 524	− 0,6	− 8,5	57,5	64,9	65,4	1,7	0,8	56 750	60 450	55 708	0,9	− 7,8
Ireland	400	300	280	− 4,0	− 6,7	52,4	67,1	60,3	3,6	− 10,1	2 095	2 016	1 686	− 0,5	− 16,4
Italia	4 616	4 009	3 844	− 2,0	− 4,1	36,2	46,5	48,0	3,6	3,2	16 687	18 620	18 460	1,6	− 0,9
Luxembourg	35	30	29	− 2,2	− 3,3	37,7	51,4	51,9	4,5	1,0	132	152	153	2,0	0,7
Nederland	184	183	187	− 0,1	2,2	61,4	75,2	80,8	2,9	7,4	1 129	1 378	1 508	2,9	9,4
Portugal	930	763	704	− 2,8	− 7,7	13,2	14,3	18,7	1,2	30,8	1 228	1 091	1 314	− 1,7	20,4
United Kingdom	4 017	3 489	3 031	− 2,0	− 13,1	56,9	63,2	64,3	1,5	1,7	22 471	22 063	19 483	− 0,3	− 11,7

(¹) Including ' triticale '.

4.1.2.1   World production of cereals and production in principal exporting countries

	%			Mio t			% TAV	
	1985	1992 ∞	1993 ∞	1985	1992 ∞	1993 ∞	1992 / 1985	1993 / 1992
1	2	3	4	5	6	7	8	9
I — *Wheat* (¹) World of which:	100,0	100,0	100,0	505,7	566,3	564,3	1,6	− 0,4
– EUR 12	14,1	15,1	14,4	71,2	85,3	81,4	2,6	− 4,6
– USA	13,1	11,8	11,7	66,0	66,9	65,9	0,2	− 1,5
– Canada	4,8	5,3	5,0	24,2	29,9	28,2	3,1	− 5,7
– Argentina	1,7	1,7	1,8	8,7	9,7	10,0	1,6	3,1
– Australia	3,2	2,9	2,7	16,2	16,2	15,3	0,0	− 5,6
– Others	63,2	63,3	64,4	319,4	358,3	363,6	1,7	1,5
II — *Other cereals* (²) World of which:	100,0	100,0	100,0	864,9	871,4	804,9	0,1	− 7,6
– EUR 12	10,3	9,5	10,4	89,1	83,1	84,1	− 1,0	1,2
– USA	31,8	31,9	25,6	275,3	278,3	206,0	0,2	− 26,0
– Canada	2,8	2,3	3,1	24,0	19,8	24,6	− 2,7	24,2
– Argentina	2,2	1,7	1,9	19,0	14,8	15,6	− 3,5	5,4
– Australia	0,9	1,0	0,9	8,1	9,1	7,6	1,7	− 16,5
– Others	52,0	53,5	58,0	449,4	466,2	467,1	0,5	0,2

Source: FAO — Production Directory + Monthly Bulletin: Economics and Statistics. Eurostat for Community figures.
(¹) Common and durum wheat.
(²) Excl. rice.

4.1.3.1   The Community's share in world cereals trade

		Mio t						% TAV	
		1990	%	1991 ∞	%	1992 ∞	%	1991 / 1990	1992 / 1991
1	2	3	4	5	6	7	8	9	10
1. *Imports* (¹) Wheat and flour (wheat equivalent)	World	107	100,0	115,4	100,0	121,5	100,0	7,9	5,3
	EUR 12	2	1,8	2,0	1,6	2,0	1,7	0,0	0,0
Other cereals (²)	World	106	100,0	105,4	100,0	114,3	100,0	− 0,6	8,4
	EUR 12	4	3,8	4,0	3,6	3,0	2,3	0,0	− 25,0
All cereals (²)	World	213	100,0	220,8	100,0	235,7	100,0	3,7	6,7
	EUR 12	6	2,8	6,0	2,6	5,0	2,0	0,0	− 16,7
2. *Exports* (¹) Wheat and flour (wheat equivalent)	World	108	100,0	119,9	100,0	122,9	100,0	11,0	2,5
	EUR 12	22	20,5	32,4	27,0	36,4	29,6	47,3	12,3
Other cereals (²)	World	105	100,0	98,5	100,0	114,8	100,0	− 6,2	16,5
	EUR 12	13	12,1	9,5	9,6	12,9	11,2	− 26,9	35,8
All cereals (²)	World	212	100,0	218,4	100,0	237,7	100,0	3,0	8,8
	EUR 12	35	16,4	42,0	19,2	49,0	20,7	20,0	16,7

Sources: FAO but Eurostat for Community figures.
(¹) Excl. intra-EC trade.
(²) Excl. rice + malt in barley equivalent.

4.1.4.1   **Supply balances   — durum wheat**
(1 July-30 June)   **— common wheat**

EUR 12

	1 000 t			% TAV	
	1985/86	1991/92 ∞	1992/93 ∞	1991/92 / 1985/86	1992/93 / 1991/92
1	2	3	4	5	6
Durum wheat					
Usable production	5 753	11 319	9 396	11,9	− 17,0
Change in stocks	− 280	2 510	− 639	×	×
Imports	555	304	101	− 9,5	− 66,8
Exports	1 872	4 006	2 555	13,5	− 36,2
Intra-EC trade (¹)	1 374	3 183	2 122	15,0	− 33,3
Internal use	4 716	5 107	7 581	1,3	48,4
of which :					
— animal feed	200	200	709	0,0	254,5
— seed	533	713	817	5,0	14,6
— industrial use	2	0	3	×	×
— losses (market)	27	26	251	− 0,6	865,4
— human consumption (grain)	3 954	4 168	5 801	0,9	39,2
Human consumption (after processing)	2 790	2 941	4 093	0,9	39,2
Human consumption (kg/head)	8,7	8,4	11,7	− 0,6	39,3
Self-sufficiency (%)	122,0	221,6	123,9	10,5	− 44,1
Common wheat					
Usable production	65 452	79 220	76 050	3,2	− 4,0
Change in stocks	− 1 656	3 806	1 901	×	− 50,1
Imports	2 629	1 343	1 045	− 10,6	− 22,2
Exports	13 490	19 526	20 004	6,4	2,4
Intra-EC trade (¹)	12 370	13 997	10 665	2,1	− 23,8
Internal use	56 247	57 231	55 190	0,3	− 3,6
of which :					
— animal feed	24 037	22 925	23 146	− 0,8	1,0
— seed	2 214	2 287	2 213	0,5	− 3,2
— industrial use	1 175	2 801	2 506	15,6	− 10,5
— losses (market)	995	1 014	71	0,3	− 93,0
— human consumption (grain)	27 826	28 204	27 254	0,2	− 3,4
Human consumption (after processing)	20 770	21 052	20 343	0,2	− 3,4
Human consumption (kg/head)	64,4	61,1	59,1	− 0,9	− 3,3
Self-sufficiency (%)	116,4	138,4	137,8	2,9	− 0,4

Source : Eurostat and EC Commission, Directorate-General for Agriculture.

(¹) Calculated on intra-import basis.

4.1.4.2 Supply balances   — barley
(1 July-30 June)   — rye

EUR 12

| | 1 000 t | | | % TAV | |
	1985/86	1991/92 ∞	1992/93 ∞	1991/92 / 1985/86	1992/93 / 1991/92
1	2	3	4	5	6
Barley					
Usable production	51 413	51 224	43 320	− 0,1	− 15,4
Change in stocks	1 248	3 309	− 206	17,6	×
Imports	168	119	85	− 5,6	− 28,6
Exports	9 218	9 509	8 327	0,5	− 12,4
Intra-EC trade (¹)	5 531	5 943	5 443	1,2	− 8,4
Internal use of which:	41 115	38 525	35 284	− 1,1	− 8,4
— animal feed	32 237	29 322	26 666	− 1,6	− 9,1
— seed	2 035	1 840	2 011	− 1,7	9,3
— industrial use	5 814	6 543	6 272	2,0	− 4,1
— losses (market)	910	734	254	− 3,5	− 65,4
— human consumption (grain)	119	86	81	− 5,3	− 5,8
Human consumption (after processing)	66	48	45	− 5,2	− 6,3
Human consumption (kg/head)	0,2	0,1	0,1	− 10,9	0,0
Self-sufficiency (%)	125,1	133,0	122,8	1,0	− 7,7
Rye					
Usable production	3 256	4 464	3 422	5,4	− 23,3
Change in stocks	273	452	− 1 353	8,8	×
Imports	58	24	20	− 13,7	− 16,7
Exports	127	715	1 808	33,4	152,9
Intra-EC trade (¹)	153	135	125	− 2,1	− 7,4
Internal use of which:	2 914	3 321	2 987	2,2	− 10,1
— animal feed	1 356	1 564	1 501	2,4	− 4,0
— seed	138	150	201	1,4	34,0
— industrial use	35	60	59	9,4	− 1,7
— losses (market)	74	113	45	7,3	− 60,2
— human consumption (grain)	1 311	1 434	1 182	1,5	− 17,6
Human consumption (after processing)	1 110	1 214	1 001	1,5	− 17,5
Human consumption (kg/head)	3,4	3,5	2,9	0,5	− 17,1
Self-sufficiency (%)	111,7	134,4	114,6	3,1	− 14,7

Source: Eurostat and EC Commission, Directorate-General for Agriculture.

(¹) Calculated on intra-import basis.

4.1.4.3   **Supply balances**   — **maize**
      (1 July-30 June)   — **oats and mixed summer cereals**   **EUR 12**

	1 000 t			% TAV	
	1985/86	1991/92 ∞	1992/93 ∞	1991/92 ── 1985/86	1992/93 ── 1991/92
1	2	3	4	5	6
Maize					
Usable production	25 752	27 281	29 822	1,0	9,3
Change in stocks	1 842	1 356	1 807	− 5,0	33,3
Imports	7 336	3 284	1 638	− 12,5	− 50,1
Exports	1 083	933	2 307	− 2,5	147,3
Intra-EC trade (¹)	8 132	10 444	6 358	4,3	− 39,1
Internal use	30 163	28 276	27 346	− 1,1	− 3,3
of which:					
— animal feed	24 045	22 195	21 452	− 1,3	− 3,3
— seed	232	208	218	− 1,8	4,8
— industrial use	2 737	2 617	3 159	− 0,7	20,7
— losses (market)	257	284	38	1,7	− 86,6
— human consumption (grain)	2 892	2 972	2 478	0,5	− 16,6
Human consumption (after processing)	2 170	2 230	1 859	0,5	− 16,6
Human consumption (kg/head)	5,9	5,6	4,7	− 0,9	− 16,1
Self-sufficiency (%)	85,4	96,5	109,1	2,1	13,1
Oats and mixed corn					
Usable production	7 841	4 950	4 569	− 7,4	− 7,7
Change in stocks	− 70	− 75	− 54	×	×
Imports	76	31	437	− 13,9	1 309,7
Exports	5	59	25	50,9	− 57,6
Intra-EC trade (¹)	409	307	112	− 4,7	− 63,5
Internal use	7 982	4 997	5 035	− 7,5	0,8
of which:					
— animal feed	7 169	4 163	4 240	− 8,7	1,8
— seed	313	266	241	− 2,7	− 9,4
— industrial use	50	2	7	− 41,5	250,0
— losses (market)	78	92	83	2,8	− 9,8
— human consumption (grain)	372	474	464	4,1	− 2,1
Human consumption (after processing)	236	301	294	4,1	− 2,3
Human consumption (kg/head)	0,7	0,9	0,8	4,3	− 11,1
Self-sufficiency (%)	98,2	99,1	90,7	0,2	− 8,5

Source: Eurostat and EC Commission, Directorate-General for Agriculture.

(¹) Calculated on intra-import basis.

4.1.4.4 **Supply balances**   **— other cereals**
          (1 July-30 June)   **— total cereals** (excl. rice)   **EUR 12**

	1 000 t			% TAV	
	1985/86	1991/92 ∞	1992/93 ∞	1991/92 / 1985/86	1992/93 / 1991/92
1	2	3	4	5	6
Other cereals ([1])					
Usable production	722	2 479	2 004	22,8	− 19,2
Change in stocks	− 3	56	− 9	×	×
Imports	232	438	87	11,2	− 80,1
Exports	1	39	1	84,2	− 97,4
Intra-EC trade ([2])	153	358	252	15,2	− 29,6
Internal use	956	2 822	2 099	19,8	− 25,6
of which:					
— animal feed	923	2 694	1 943	19,5	− 27,9
— seed	22	97	109	28,1	12,4
— industrial use	4	4	0	0,0	×
— losses (market)	5	17	36	22,6	111,8
— human consumption (grain)	2	10	11	30,8	10,0
Human consumption (after processing)	2	10	11	30,8	10,0
Human consumption (kg/head)	0,0	0,0	0,0	×	×
Self-sufficiency (%)	75,5	87,8	95,5	2,5	8,8
Total cereals (excl. rice)					
Usable production	160 189	180 937	168 582	2,1	− 6,8
Change in stocks	1 354	11 414	1 447	42,7	− 87,3
Imports	11 054	5 543	3 413	− 10,9	− 38,4
Exports	25 796	34 787	35 027	5,1	0,7
Intra-EC trade ([2])	28 122	34 367	25 077	3,4	− 27,0
Internal use	144 093	140 279	135 521	− 0,4	− 3,4
of which:					
— animal feed	89 967	83 063	79 656	− 1,3	− 4,1
— seed	5 487	5 561	5 810	0,2	4,5
— industrial use	9 817	12 027	12 005	3,4	− 0,2
— losses (market)	2 346	2 280	779	− 0,5	− 65,8
— human consumption (grain)	36 476	37 348	37 271	0,4	− 0,2
Human consumption (after processing)	27 144	27 793	27 736	0,4	− 0,2
Human consumption (kg/head)	83,4	79,6	79,6	− 0,8	0,0
Self-sufficiency (%)	111,2	129,0	124,4	2,5	− 3,6

Source: Eurostat and EC Commission, Directorate-General for Agriculture.

([1]) Including 'triticale'.
([2]) Calculated on intra-import basis.

(NC/100 kg)

4.1.5.1   Producer prices of certain cereals

1	2	Belgique/ Belgïe (BFR)	Danmark (DKR)	BR Deutschland (DM)	Elláda (DR)	España (PTA)	France (FF)	Ireland (IRL)	Italia (LIT)	Luxembourg (LFR)	Nederland (HFL)	Portugal (ESC)	United Kingdom (UKL)
		3	4	5	6	7	8	9	10	11	12	13	14
Common wheat	1985	791,3	152,46	42,05	1 947	2 595	110,83	9,11	31 301	770	45,65	4 050	11,18
	1991	653,3	128,01	32,91	3 986	2 765	106,63	11,47	32 425	614	36,15	4 399	11,66
	1992	648,4	131,03	33,60	4 187	2 675	99,85	10,92	31 404	604	35,85	3 979	12,17
% TAV	1991/1985	−3,1	−2,9	−4,0	12,7	1,1	−0,6	3,9	0,6	−3,7	−3,8	1,4	0,7
% TAV	1992/1991	−0,8	2,4	2,1	5,0	−3,3	−6,4	−4,8	−3,1	−1,6	−0,8	−9,5	4,4
Rye	1985	775,3	138,59	41,91	—	2 339	95,64	—	30 594	780	43,60	3 730	—
	1991	615,1	119,91	30,55	—	2 276	91,86	—	29 123	587	33,50	3 342	—
	1992	603,7	124,60	31,50	—	2 392	96,57	—	29 204	563	34,90	3 205	—
% TAV	1991/1985	−3,8	−2,4	−5,1	x	−0,5	−0,7	x	−0,8	−4,6	−4,3	−1,8	x
% TAV	1992/1991	−1,9	3,9	3,1	x	5,1	5,1	x	0,3	−4,1	4,2	−4,1	x
Barley	1985	759,9	143,24	39,85	1 919	2 174	104,09	9,25	30 380	740	45,90	3 650	10,66
	1991	599,5	121,96	29,59	3 902	2 255	93,28	10,81	30 491	519	35,35	3 342	11,23
	1992	600,9	126,55	30,14	4 099	2 188	87,71	10,24	29 025	533	38,85	3 205	11,82
% TAV	1991/1985	−3,9	−2,6	−4,8	12,6	0,6	−1,8	2,6	0,1	−5,7	−4,3	−1,5	0,9
% TAV	1992/1991	0,2	3,8	1,9	5,0	−3,0	−6,0	−5,3	−4,8	2,7	9,9	−4,1	5,3
Oats	1985	694,6	133,38	38,62	2 597	2 025	88,18	7,77	38 289	700	42,50	3 040	10,03
	1991	628,9	120,68	28,96	4 630	2 130	92,88	10,50	33 413	544	34,65	3 520	10,68
	1992	696,6	138,44	32,35	4 771	2 316	110,30	10,98	36 691	621	38,30	3 495	11,80
% TAV	1991/1985	−1,6	−1,7	−4,7	10,1	0,8	0,9	5,1	−2,2	−4,1	−3,3	2,5	1,1
% TAV	1992/1991	10,8	14,7	11,7	3,0	8,7	18,8	4,6	9,8	14,2	10,5	−0,7	10,5
Maize	1985	—	—	47,59	1 818	2 619	120,52	—	33 957	—	—	3 850	—
	1991	—	—	37,57	3 876	2 742	116,94	—	37 587	—	—	3 517	—
	1992	—	—	33,12	4 410	2 564	94,43	—	33163	—	—	3 372	—
% TAV	1991/1985	x	x	−3,9	13,4	0,8	−0,5	x	1,7	x	x	−1,5	x
% TAV	1992/1991	x	x	−11,8	13,8	−6,5	−19,2	x	−11,8	x	x	−4,1	x

Source : Eurostat.

4.1.5.4 Consumer price indices — bread and cereals
(in nominal and real terms)

1	1985 = 100			% TAV	
	1991	1992	1993	1992 / 1991	1993 / 1992
1	2	3	4	5	6
Nominal terms					
Belgique/België	117,1	120,7	123,4	3,1	2,2
Danmark	130,4	134,8	138,1	3,4	2,4
BR Deutschland	113,4	119,4	123,5	5,3	3,4
Elláda	261,4	315,3	360,3	20,6	14,3
España	163,1	178,5	190,8	9,4	6,9
France	122,7	126,8	129,6	3,3	2,2
Ireland	122,6	126,1	124,9	2,9	− 1,0
Italia	139,3	147,0	156,8	5,5	6,7
Luxembourg	122,6	124,3	126,1	1,4	1,4
Nederland	107,8	112,6	115,0	4,5	2,1
Portugal	203,4	229,5	239,4	12,8	4,3
United Kingdom	140,8	145,4	145,7	3,3	0,2
Real terms					
Belgique/België	98,7	97,5	97,1	− 1,2	− 0,4
Danmark	105,0	106,5	107,2	1,4	0,7
BR Deutschland	96,8	97,6	97,8	0,8	0,2
Elláda	105,4	110,6	111,1	4,9	0,5
España	106,7	109,7	112,1	2,8	2,2
France	100,6	101,6	105,5	1,0	3,8
Ireland	106,8	108,6	104,1	1,7	− 4,1
Italia	92,6	93,8	95,8	1,3	2,1
Luxembourg	102,1	99,0	98,2	− 3,0	− 0,8
Nederland	100,4	102,3	103,3	1,9	1,0
Portugal	91,9	91,4	88,8	− 0,5	− 2,8
United Kingdom	100,9	99,9	96,9	− 1,0	− 3,0

Sources: Eurostat and EC Commission, Directorate-General for Agriculture.

4.1.5.5 Cif Rotterdam prices for cereals

(ECU/t)

	Year	Months												Ø	% TAV compared with previous year
		I	II	III	IV	V	VI	VII	VIII	IX	X	XI	XII		
1	2	3	4	5	6	7	8	9	10	11	12	13	14	15	16
Common wheat	1991	47,45	44,85	59,16	72,38	84,94	87,49	77,14	73,72	73,69	81,72	79,94	76,55	71,73	− 20,4
	1992	80,20	90,68	97,89	88,40	83,12	91,36	85,15	77,12	81,32	86,26	97,33	83,45	86,86	21,3
	1993	89,62	93,51	93,08	89,96	92,49	85,27	90,46	93,21	88,82	92,63	101,91	91,73	91,89	5,8
Rye	1991	61,22	59,77	64,33	68,60	70,01	72,39	73,17	64,81	50,78	49,69	50,94	50,55	61,38	− 25,0
	1992	50,27	53,82	56,77	56,24	55,28	52,89	49,23	48,19	48,07	49,83	53,12	51,88	52,13	− 15,0
	1993	53,24	60,39	63,25	61,45	61,24	62,73	65,28	65,86	63,34	63,88	65,94	66,02	62,72	20,3
Barley	1991	67,50	57,06	68,45	72,99	76,45	76,21	70,77	67,81	70,80	74,11	73,19	72,79	70,76	− 18,7
	1992	72,86	74,85	77,68	77,29	74,81	73,38	77,31	75,73	75,53	80,55	85,42	83,59	77,42	9,5
	1993	85,57	86,72	82,75	76,98	75,27	76,67	75,32	74,86	66,24	56,89	59,30	61,79	73,2	− 5,5
Maize	1991	76,86	77,39	85,68	88,63	88,88	93,17	92,35	93,73	89,83	89,78	86,69	83,80	87,29	1,5
	1992	83,91	88,18	91,38	80,33	80,37	82,05	74,65	68,84	68,96	68,73	74,99	75,10	78,12	− 10,4
	1993	76,56	77,13	76,68	75,67	77,37	76,87	84,03	85,95	79,53	84,06	94,42	98,85	82,26	5,3

Source: EC Commission, Directorate-General for Agriculture.

4.1.6.2 Market prices for cereals as a percentage of the intervention price ([1])

		1993				
		VII	VIII	IX	X	XI
1	2	3	4	5	6	7
Common wheat of breadmaking quality	Belgique/België	:	102,51	:	:	:
	Danmark	:	:	95,12	94,70	102,84
	BR Deutschland	125,14	100,62	103,83	105,36	106,43
	Elláda	122,16	123,48	129,98	137,06	132,11
	España	129,04	120,46	125,48	113,39	121,13
	France	100,83	101,54	101,86	101,06	110,00
	Italia	125,09	129,10	131,38	131,20	131,16
	Nederland	:	101,37	103,01	103,93	104,80
	Portugal	111,36	102,46	104,96	102,33	:
	United Kingdom	157,56	120,69	122,47	121,74	118,73
Common feed wheat ([2])	Belgique/België	:	96,74	:	:	:
	BR Deutschland	121,77	97,93	98,73	96,96	96,79
	Nederland	:	99,52	100,21	100,98	101,53
	Portugal	120,01	97,47	98,67	:	:
	United Kingdom	122,56	100,83	98,81	98,36	96,46
Durum wheat	Elláda	123,06	131,13	129,42	134,86	117,14
	España	141,62	138,56	138,70	142,68	123,72
	France	133,79	140,01	140,78	143,26	143,35
	Italia	142,80	144,41	143,36	143,30	135,49
Barley ([3])	Belgique/België	:	96,00	:	:	:
	Danmark	:	:	96,81	98,92	99,11
	BR Deutschland	98,74	93,24	98,45	97,29	98,39
	Elláda	118,12	123,32	128,45	128,12	141,87
	España	90,96	87,69	83,51	92,92	101,38
	France	93,18	95,94	96,00	98,47	99,41
	Italia	114,12	114,86	119,93	118,50	114,64
	Nederland	:	99,44	100,84	101,81	103,86
	Portugal	101,06	95,66	101,11	:	:
	United Kingdom	100,02	93,56	97,03	94,66	95,89
Rye ([3])	Belgique/België	:	93,65	:	:	:
	Danmark	:	:	:	94,25	92,73
	BR Deutschland ([4])	106,66	99,47	104,12	109,08	107,34
	Portugal	:	90,25	91,36	91,36	:
Maize ([3])	Belgique/België	134,88	191,39	:	:	:
	BR Deutschland	102,37	112,63	77,07	100,77	99,72
	Elláda	118,36	174,25	91,55	120,94	122,35
	España	112,92	162,71	107,07	123,06	114,04
	France	95,35	146,71	92,93	100,31	103,47
	Italia	109,03	146,51	86,01	115,54	113,61
	Portugal	:	:	101,54	122,43	:

Source: EC Commission, Directorate-General for Agriculture.
([1]) Average prices at certain representative marketing centres adjusted to the standard quality.
([2]) Figures based on intervention price for common wheat of breadmaking quality reduced by 5%.
([3]) Feed grains.
([4]) Rye of breadmaking quality.

(%)

	1994					
XII	I	II	III	IV	V	VI
8	9	10	11	12	13	14
103,69	103,13	102,72	101,79	101,63	102,14	102,99
101,24	102,28	102,74	103,09	104,44	104,87	104,87
104,63	104,21	103,21	100,92	104,42	106,05	106,81
130,98	133,53	136,34	141,36	136,47	132,83	103,72
128,09	126,03	123,32	120,75	111,95	116,77	95,63
106,61	102,96	99,57	100,81	103,16	102,94	97,87
126,72	127,74	122,87	119,72	117,85	113,71	104,57
105,27	105,45	104,31	103,35	107,01	108,27	:
:	:	:	:	:	:	:
116,02	110,19	106,30	104,48	109,27	110,31	110,35
95,97	94,68	91,30	91,42	94,74	97,58	99,51
96,48	102,76	96,12	97,42	101,20	103,79	104,81
100,87	100,81	98,79	100,17	105,26	106,26	106,50
:	:	:	:		:	
92,57	94,68	93,51	95,39	100,94	104,49	104,75
110,43	113,76	108,25	110,68	107,29	102,54	82,74
137,22	129,02	125,72	119,83	119,10	112,88	92,62
133,95	136,46	136,85	:	:	:	:
137,11	136,81	134,34	125,67	118,53	105,53	96,15
99,40	98,27	97,21	96,57	98,38	97,46	96,44
99,28	100,25	99,44	98,80	98,17	99,33	99,33
98,46	100,39	101,39	100,50	101,93	103,25	100,09
139,97	136,81	134,55	132,92	130,41	:	95,80
102,17	100,53	98,37	95,17	95,23	95,01	78,38
102,39	100,92	99,72	100,66	100,13	97,40	84,88
122,61	115,71	116,45	114,40	112,30	111,88	98,94
105,15	105,62	104,74	104,21	105,71	105,29	100,61
:	:	:	:	:	:	:
98,42	98,44	97,42	97,90	98,46	100,22	101,71
:	90,29	87,92	88,15	92,74	93,90	96,11
92,50	94,64	93,87	93,77	96,47	100,22	76,74
106,55	108,15	106,07	105,69	105,78	102,66	94,08
:	:	:	:	:	:	:
156,29	153,91	154,65	154,20	152,57	151,19	155,18
101,67	106,82	102,68	104,07	108,17	111,04	108,88
121,56	125,78	120,19	121,34	119,62	116,39	108,35
117,99	117,49	115,71	112,97	114,69	115,11	111,02
104,06	102,76	102,76	104,62	108,78	108,79	109,16
115,09	115,63	106,98	105,19	103,07	105,93	104,19
:	:	:	:	:	:	:

4.1.6.3   Intervention stocks in the EC at the end of the marketing year

(1 000 t)

Products	1985/86	1990/91	1991/92 ∞	1992/93 ∞	1993/94 ∞
1	2	3	4	5	6
Common wheat	10 312	8 520	10 943	14 974	6 480
— common wheat of breadmaking quality	2 917	8 375	10 754	14 489	6 316
— common feed wheat	7 395	145	189	485	164
Rye	1 161	3 163	3 552	2 458	2 545
Barley	5 296	5 538	7 418	8 694	6 526
Durum wheat	887	1 528	4 168	3 392	1 152
Maize	392	1	301	3 670	1 130
Sorghum	454	–	0	151	160
Total	18 502	18 750	26 383	33 339	17 993

Source : EC Commission, Directorate-General for Agriculture.

4.2.1.1   Area, yield and production of rice (paddy)

	Area					Yield					Production				
	1 000 ha			% TAV		100 kg/ha			% TAV		1 000 t			% TAV	
1	1991	1992	1993	1992/1991	1993/1992	1991	1992	1993	1992/1991	1993/1992	1991	1992	1993	1992/1991	1993/1992
	2	3	4	5	6	7	8	9	10	11	12	13	14	15	16
EUR 12	367	353	336	– 3,8	– 4,8	61,3	61,7	58,9	0,7	– 4,5	2 225	2 173	1 979	– 2,3	– 8,9
Elláda	15	15	19	0,0	26,7	59,7	75,4	72,5	26,3	– 3,8	89	110	137	23,6	24,5
España	94	83	48	– 11,7	– 42,2	58,0	67,6	64,6	16,6	– 4,4	582	528	310	– 9,3	– 41,3
France	20	22	24	10,0	9,1	61,0	55,5	52,0	– 9,0	– 6,3	116	122	125	5,2	2,5
Italia	206	216	232	4,9	7,4	63,6	59,8	57,4	– 6,0	– 4,0	1 278	1 228	1 331	– 3,9	8,4
Portugal	32	17	13	– 46,9	– 23,5	56,0	58,5	56,3	4,5	– 3,8	160	97	76	– 39,4	– 21,6

Source : Eurostat and reports from Member States.

4.2.4.1   Supply balance — rice ([1])

EUR 12

1	1 000 t wholly milled rice			% TAV	
	1985/86	1991/92	1992/93	1991/92 ÷ 1985/86	1992/93 ÷ 1991/92
1	2	3	4	5	6
Usable production	1 115	1 721	1 505	7,5	− 12,6
Changes in stock	16	− 147	− 54	×	− 63,3
Imports	1 031	329	374	− 17,3	13,7
Exports	489	375	190	− 4,3	− 49,3
Intra-Community trade ([2])	689	699	704	0,2	0,7
Internal use of which :	1 641	1 822	1 743	1,8	− 4,3
— animal feed	116	120	120	0,6	0,0
— seed	47	47	47	0,0	0,0
— industrial use	35	110	74	21,0	− 32,7
— losses (market)	10	25	30	16,5	20,0
— gross human consumption	1 433	1 520	1 472	1,0	− 3,2
Self-sufficiency (%)	67,9	94,4	86,3	5,6	− 8,6

Source : Eurostat.

([1]) Broken rice included.
([2]) Calculated on intra-import basis.

4.2.5.1   Cif Rotterdam prices (¹) for husked rice

(ECU/t)

1	IX	X	XI	XII	I	II	III	IV	V	VI	VII	VIII	Ø	% TAV compared with previous year
	2	3	4	5	6	7	8	9	10	11	12	13	14	15
Round-grain rice (²)														
1985/86	219,0	205,0	200,4	192,9	189,5	183,0	172,8	174,9	165,4	170,0	165,0	158,5	183,0	− 34,4
1986/87	156,5	162,6	174,1	171,5	158,6	143,9	139,2	134,3	132,2	133,3	134,2	136,6	148,3	− 19,0
1987/88	133,0	133,4	126,9	157,8	158,1	167,4	167,8	166,8	167,1	170,5	180,4	186,1	159,7	7,6
1988/89	186,2	185,1	176,6	174,4	183,3	186,3	184,6	187,5	189,3	198,1	193,3	190,1	186,1	16,5
1989/90	191,8	188,2	184,9	178,1	169,7	167,7	167,7	167,2	164,5	165,4	163,9	157,8	172,2	− 7 5
1990/91	153,1	151,5	147,6	146,4	149,9	146,0	152,1	165,5	169,3	171,6	178,1	172,1	158,6	− 7,9
1991/92	170,2	166,6	166,0	163,5	160,9	163,7	166,2	165,8	164,7	162,9	159,4	157,3	163,9	3,3
1992/93	156,5	158,3	163,8	162,5	164,4	165,5	165,9	163,7	162,6	164,9	168,1	169,0	163,8	− 0,1
1993/94	163,4	164,5	169,9	170,6	170,1	170,2	168,7	168,3	166,1	165,8	162,2	162,7	166,9	1,9
Long-grain rice (³)														
1985/86	230,8	213,8	213,5	206,7	194,1	180,2	168,7	161,3	146,1	144,7	137,4	134,1	177,7	− 35,0
1986/87	130,4	122,3	124,8	122,5	112,3	101,5	100,7	99,0	99,6	111,2	114,3	121,1	113,6	− 36,1
1987/88	139,3	187,8	175,2	169,2	182,5	209,6	207,7	201,3	193,2	198,1	220,6	220,1	192,1	69,1
1988/89	205,9	199,6	182,2	175,5	181,9	185,2	183,1	193,6	218,0	239,2	247,5	248,4	205,2	6,8
1989/90	231,7	220,3	190,7	175,8	167,9	189,1	191,5	181,9	175,8	180,3	177,4	166,8	187,4	− 8,7
1990/91	147,7	141,2	134,7	137,2	147,4	159,4	178,4	191,7	198,9	213,5	226,3	214,4	174,2	− 7,0
1991/92	207,7	197,6	187,7	181,8	186,4	192,3	193,0	185,8	182,3	176,0	169,2	168,0	185,7	6,6
1992/93	170,5	176,3	181,2	174,2	171,6	166,9	165,6	160,8	156,2	156,5	162,8	172,9	168,0	− 9,5
1993/94	150,6	155,8	205,1	219,6	212,0	222,0	208,3	195,5	190,4	185,5	173,6	169,0	190,6	13,5

Source: EC Commission, Directorate-General for Agriculture.

(¹) Monthly averages.
(²) Round-grain rice of standard quality.
(³) Rice equivalent to Community-produced long-grain standard (Ribe).

4.2.6.1 Average market prices (¹) for paddy rice in surplus areas (²) compared with intervention prices

Month	Italy						España	
	Balilla round-grain rice Community origin		Ribe long-grain rice		Lido medium-grain rice		Bahia rice	
	LIT/100 kg	% of intervention price	LIT/100 kg	% of intervention price	LIT/100 kg	% of intervention price	PTA/100 kg	% of intervention price
1	2	3	4	5	6	7	8	9
IX. 1992	49 000	86,53	55 500	98,18	47 363	83,74	4 260	89,85
X.	49 225	86,89	51 416	90,80	49 919	88,15	4 158	85,97
XI.	50 833	85,25	53 283	89,41	51 050	89,82	4 645	96,00
XII.	52 693	87,27	54 556	90,15	52 217	85,87	4 833	99,16
I. 1993	59 145	88,58	59 830	89,77	57 492	86,12	5 008	96,06
II.	60 785	88,78	61 142	89,30	60 089	87,76	5 219	98,90
III.	63 548	90,22	66 072	93,81	64 733	91,90	5 392	101,50
IV.	71 066	96,88	75 016	102,27	70 966	96,75	5 510	102,72
V.	75 000	104,54	84 588	117,91	74 750	104,19	5 616	100,12
VI.	75 000	104,83	84 750	118,46	74 750	104,48	5 903	99,15
VII.	:	—	:	—	:	—	5 559	92,59
VIII.	:	—	:	—	:	—	5 200	83,23
IX.	59 750	89,08	63 500	94,66	59 541	88,76	5 518	93,61
X.	65 787	96,62	66 875	99,88	65 784	96,61	6 163	104,56
XI.	69 330	100,73	71 100	103,30	69 643	101,18	6 404	108,65
XII.	74 900	108,82	75 100	109,19	71 100	103,84	6 540	110,95
I. 1994	74 516	105,48	75 379	106,70	73 193	103,6	6 867	114,93
II.	72 975	102,24	73 178	102,52	72 428	101,47	7 226	119,75
III.	72 854	101,4	72 669	101,14	72 000	100,21	7 140	117,55
IV.	69 873	96,62	69 800	96,52	69 900	97,29	7 309	119,57
V.	65 800	90,4	66 653	91,57	66 903	91,91	7 332	119,15
VI.	65 800	89,82	64 905	88,59	65 483	89,38	7 077	114,26
VII.	66 476	90,16	64 950	88,09	65 700	89,11	7 040	112,94
VIII.	68 700	93,17	67 250	91,21	68 000	92,23	:	—

Source : Camera di commercio di Vercelli.

(¹) Monthly averages.
(²) There are no regular market prices for paddy rice in France, as rice is usually sold in its husked form, for which no intervention price is quoted.

4.2.6.2   Aid scheme for producers of certain arable crops (Regulation (EEC) No 1765/92) (¹)

Marketing year 1993/94

EUR 12

Crop	No	General scheme						Simplified scheme					
		Yield (t/ha)			Area (ha)			Yield (t/ha)			Area (ha)		
		Total number of applications: 393 362						Total number of applications: 1 319 314					
		Total	Not irrigated	Irrigated (²)	Total	Not irrigated	Irrigated (²)	Total	Not irrigated	Irrigated (²)	Total	Not irrigated	Irrigated (²)
1	2	3	4	5	6	7	8	9	10	11	12	13	14
Durum wheat in traditional areas	1	2,49	2,39	5,09	792 235,70	762 548,50	29 687,20	2,26	2,22	6,49	1 722 566,20	1 708 151,20	14 415,00
Maize in separate yield areas	2	7,79	7,84	7,66	837 101,83	607 507,83	229 594,00	6,01	6,04	5,58	1 285 082,46	1 210 469,46	74 613,00
Other cereals	3	4,78	4,76	5,17	17 575 534,51	16 854 983,61	720 550,90	4,24	4,09	6,97	8 527 113,14	8 093 629,05	433 484,09
Total for all cereals	4	4,82	4,77	5,75	19 204 872,04	18 225 039,94	979 832,10	4,14	4,02	6,76	11 534 761,80	11 012 249,71	522 512,09
of which silage	5	0,00	0,00	0,00	802 763,00	775 313,00	27 450,00	0,00	0,00	0,00	608 089,00	606 089,00	0,00
Soya	6	7,69	8,12	6,04	206 561,30	163 861,30	42 700,00	3,99	3,96	6,38	12 084,79	11 975,39	109,40
Rape	7	5,94	5,94	5,52	2 040 367,55	2 038 433,55	1 934,00	5,27	5,27	5,51	12 901,67	12 815,67	86,00
Sunflower	8	3,47	2,80	6,54	2 709 623,75	2 225 365,75	484 258,00	3,38	2,37	6,21	485 854,25	357 713,15	128 141,10
Total for oilseeds	9	4,67	4,45	6,49	4 956 552,60	4 427 660,60	528 892,00	3,45	2,52	6,21	510 851,04	382 514,23	128 336,81
Total for protein plants	10	5,54	5,57	5,16	1 270 257,83	1 170 890,83	99 367,00	4,06	4,02	5,13	47 400,05	45 610,05	1 790,00
Total for flax other than fibre flax	11	5,81	5,81	0,00	204 189,83	204 189,83	0,00	4,52	4,52	0,00	80,00	80,00	0,00
Rotational set-aside	12	–	–	–	4 604 963,28	4 472 637,68	132 325,60	–	–	–	0,00	0,00	0,00
of which non-food	13	–	–	–	253 142,51	252 136,51	1 006,00	–	–	–	0,00	0,00	0,00
Non-rotational set-aside	14	–	–	–	0,00	0,00	0,00	–	–	–	0,00	0,00	0,00
of which non-food	15	–	–	–	0,00	0,00	0,00	–	–	–	0,00	0,00	0,00
Five-year set-aside (Regulation (EEC) No 2328/91)	16	–	–	–	1 655 000,00	–	–	–	–	–	0,00	0,00	0,00
Total for set-aside (³)	17	–	–	–	4 604 963,28	4 472 637,68	132 325,60	–	–	–	0,00	0,00	0,00
Arable crops declared as feed areas for cattle and sheep premiums	18	–	–	–	556 933,95	556 117,95	816,00	–	–	–	461 948,40	459 384,40	2 564,00
Peas (Article 3 of Regulation (EEC) No 3738/92)	19	–	–	–	75 116,00	74 361,00	755,00	–	–	–	3 316,00	2 931,00	385,00
Total (⁴)	20	–	–	–	30 240 835,58	28 500 418,88	1 740 416,70	–	–	–	12 093 092,89	11 440 453,99	652 638,90

Source: EC Commission, Directorate-General for Agriculture.

(¹) Provisional data.
(²) Including 'regadio'.
(³) Five-year set-aside excluded.
(⁴) Five-year set-aside area excluded as well as arable crops declared as feed areas.

4.3.1.1 Area under sugarbeet, (¹) yield (²) and production (²) of sugar

	Area					Yield					Production				
	1 000 ha			% TAV		t/ha			% TAV		1 000 t			% TAV	
1	1985/86	1993/94 ∞	1994/95 ∞	1993/94 / 1985/86	1994/95 / 1993/94	1985/86	1993/94 ∞	1994/95 ∞	1993/94 / 1985/86	1994/95 / 1993/94	1985/86	1993/94 ∞	1994/95 ∞	1993/94 / 1985/86	1994/95 / 1993/94
	2	3	4	5	6	7	8	9	10	11	12	13	14	15	16
EUR 12	1 886	1 921	1 892	0,2	– 1,5	7,05	8,32	7,57	2,1	– 9,0	13 626	16 235	14 644	2,2	– 9,8
Belgique/België	125	104	102	– 2,3	– 1,9	7,55	10,03	8,33	3,6	– 16,9	944	1 043	850	1,3	– 18,5
Danmark	73	67	67	– 1,1	0,0	7,26	7,78	7,36	0,9	– 5,4	530	521	493	– 0,2	– 5,4
BR Deutschland (³)	415	530	507	3,1	– 4,3	7,56	8,18	7,2	1,0	– 12,0	3 155	4 352	3 674	4,1	– 15,6
Elláda	43	45	40	0,6	– 11,1	7,37	6,82	7,25	– 1,0	6,3	317	307	290	– 0,4	– 5,5
España (⁵)	178	178	170	0,0	– 4,5	4,99	6,85	5,88	4,0	– 14,2	900	1 213	1 015	3,8	– 16,3
France (⁴)	464	420	405	– 1,2	– 3,6	8,52	10,45	9,94	2,6	– 4,9	4 249	4 633	4 305	1,1	– 7,1
Ireland	34	32	36	– 0,8	12,5	5,12	5,53	5,97	1,0	8,0	174	177	215	0,2	21,5
Italia	221	260	280	2,1	7,7	5,63	5,48	5,54	– 0,3	1,1	1 244	1 419	1 550	1,7	9,2
Nederland	130	116	115	– 1,4	– 0,9	7,08	9,76	8,69	4,1	– 11,0	897	1 133	1 000	3,0	– 11,7
Portugal (⁵)	1	0	0	x	x	–	–	–	x	x	6	3	2	– 8,3	– 33,3
United Kingdom	202	169	170	– 2,2	0,6	6,00	8,49	7,35	4,4	– 13,4	1 210	1 434	1 250	2,1	– 12,8

Source : EC Commission, Directorate-General for Agriculture.

(¹) Area planted with sugarbeet exclusive of area planted for distillery supply.
(²) In terms of white-sugar value.
(³) Including production of molasses.
(⁴) Area and yield, metropolitan France only; production, including the French overseas departments.
(⁵) Including production of sugar from sugar cane.

4.3.2.1   World production of sugar and production of the main producing and/or exporting countries

	Raw sugar						% TAV	
	%			1 000 t				
	1985	1992	1993	1985	1992	1993	1992/1985	1993/1992
1	2	3	4	5	6	7	8	9
World	100,0	100,0	100,0	98 155	116 901	113 616	2,5	− 2,8
of which :								
Europe								
EUR 12	15,0	14,6∞	15,4∞	14 680	17 102∞	17 500∞	2,2	2,3
America								
USA	5,5	5,8	6,1	5 415	6 805	6 875	3,3	1,0
Cuba	8,0	6,2	4,0	7 889	7 219	4 550	− 1,3	− 37,0
Dominican Rep.	0,9	0,5	0,5	921	593	610	− 6,1	2,9
Mexico	3,6	3,2	3,6	3 492	3 745	4 100	1,0	9,5
Argentina	1,2	1,2	1,0	1 188	1 379	1 175	2,2	− 14,8
Brazil	8,6	8,5	8,2	8 455	9 925	9 375	2,3	− 5,5
Asia								
India	7,2	11,9	9,6	7 016	13 873	10 900	10,2	− 21,4
Peop. Rep. China	4,9	7,6	6,6	4 800	8 864	7 450	9,2	− 16,0
Pakistan	1,4	2,2	2,8	1 410	2 530	3 175	8,7	25,5
Philippines	1,7	1,6	1,8	1 665	1 919	2 025	2,0	5,5
Thailand	2,4	4,3	3,5	2 393	5 078	3 975	11,3	− 21,7
Africa								
South Africa	2,6	1,5	1,2	2 540	1 715	1 325	− 5,5	− 22,7
Oceania								
Australia	3,5	3,7	4,0	3 439	4 363	4 540	3,5	4,1

Source : Statistical Bulletin of the International Sugar Organization (ISO).

4.3.3.1 World supply balance and international trade in sugar

	1 000 t raw sugar			% TAV	
	1985/86	1992/93 ∞	1993/94 ∞	1992/93 / 1985/86	1993/94 / 1992/93
1	2	3	4	5	6
(I) *Supply balance (marketing year Sept./August)*					
Initial stock	41 406	39 549	39 098	− 0,7	− 1,1
Production	98 578	112 608	109 608	1,9	− 2,7
Imports	32 442	31 212	30 184	− 0,6	− 3,3
Availability	172 426	183 369	178 890	0,9	− 2,4
Exports	33 430	32 071	31 034	− 0,6	− 3,2
Consumption	100 467	112 200	112 955	1,6	0,7
Final stock	38 529	39 098	34 901	0,2	− 10,7
of which : as % of consumption	38,4	34,9	30,9	− 1,4	− 11,5
	1985	1992 ∞	1993 ∞	1992 / 1985	1993 / 1992
(II) *International trade*					
Imports/world	26 565	29 471	26 870	1,5	− 8,8
of which : EUR 12	1 946	1 786	1 680	− 1,2	− 5,9
%	7,3	6,1	6,3	− 2,5	3,3
Exports/world	27 750	30 362	28 097	1,3	− 7,5
of which : EUR 12	4 280	4 983	5 820	2,2	16,8
%	15,4	16,4	20,7	0,9	26,2

Sources : (I) FO Licht — European Sugar Journal (for the supply balance). (II) International Sugar Organization (for international trade).

4.3.4.1 Sugar supply balance
(October/September)

1	1 000 t white sugar			% TAV	
	1986/87	1992/93 ∞	1993/94 ∞	1992/93 / 1986/87	1993/94 / 1992/93
	2	3	4	5	6
Total production	14 096	16 012	16 235	2,1	1,4
of which: C sugar production for export	1 312	2 265	2 940	9,5	29,8
Usable production (¹)	12 784	13 747	13 295	1,2	− 3,3
Change in stocks	89	272	− 38	20,5	×
Imports (²)	1 769	1 977	2 014	1,9	1,9
Exports (¹) (²)	3 557	3 545	3 472	− 0,1	− 2,1
of which: Intra-Community trade	818	1 240	1 300	7,2	4,8
Internal use					
of which:	10 907	11 907	11 875	1,5	− 0,3
— animal feed	11	10	11	− 1,6	10,0
— industrial use	170	174	215	0,4	23,6
— human consumption	10 726	11 723	11 649	1,5	− 0,6
Human consumption (kg/head) (³)	33,2	:	:	×	×
Self-sufficiency (%) (¹)	117,2	115,5	112,0	− 0,2	− 3,0

Source: EC Commission, Directorate-General for Agriculture.

(¹) Excl. C sugar.
(²) Excl. sugar traded for processing.
(³) Ratio of human consumption to resident population at 1 January.

4.3.5.1 Average world sugar prices (¹)

1	ECU/100 kg			% TAV (⁵)	
	1985/86	1992/93	1993/94	1992/93 / 1985/86	1993/94 / 1992/93
	2	3	4	5	6
Paris Exchange (²)	19,88	21,91	26,79	2,3	10,4
London Exchange (³)	16,40	19,82	23,59	7,1	5,9
New York Exchange (⁴)	14,03	16,95	20,37	7,1	15,4

Source: EC Commission, Directorate-General for Agriculture.

(¹) Arithmetic mean of spot prices (June/July).
(²) White sugar, loaded fob designated European ports, in new bags.
(³) Raw sugar, 96°, cif — United Kingdom, ex. hold.
(⁴) Raw sugar, 96°, loaded fob Caribbean — Contract No 11.
(⁵) Calculated on the basis of prices in national currencies.

4.3.5.2   Consumer prices for refined sugar

| | 1985 | 1991 | 1992 | % TAV | |
| | | | | 1991 / 1985 | 1992 / 1991 |
1	2	3	4	5	6
Belgique/België	0,93	0,92	0,94	0,0	2,2
Danmark	1,73	1,13	:	− 6,9	×
BR Deutschland	0,87	0,92	0,96	0,9	4,3
Elláda	0,60	0,76	0,81	4,0	6,6
España	:	:	:	:	×
France	0,86	0,94	1,03	1,5	9,6
Ireland	0,86	1,06	1,08	3,5	1,9
Italia	0,90	1,05	1,03	2,6	− 0,2
Nederland	0,93	0,88	0,90	− 0,1	2,3
Portugal	:	:	:	×	×
United Kingdom	0,81	0,94	0,88	2,5	− 0,6

Source : Eurostat.

4.3.6.1 Sugar and isoglucose production, by quota

	Sugar (1 000 t white sugar)								Isoglucose (1 000 t dry matter)				
	Basic quantity		Carry-over and production (p)						Basic quantity		Production (p)		
	A sugar	B sugar	Quantity of sugar carried over from 1992/93	1993/94 crop	Production of A sugar	Production of B sugar not carried over	Production of C sugar not carried over	Quantity of sugar carried over into 1994/95	Iso-glucose A	Iso-glucose B	Total	of which: A+B	Isoglucose C
1	2	3	4	5	6	7	8	9	10	11	12	13	14
EUR 12 ∞	11 187	2 482	1 314	16 258	10 935	2 422	2 955	1 258	241	49	289	288	–
Belgique/België	680	146	64	1 043	680	146	203	78	57	15	72	72	–
Danmark	328	97	–	521	328	97	96	–	–	–	–	–	–
BR Deutschland ∞	2 637	812	195	4 354	2 638	812	887	211	29	7	36	36	–
Elláda	290	29	35	307	290	29	–	22	11	2	12	12	–
España	960	40	28	1 234	960	40	81	181	75	8	83	83	–
France (¹)	2 996	800	432	4 633	2 803	759	1 082	421	16	4	20	20	–
Ireland	182	18	10	177	182	5	–	–	–	–	–	–	–
Italia	1 320	248	262	1 419	1 320	248	–	113	16	4	20	20	–
Luxembourg	–	–	–	–	–	–	–	–	–	–	–	–	–
Nederland	690	182	80	1 133	690	182	317	24	7	2	9	9	–
Portugal	64	6	–	4	4	–	–	–	8	2	10	9	–
United Kingdom	1 040	104	208	1 433	1 040	104	289	208	22	5	27	27	–

Source: EC Commission, Directorate-General for Agriculture.

(¹) Incl. French overseas departments.

4.4.1.1   Area, yield and production of: (a) rapeseed, (b) sunflower seed and (c) soya beans

		Area 1 000 ha			% TAV		Yield 100 kg/ha			% TAV		Production 1 000 t			% TAV	
		1985	1992∞	1993∞	1992/1985	1993/1992	1985	1992∞	1993∞	1992/1985	1993/1992	1985	1992∞	1993∞	1992/1985	1993/1992
1	2	3	4	5	6	7	8	9	10	11	12	13	14	15	16	17
	EUR 12	1 287	2 380	2 227	9,2	−6,4	29,0	26,1	28,5	−1,5	9,2	3 738	6 269	6 297	7,7	0,4
Rapeseed	BLEU/UEBL	2	8	8	21,9	0,0	25,4	30,0	30,0	2,4	0,0	6	23	23	21,2	0,0
	Danmark	218	189	161	−2,0	−14,8	25,0	23,8	25,4	−0,7	6,7	544	450	410	−2,7	−8,9
	BR Deutschland	266	1 050	1 061	21,7	1,0	30,2	25,9	28,3	−2,2	9,3	803	2 728	3 002	19,1	10,0
	España	10	9	10	−1,5	11,1	12,2	15,1	12,3	3,1	−18,5	12	13	12	1,2	−7,7
	France	474	686	558	5,4	−18,7	29,9	27,0	28,5	−1,4	5,6	1 419	1 853	1 571	3,9	−15,2
	Ireland	4	5	2	3,2	−60,0	31,1	33,0	33,0	0,9	0,0	14	16	7	1,9	−56,3
	Italia	6	9	4	6,0	−55,6	21,4	25,0	20,0	2,2	−20,0	13	22	8	7,8	−63,6
	Nederland	10	4	2	−12,3	−50,0	30,3	30,0	33,0	−0,1	10,0	31	13	7	−11,7	−46,2
	United Kingdom	296	420	421	5,1	0,2	30,2	27,6	29,8	−1,3	8,0	895	1 159	1 256	3,8	8,4
	EUR 12	1 813	2 712	3 173	5,9	17,0	14,9	14,6	10,8	−0,3	−26,0	2 703	3 947	3 540	5,6	−10,3
Sunflower seed	BR Deutschland	0	39	88	x	125,6	−	24,6	26,2	x	6,5	0	183	232	x	26,8
	Elláda	50	27	17	−8,4	−37,0	17,0	16,3	13,2	−0,6	−19,0	85	44	23	−9,0	−47,7
	España	989	1 456	2 098	5,7	44,1	9,3	9,3	5,9	0,0	−36,6	915	1 343	1 218	5,6	−9,3
	France	639	986	760	6,4	−22,9	23,7	21,6	20,9	−1,3	−3,2	1 513	2 129	1 643	5,0	−22,8
	Italia	95	127	115	4,2	−9,4	17,2	20,5	22,0	2,5	7,3	162	260	249	7,0	−4,2
	Portugal	40	77	98	9,8	27,3	6,3	8,0	8,0	3,5	0,0	28	61	77	11,8	26,2
	EUR 12	123	421	219	19,2	−48,0	28,2	28,0	30,9	0,1	10,4	347	1 231	691	19,8	−43,9
Soya beans	BR Deutschland	0	1	1	x	0,0	−	34,9	34,5	x	−1,1	0	3	2	x	−33,3
	Elláda	0	1	0	x	x	−	29,5	−	x	x	0	3	0	x	x
	España	2	7	11	19,6	57,1	22,5	20,2	20,0	1,5	−1,0	5	33	3	30,9	−90,9
	France	27	41	50	6,1	22,0	20,7	16,1	24,6	−3,5	52,8	56	66	140	2,4	112,1
	Italia	94	371	157	21,7	−57,7	30,5	29,6	33,7	−0,4	13,9	286	1 099	536	21,2	−51,2

Source: EC Commission, Directorate-General for Agriculture.

4.4.3.1   Internal and external trade: (a) rapeseed, (b) sunflower seed, (c) soya beans and (d) flax seed

(1 000 t)

			EUR 12	BLEU/UEBL	Danmark	BR Deutschland	Elláda	España	France	Ireland	Italia	Nederland	Portugal	United Kingdom
1	2	3	4	5	6	7	8	9	10	11	12	13	14	15
Intra-EC trade (¹)	Rapeseed	1991	1 864	712	3	544	1	6	146	1	12	350	:	88
		1992	1 885	660	69	545	0	21	57	0	0	345	6	179
		1993	:	:	:	:	:	:	:	:	:	:	:	:
	Sunflower seed	1991	981	174	2	214	0	10	2	0	43	273	223	40
		1992	1 226	230	1	231	0	60	21	0	40	326	232	85
		1993	:	:	:	:	:	:	:	:	:	:	:	:
	Soya beans	1991	435	77	2	171	26	50	78	2	1	1	7	20
		1992	511	171	1	196	7	15	53	11	1	19	11	25
		1993	:	:	:	:	:	:	:	:	:	:	:	:
	Flax seed	1991	115	41	2	43	0	1	8	0	1	18	0	1
		1992	145	43	2	56	0	—	8	0	0	33	0	—
		1993	:	:	:	:	:	:	:	:	:	:	:	:
Imports	Rapeseed	1991	301	41	0	124	0	3	4	0	37	39	0	53
		1992	155	6	0	40	0	3	22	0	15	35	0	33
		1993	469	95	9	136	0	7	103	0	11	55	3	53
	Sunflower seed	1991	428	31	5	66	20	36	4	0	156	101	3	6
		1992	644	17	8	90	27	66	9	0	249	120	40	18
		1993	840	8	7	182	27	101	22	0	153	266	68	8
	Soya beans	1991	12 748	983	31	2 722	223	2 366	261	11	713	3 677	877	610
		1992	14 214	1 199	75	3 012	320	2 541	505	10	1 068	4 292	571	622
		1993	12 833	1 034	59	2 943	309	2 104	478	13	1 460	3 226	577	630
	Flax seed	1991	224	60	1	140	4	0	0	0	4	3	0	11
		1992	239	93	2	120	0	0	0	0	6	12	0	5
		1993	245	126	1	106	0	0	0	0	4	5	0	1
Exports	Rapeseed	1991	1	0	0	0	0	0	0	0	0	0	0	1
		1992	273	17	73	81	0	0	72	0	0	0	0	46
		1993	553	—	45	173	0	0	294	0	3	3	0	20
	Sunflower seed	1991	3	0	1	1	0	0	0	0	0	1	0	0
		1992	49	0	2	1	0	44	0	0	0	1	0	0
		1993	14	0	2	1	0	4	5	0	0	1	0	0
	Soya beans	1991	22	0	0	2	0	0	0	0	1	19	0	0
		1992	16	0	0	2	0	0	0	0	0	13	0	0
		1993	18	2	2	3	3	0	0	0	0	8	0	0
	Flax seed	1991	4	1	0	1	0	0	0	0	0	1	0	1
		1992	3	1	0	0	0	0	0	0	0	1	0	0
		1993	4	1	0	1	0	0	0	0	0	1	0	0

Source: Eurostat.
(¹) Based on quantities entering.

4.4.4.1 **Supplies of rape and colza** (seed, oil, cake) EUR 12
(July/June)

	1 000 t			% TAV	
	1986/87	1992/93 ∞	1993/94 ∞	$\dfrac{1992/93}{1986/87}$	$\dfrac{1993/94}{1992/93}$
1	2	3	4	5	6
Seed					
EC production	3 682	6 209	6 297	9,1	1,4
Imports (extra-EC)	569	375	528	− 6,7	40,8
Exports (extra-EC)	0	602	533	×	− 11,5
Change in stocks	100	− 19	− 38	×	×
Availabilities	4 151	6 001	6 330	6,3	5,5
Self-sufficiency (%)	89	103	100	2,5	− 2,9
Oil and oil equivalent					
EC total production:					
— from Community seed	1 436	2 486	2 517	9,6	1,2
— from imported seed	222	150	211	− 6,3	40,7
Imports (extra-EC)	30	27	29	− 1,7	7,4
Exports (extra-EC)	441	538	664	3,4	23,4
Change in stocks	134	29	− 25	− 22,5	×
Availabilities	1 113	2 096	2 118	11,1	1,0
Self-sufficiency (%)	129	118	118	− 1,5	0,0
Cake and cake equivalent					
EC total production					
— from Community seed	2 062	3 477	3 524	9,1	1,4
— from imported seed	319	210	296	− 6,7	41,0
Imports (extra-EC)	836	927	996	1,7	7,4
Exports (extra-EC)	48	25	63	− 10,3	152,0
Change in stocks	98	5	− 2	− 39,1	×
Availabilities	3 071	4 584	4 755	6,9	3,7
Self-sufficiency (%)	67	76	74	2,1	− 2,6

Source: Eurostat and EC Commission, Directorate-General for Agriculture.

4.4.4.2   **Supplies of sunflower** (seed, oil, cake)   **EUR 12**
(July/June)

	1 000 t			% TAV	
	1986/87	1992/93 ∞	1993/94 ∞	$\dfrac{1992/93}{1986/87}$	$\dfrac{1993/94}{1992/93}$
1	2	3	4	5	6
Seed					
EC production	3 160	4 089	3 173	4,4	− 22,4
Imports (extra-EC)	291	786	798	18,0	1,5
Exports (extra-EC)	3	54	13	61,9	− 75,9
Change in stocks	− 14	102	− 161	×	×
Availabilities	3 462	4 719	4 119	5,3	− 12,7
Self-sufficiency (%)	89	87	79	− 0,4	− 9,2
Oil and oil equivalent					
EC total production :					
— from Community seed	1 327	1 717	1 338	4,4	− 22,4
— from imported seed	122	330	335	18,0	1,5
Imports (extra-EC)	154	129	144	− 2,9	11,6
Exports (extra-EC)	115	164	123	6,1	− 25,0
Change in stocks	86	− 38	− 27	×	×
Availabilities	1 402	2 050	1 721	6,5	− 16,0
Self-sufficiency (%)	95	84	79	− 2,0	− 6,0
Cake and cake equivalent					
EC total production					
— from Community seed	1 738	2 290	1 783	4,7	− 22,1
— from imported seed	160	440	447	18,4	1,6
Imports (extra-EC)	1 283	1 053	833	− 3,2	− 20,9
Exports (extra-EC)	12	6	14	− 10,9	133,3
Change in stocks	− 15	− 1	1	×	×
Availabilities	3 182	3 778	3 048	2,9	− 19,3
Self-sufficiency (%)	55	60	61	1,5	1,7

Source: Eurostat and EC Commission, Directorate-General for Agriculture.

4.4.4.3 **Supplies of soya** (seed, oil, cake)   **EUR 12**
         (July/June)

	1 000 t			% TAV	
	1986/87	1992/93 ∞	1993/94 ∞	1992/93 / 1986/87	1993/94 / 1992/93
1	2	3	4	5	6
Seed					
EC production	905	1 231	681	5,3	− 44,7
Imports (extra-EC)	12 949	14 799	12 811	2,3	− 13,4
Exports (extra-EC)	9	14	10	7,6	− 28,6
Change in stocks	0	183	− 529	×	×
Availabilities	13 845	15 861	14 011	2,3	− 11,7
Self-sufficiency (%)	7	8	5	2,3	− 37,5
Oil and oil equivalent					
EC total production:					
— from Community seed	158	222	122	5,8	− 45,0
— from imported seed	2 266	2 664	2 306	2,7	− 13,4
Imports (extra-EC)	5	4	3	− 3,7	− 25,0
Exports (extra-EC)	782	644	499	− 3,2	− 22,5
Change in stocks	0	1	− 32	×	×
Availabilities	1 647	2 245	1 964	5,3	− 12,5
Self-sufficiency (%)	10	10	6	0,0	− 40,0
Cake and cake equivalent					
EC total production					
— from Community seed	724	960	531	4,8	− 44,7
— from imported seed	10 359	11 543	9 993	1,8	− 13,4
Imports (extra-EC)	10 906	10 902	11 249	− 0,0	3,2
Exports (extra-EC)	960	1 084	1 054	2,0	− 2,8
Change in stocks	0	12	− 2	×	×
Availabilities	21 029	22 309	20 721	1,0	− 7,1
Self-sufficiency (%)	3	4	3	4,9	− 25,0

Source: Eurostat and EC Commission, Directorate-General for Agriculture.

4.4.4.4 **Supplies of olive oil**   **EUR 12**

	1 000 t			% TAV	
	1986/87	1991/92 ∞	1992/93 ∞	1991/92 / 1986/87	1992/93 / 1991/92
1	2	3	4	5	6
EC production	1 204	1 729	1 379	7,5	− 20,2
Oil imports	20	40	57	14,9	42,5
Intra-EC trade	490	328	357	− 7,7	8,8
Oil exports	154	162	162	1,0	0,0
Intra-EC trade	457	308	359	− 7,6	16,6
Change in stocks	− 288	135	− 215	×	×
Internal use	1 409	1 472	1 489	0,9	1,2
of which:					
— industrial use	26	25	26	− 0,8	4,0
— human consumption	1 383	1 447	1 463	0,9	1,1
Human consumption (kg/head) [1]	4,3	4,4	4,2	0,5	− 4,5
Self-sufficiency (%) [2]	85,5	117,5	92,6	6,6	− 21,2

Source: EC Commission, Directorate-General for Agriculture.

[1] Ratio of human consumption to resident population at 1 January.
[2] Ratio of total production to domestic use.

4.4.5.1 Prices fixed ([1]) and market prices on the Bari market for: — virgin olive oil
— lampante grade olive oil 3°

1	2	3	XI 4	XII 5	I 6	II 7
Virgin olive oil	Market price	1985/86	249,24	242,48	248,25	244,78
	Intervention price	1985/86	239,71	239,71	241,50	243,29
	Market price	1992/93	205,78	193,07	175,52	174,59
	Intervention price	1992/93	207,82	207,82	207,82	207,82
	Market price	1993/94	195,68	193,80	202,11	213,04
	Intervention price	1993/94	197,90	197,90	197,90	197,90
Lampante grade olive oil 3°	Market price ([2])	1985/86	198,70	202,76	210,50	215,15
	Intervention price ([2])	1985/86	206,68	206,68	208,47	210,26
	Market price	1992/93	188,95	179,12	165,16	165,43
	Intervention price	1992/93	185,47	185,47	185,47	185,47
	Market price	1993/94	175,59	177,69	181,53	190,04
	Intervention price	1993/94	175,71	175,71	175,71	175,71

Source: EC Commission, Directorate-General for Agriculture, and Bari Chamber of Commerce.

([1]) Calculated prices allow for monthly increments.
([2]) For 5° of acidity.

4.4.5.2 Wholesale prices: — on the Bari market for refined olive oil
— on the Milan market for refined olive oil, edible seed oils

1	2	XI 3	XII 4	I 5	II 6
Bari — refined olive oil	1985/86	231,66	233,88	243,35	247,59
	1992/93	212,77	204,30	185,52	182,02
	1993/94	194,32	200,33	206,97	215,01
Milan — refined olive oil	1985/86	234,29	235,33	248,10	255,63
	1992/93	220,70	214,78	195,23	187,93
	1993/94	202,54	204,73	195,57	219,85
Milan — edible seed oils	1985/86	67,83	67,75	64,88	55,45
	1992/93	40,72	42,17	40,26	39,21
	1993/94	46,09	51,98	54,54	53,20
Ratio: olive-oil (Bari)/edible seed oils (Milan)	1985/86	3,41	3,45	3,75	4,47
	1992/93	5,23	4,84	4,61	4,64
	1993/94	4,22	3,85	3,80	4,04

NB: The ratio olive-oil/seed oils is based on wholesale prices and excludes the consumption aid effective from 1 April 1979.
Source: Bari and Milan Chambers of Commerce.

(ECU/100 kg)

III	IV	V	VI	VII	VIII	IX	X	∅
8	9	10	11	12	13	14	15	16
242,97	244,60	242,43	241,23	241,23	241,23	251,35	262,82	246,05
245,08	246,87	248,66	250,45	252,24	252,24	252,24	239,71	245,98
177,27	179,89	185,36	187,51	194,92	198,66	205,78	199,85	189,85
207,82	207,82	207,82	207,82	207,82	207,82	207,82	207,82	207,82
212,86	206,60	206,60	206,60	206,60	206,52	212,86	:	204,75
197,90	197,90	197,90	197,90	196,98	196,98	196,98	196,98	197,59
220,98	223,01	221,74	216,42	216,77	220,65	223,85	:	215,50
212,05	213,84	215,63	217,42	219,21	219,21	219,21	206,68	212,95
168,02	169,51	173,38	173,82	181,31	184,62	184,62	180,89	173,23
185,47	185,47	185,47	185,47	185,47	185,47	185,47	185,47	185,47
186,04	186,15	191,20	193,59	194,29	195,34	199,15	:	188,24
175,71	175,71	175,71	175,71	176,08	176,08	176,08	176,08	175,83

(ECU/100 kg)

III	IV	V	VI	VII	VIII	IX	X	∅
7	8	9	10	11	12	13	14	15
252,75	255,55	254,99	252,45	249,83	251,01	252,56	252,80	248,20
190,30	192,06	197,28	195,46	202,15	208,74	217,59	206,67	199,57
209,44	208,71	213,78	214,07	214,07	218,32	226,34	:	213,80
259,47	260,68	261,06	258,96	255,07	256,82	259,22	261,40	253,84
194,87	196,36	200,52	199,88	208,39	214,24	221,86	220,39	206,26
215,94	214,55	219,20	219,48	220,48	218,95	230,63	:	216,88
49,29	50,60	47,15	45,25	43,13	39,23	36,34	38,31	50,43
38,99	36,92	37,39	38,95	45,45	46,23	43,56	43,18	41,01
51,42	50,48	50,59	50,57	47,57	46,47	53,36	:	50,85
5,13	5,05	5,41	5,58	5,79	6,40	6,95	6,60	5,17
4,88	5,20	5,28	5,02	4,45	4,52	5,00	4,74	4,87
4,07	4,13	4,23	4,23	4,50	4,70	4,24	:	4,20

4.4.5.3   Average monthly prices for oilseed products

(ECU/100 kg)

| | 1993 | | | | | | 1994 | | | | | |
	VII	VIII	IX	X	XI	XII	I	II	III	IV	V	VI
1	2	3	4	5	6	7	8	9	10	11	12	13
Soya beans (1)	19,14	19,17	18,12	17,29	19,44	19,76	20,77	18,38	18,30	18,57	18,42	18,54
Oils (2)												
Soya oil	41,80	34,29	33,37	34,38	40,79	43,71	45,85	42,47	42,52	42,49	42,02	40,70
Rapeseed oil	33,11	32,28	31,74	32,64	40,79	42,85	45,47	43,10	42,05	40,17	43,55	40,52
Sunflower oil	39,23	44,09	40,62	45,03	50,72	57,84	54,04	50,25	47,31	46,68	47,02	47,44
Oil cake (2)												
Soya cake	15,98	15,51	14,07	14,03	15,37	14,75	15,72	16,40	16,27	15,63	14,06	14,03
Rapeseed cake	10,65	9,45	9,92	10,80	11,29	11,81	12,07	11,34	11,13	11,66	11,65	11,04
Sunflower cake	11,96	9,56	8,62	7,81	8,20	9,67	9,83	9,37	9,31	9,51	8,04	7,71

Source : EC Commission, Directorate-General for Agriculture.

(1) Unloaded at Rotterdam.
(2) Ex-EEC factory.

(1992)

4.4.9.1 Apparent human consumption of fats, subdivided by : — base materials (pure fat) — processed products consumed (pure fat)

		Base materials					Processed products consumed							
		Vegetable oils and fats	Oils and fats of land animals	Oils and fats of marine animals	Total (without butter)	Butter	Total	Margarine	Other prepared oils and fats	Other oils and fats of land animals	Other oils and fats of marine animals	Edible oils	Total (without butter)	Butter
1	2	3	4	5	6	7	8	9	10	11	12	13	14	15
1 000 t	EUR 12	:	:	:	:	1 288	:	:	:	:	:	:	:	1 288
	BLEU/UEBL	177	114	9	300	60	360	92	53	94	–	61	300	60
	Danmark	:	75	23	:	8	:	75	4	66	–	:	:	8
	BR Deutschland	:	:	:	:	456	:	:	:	:	:	:	:	456
	Elláda	:	:	:	:	8	:	:	:	:	:	:	:	8
	España	1 041	162	–	1 203	13	1 216	82	95	125	–	901	1 203	13
	France*	:	:	:	:	422	:	:	:	:	:	:	:	422
	Ireland*	:	:	:	:	10	:	:	:	:	:	:	:	10
	Italia	1 319	283	3	1 605	112	1 717	54	28	279	3	1 241	1 605	112
	Nederland	246	:	:	:	32	:	56	:	:	:	192	:	32
	Portugal	:	:	:	:	11	:	:	:	:	:	:	:	11
	United Kingdom*	923	:	121	:	151	:	475	138	:	24	552	:	151
%	EUR 12	:	:	:	:	:	:	:	:	:	:	:	:	:
	BLEU/UEBL	49	31	3	83	17	100	25	15	26	–	17	83	17
	Danmark	:	:	:	:	:	:	:	:	:	:	:	:	:
	BR Deutschland	:	:	:	:	:	:	:	:	:	:	:	:	:
	Elláda	:	:	:	:	:	:	:	:	:	:	:	:	:
	España	86	13	–	99	1	100	7	8	10	–	74	99	1
	France	:	:	:	:	:	:	:	:	:	:	:	:	:
	Ireland	:	:	:	:	:	:	:	:	:	:	:	:	:
	Italia	77	16	0	93	7	100	3	2	16	0	72	93	7
	Nederland	:	:	:	:	:	:	:	:	:	:	:	:	:
	Portugal	:	:	:	:	:	:	:	:	:	:	:	:	:
	United Kingdom	:	:	:	:	:	:	:	:	:	:	:	:	:
kg/head	EUR 12	:	:	:	:	4	:	:	:	:	:	:	:	4
	BLEU/UEBL	17	11	0	28	7	35	8	5	9	–	6	28	7
	Danmark	:	15	4	:	2	:	15	0	13	–	:	:	2
	BR Deutschland	:	:	:	:	6	:	:	:	:	:	:	:	6
	Elláda	:	:	:	:	0	:	:	:	:	:	:	:	0
	España	27	4	–	31	0	31	2	2	4	–	23	31	0
	France	:	:	:	:	7	:	:	:	:	:	:	:	7
	Ireland	:	:	:	:	3	:	:	:	:	:	:	:	3
	Italia	23	5	0	28	2	30	1	0	5	0	22	28	2
	Nederland	25	:	:	:	2	:	6	:	:	:	20	:	2
	Portugal	:	:	:	:	1	:	:	:	:	:	:	:	1
	United Kingdom	16	:	2	:	3	:	8	2	:	0	10	:	3

Source : Eurostat.

4.5.1.1   Area, yield and harvested production of (a) fruit, (b) citrus fruit and (c) vegetables

	Area 1 000 ha					Yield 100 kg/ha					Harvested production 1 000 t				
	1985	1992 ∞	1993 ∞	% TAV 1992/1985	% TAV 1993/1992	1985	1992 ∞	1993 ∞	% TAV 1992/1985	% TAV 1993/1992	1985	1992 ∞	1993 ∞	% TAV 1992/1985	% TAV 1993/1992
1	2	3	4	5	6	7	8	9	10	11	12	13	14	15	16
A. Fruit (excl. citrus)															
A.1 All fruit															
EUR 12	2 899	2 973**	2 961**	0.4	−0.4	75	88	77	2.3	−12.5	21 689	26 089**	22 676**	2.7	−13.1
Belgique/België	11	15	15**	4.5	0.0	315	434	473	4.7	9.0	346	651	709	9.4	8.9
Danmark	9	8	4	−1.7	−50.0	81	61	145	−4.0	137.7	73	49	58	−5.5	18.4
BR Deutschland	54	56	56**	−0.5	0.0	499	824	485	7.4	−41.1	2 694	4 615	2 717	8.0	−41.1
Elláda	288	282**	282**	−0.3	0.0	87	94	87	1.1	−7.4	2 500	2 662	2 440**	0.9	−8.3
España	1 119	1 178	1 180**	−0.7	0.2	137	28	34	−4.6	21.4	4 349	3 260**	3 969**	−4.0	21.7
France	250	242**	242**	−0.5	0.0	174	174	138	3.5	−20.7	3 434	4 201	3 328	2.9	−20.8
Ireland	2	1**	1**	0.0	0.0	72	80	85	0.9	6.3	15	16	17**	0.9	6.3
Italia	952	957	949	0.1	−0.8	72	90	81	3.2	−10.0	6 898	8 617	7 678	3.2	−10.9
Luxembourg	0	0	0	x	x	x	x	x	x	x	7	15	8	11.5	−46.7
Nederland	25	26	26	0.6	0.0	176	304	212	8.1	−30.3	439	790	550**	8.8	−30.4
Portugal	139	163	162**	2.3	−0.6	45	41	40	−1.3	−2.4	624	668	650**	1.0	−2.7
United Kingdom	50	44	43	−1.8	−2.3	98	124	128	3.4	3.2	491	545	552	1.5	1.3
A.2 Apples															
EUR 12	323	325**	323**	0.1	−0.6	230	339	268	5.7	−20.9	7 433	11 016	8 651	5.8	−21.5
Belgique/België	6	9	9**	6.0	0.0	360	536	548	5.9	2.2	216	482	493	12.1	2.3
Danmark	4	3	3	−4.0	0.0	113	143	133	3.4	−7.0	45	43	40	−0.6	−7.0
BR Deutschland	24	36	36	6.0	0.0	576	815	409	5.1	−49.8	1 383	2 933	1 474	11.3	−49.7
Elláda	18	17**	17**	−0.8	0.0	148	226	208	6.2	−8.0	267	385	353	5.4	−8.3
España	57	46	46**	−3.0	0.0	176	232	178	4.0	−23.3	1 004	1 069	821	0.9	−23.2
France	66	69	70	0.6	1.4	272	348	290	3.6	−16.7	1 793	2 398	2 027	4.2	−15.5
Ireland	1	1**	0	0.0	−3.7	90	80	80	−1.7	0.0	9	8	8	−1.7	0.0
Italia	85	82	79	−0.5	0.0	237	292	271	3.0	−7.2	2 014	2 394	2 144	2.5	−10.4
Luxembourg	0	0	0	x	x	x	x	x	x	x	6	10	5	7.6	−50.0
Nederland	16	17	17	0.9	0.0	188	376	394	10.4	4.8	300	640	670	11.4	4.7
Portugal	22	25	25**	1.8	0.0	43	111	97	14.5	−12.6	95	277	243	16.5	−12.3
United Kingdom	24	20	20	−2.6	0.0	125	189	187	6.1	−1.1	301	377	373	3.3	−1.1
A.3 Pears															
EUR 12	137	138**	140**	0.1	1.4	188	226	180	2.7	−20.4	2 579	3 117	2 520	2.7	−19.2
Belgique/België	3	3	4**	0.0	33.3	260	370	380	5.2	2.7	78	111	152	5.2	36.9
Danmark	0	0	0	x	0.0	x	x	x	x	x	4	6	6	6.0	0.0
BR Deutschland	2	2	2	0.0	0.0	1 620	2 670	1 760	7.4	−34.1	324	534	352	7.4	−34.1
Elláda	7	5**	5**	−4.7	0.0	200	182	190	−1.3	4.4	140	91	95	−6.0	4.4
España	36	37	37**	0.4	0.0	165	163	124	−0.2	−23.9	595	602	459	0.2	−23.8
France	20	16	16	−3.1	0.0	209	246	148	2.4	−39.8	417	394	237	−0.8	−39.8
Italia	49	52	52	0.9	0.0	164	219	176	4.2	−19.6	806	1 138	916	5.1	−19.5
Nederland	6	5	6	−2.6	20.0	178	230	283	3.7	23.0	107	115	170	1.0	47.8
Portugal	10	14	14**	4.9	0.0	57	71	62	3.2	−12.7	57	100	87	8.4	−13.0
United Kingdom	4	4	4	0.0	0.0	128	65	115	−9.2	76.9	51	26	46	−9.2	76.9

A.4 Peaches

	EUR 12														
	210	228**	225**	1,2	−1,3	128	158	142	3,1	−10,1	2 682	3 594	3 186	4,3	−11,4
BR Deutschland	0	0	0	x	x	x	x	x	x	x	18	32	20	8,6	−37,5
Elláda	32	42	42**	4,0	0,0	164	231	228	5,0	−1,3	524	972	959	9,2	−1,3
España	58	66	66**	1,9	0,0	92	131	106	5,2	−19,1	532	867	699	7,2	−19,4
France	29	24	22	−2,7	8,3	135	145	126	1,0	−13,1	391	348	277	−1,7	−20,4
Italia	83	79	78**	−0,7	−1,3	143	163	150	1,9	−8,0	1 191	1 289	1 167	1,1	−9,5
Portugal	8	17	17**	11,4	0,0	33	51	38	6,4	−25,5	26	86	64	18,6	−25,6

A.5 Nectarines

	EUR 12														
	35**	67**	67**	9,7	0,0	107	166	134	6,5	−19,3	375	1 113	896	16,8	−19,5
Elláda	2	5**	5**	14,0	0,0	120	300	244	14,0	−18,7	24	150	122	29,9	−18,7
España	3	12	12**	21,9	0,0	53	131	130	13,8	−0,8	16	157	156	38,6	−0,6
France	8	13	13	7,2	0,0	123	139	97	1,8	−30,2	98	181	126	9,2	−30,4
Italia	22	35	35	6,9	−5,9	106	172	135	7,2	−21,5	234	603	473	14,5	−21,6
Portugal	0**	2**	2**	x	−5,7	x	110	95	x	−13,6	3	22	19	32,9	−13,6

A.6 Table grapes

	EUR 12														
	207	190**	184**	−1,2	−3,2	118	135	130	1,9	−3,7	2 433	2 571	2 389**	0,8	−7,1
Belgique/België	0	0**	0**	0,7	0,0	131	165	160	3,4	−3,0	3	1	1	−14,5	0,0
Elláda	19	20	20**	−4,5	0,0	73	78	63	1,0	−19,2	249	330	320**	4,1	−3,0
España	76	55	55**	−5,9	−5,9	60	47	64	−3,4	36,2	558	429	345	−3,7	−19,6
France	26	17	16	1,4	−5,7	175	191	190	1,3	−0,5	156	80	103	−9,1	28,8
Italia	80	88	83	x	x	107	53	47	−9,5	x	1 402	1 678	1 573	2,6	−6,3
Nederland	0	0	0	x	x	x	x	x	x	x	1	1	0**	x	x
Portugal	6	10	10**	7,6	0,0	107	53	47	−11,3	−11,3	64	53	47	−2,7	−11,3

A.7 Apricots

	EUR 12														
	60	69**	69**	2,0	0,0	98	92	77	−0,9	−16,3	587	634	529	1,1	−16,6
BR Deutschland	0	0	0	x	x	x	x	x	x	x	2	2	1	0,0	−50,0
Elláda	7	7**	7**	0,0	0,0	187	123	123	−5,8	0,0	131	86	86	−5,8	0,0
España	21	26	26	3,1	0,0	72	77	76	1,0	−1,3	151	199	197	4,0	−1,0
France	14	19	19	4,5	0,0	73	88	41	2,7	−53,4	102	167	78	7,3	−53,3
Italia	16	16	16	0,0	0,0	123	109	102	−1,7	−6,4	196	175	163	−1,6	−6,9
Portugal	2	1	1**	−9,4	0,0	25	50	40	10,4	−20,0	5	5	4	0,0	−20,0

B. Citrus fruit
B.1 All citrus fruit

	EUR 12														
	520	537**	507**	0,5	−5,6	155	185	172	2,6	−7,0	8 082	9 955	8 738**	3,0	−12,2
Elláda	52	58	58**	1,6	0,0	175	186	183	0,9	−1,6	911	1 079	1 060**	2,4	−1,8
España	252	268	270**	0,9	0,7	139	196	175	5,0	−10,7	3 514	5 240	4 731	5,9	−9,7
France	2	3	3	6,0	6,0	185	80	107	−11,3	33,8	37	24	32	−6,0	33,3
Italia	184	182	150	−0,2	−17,6	189	187	180	−0,2	−3,7	3 484	3 400	2 705**	−0,3	−20,4
Portugal	30	26	26**	−2,0	0,0	45	82	81	9,0	−1,2	136	212	210**	6,5	−0,9

B.2 Oranges

	EUR 12														
	295	308**	288**	0,6	−6,5	164	197	171	2,7	−13,2	4 836	6 080	4 934**	3,3	−18,8
Elláda	34	37	37**	1,2	0,0	185	236	230	3,5	−2,5	630	872	850**	4,8	−2,5
España	132	143	143**	1,2	0,0	147	205	162	4,9	−21,0	1 945	2 926	2 320	6,0	−20,7
France	0	0	0	x	x	x	x	x	x	x	3	3	3	−5,6	50,0
Italia	109	109	89	0,0	−18,3	198	194	179	−0,3	−7,7	2 162	2 112	1 591	−0,3	−24,7
Portugal	20	19	19**	−0,7	0,0	48	88	89	9,0	1,1	96	168	170**	8,3	1,2

4.5.1.1   *(cont.)*

		Area 1000 ha			% TAV		Yields 100 kg/ha			% TAV		Harvested production 1000 t			% TAV	
		1985	1992 ∞	1993 ∞	1992/1985	1993/1992	1985	1992 ∞	1993 ∞	1992/1985	1993/1992	1985	1992 ∞	1993 ∞	1992/1985	1993/1992
1		2	3	4	5	6	7	8	9	10	11	12	13	14	15	16
B.3 Lemons	EUR 12	112	100**	98**	-1,6	-2,0	129	162	149	3,3	-8,0	1 442	1 619	1 459***	1,7	-9,9
Elláda		13	13**	13**	0,0	0,0	158	92	100	-7,4	8,7	205	119	130**	-7,5	9,2
España		55	46	46**	-2,5	0,0	88	162	120	9,1	-25,9	482	743	553	6,4	-25,6
France		0	0	0	x	x	x	x	x	x	x	0	0	0	x	x
Italia		40	39	37	-0,4	-5,1	185	191	207	0,5	8,4	738	746	765	0,2	2,5
Portugal		4	2	2**	-9,4	0,0	23	55	55	13,3	0,0	9	11	11**	2,9	0,0
B.4 Mandarins	EUR 12	29	32**	30**	1,4	-6,3	142	124	132	-1,9	6,5	411	398	396**	-0,5	-0,5
Elláda		4	6**	6**	6,0	0,0	125	85	85	-5,4	0,0	50	51	51**	0,3	0,0
España		3	8	8**	15,0	0,0	160	151	181	-0,8	19,9	48	121	145	14,1	19,8
France		0	0	0	x	x	x	x	x	x	x	0	0	0	x	x
Italia		16	13	11	-2,9	-15,4	182	149	152	-2,8	2,0	291	194	167	-5,6	-13,9
Portugal		6	5	5**	-2,6	0,0	37	64	66	8,1	3,1	22	32	33**	5,5	3,1
B.5 Clementines	EUR 12	58	75**	69**	3,7	-8,0	150	183	211	2,9	15,3	872	1 374	1 454**	6,7	5,8
Elláda		0	2**	2**	x	0,0	x	110	120	x	9,1	14	22	24**	6,7	9,1
España		40	52	52**	3,8	0,0	145	198	237	4,6	19,7	579	1 027	1 230	8,5	19,8
France		2	2	2	0,0	0,0	165	100	130	-6,9	30,0	33	20	26	-6,9	30,0
Italia		16	19	13	2,5	-31,6	154	161	134	0,6	-16,8	246	305	174	3,1	-43,0
B.6 Satsumas	EUR 12	18	16**	16**	-1,7	0,0	237	236	283	-0,1	19,9	427	378	452**	-1,7	19,6
Elláda		0	0**	0**	x	x	x	x	x	x	x	3	4	4**	4,2	0,0
España		18	16	16**	-1,7	0,0	236	234	280	-0,1	19,7	424	374	448	-1,8	19,8
B.7 Other citrus fruit	EUR 12	7	3**	1**	-11,4	-66,7	99	187	180	9,5	-3,7	69	56	18**	-2,9	-67,9
Elláda		0	0**	0**	x	0,0	x	x	x	x	x	5	3	3**	-7,0	0,0
España		4	1	1**	-18,0	0,0	60	160	130	15,0	-18,8	24	16	13	-5,6	-18,8
Italia		3	2	0	-5,6	x	133	185	x	4,8	x	40	37	2	-1,1	-94,6

C. Vegetables

C.1 All vegetables

EUR 12	1 962	1 915**	1 892**	−0,3	−1,2	223	246	251	1,4	2,0	43 712	47 017**	47 539**	1,0	1,1
Belgique/België	50	55	56	1,4	1,8	204	249	272	2,9	9,2	1 021	1 372	1 525	4,3	11,2
Danmark	19	9**	9**	−10,1	0,0	146	22	22	−23,7	0,0	278	20**	20**	−31,3	0,0
BR Deutschland	55	82	77	5,9	−6,1	277	276	304	−0,1	10,1	1 526	2 264	2 344	5,8	3,5
Elláda	155	134	135**	−2,1	0,7	278	312	304	1,7	−2,6	4 305	4 177	4 100**	−0,4	−1,8
España	472	454	460**	−0,6	1,3	201	240	250	2,6	4,2	9 500	10 900**	11 500**	2,0	5,5
France	328	332	320	0,2	−3,6	169	169	181	0,0	7,1	5 538	5 612	5 785	0,2	3,1
Ireland	7	7	7**	−0,9	0,0	314	321	329	0,3	2,5	220	225	230**	0,3	2,2
Italia	533	501	491	−0,9	−2,0	245	251	247	0,3	−1,6	13 061	12 569	12 133	−0,5	−3,5
Luxembourg	0	0	0	x	x	x	x	x	x	x	3	2	2	−5,6	0,0
Nederland	77	80**	76	0,5	−5,0	356	485	513	4,5	5,8	2 742	3 881	3 900**	5,1	0,5
Portugal	94	80**	80**	−2,3	0,0	211	247	250	2,3	1,2	1 984	1 976**	2 000**	−0,1	1,2
United Kingdom	172	181	181**	0,7	0,0	205	222	221	1,1	−0,5	3 534	4 019	4 000**	1,9	−0,5

C.2 Cauliflowers

EUR 12	123	142**	139**	2,1	−2,1	155	159	169	0,4	6,3	1 903	2 261**	2 344**	2,5	3,7
Belgique/België	4	5	5	3,2	0,0	155	194	238	3,3	22,7	62	97	119	6,6	22,7
Danmark	1	1**	1**	0,0	0,0	110	70	70	−6,3	0,0	11	7**	7**	−6,3	0,0
BR Deutschland	3	6	6	10,4	0,0	263	250	252	−0,7	0,8	79	150	151	9,6	0,7
Elláda	3	3	3**	0,0	0,0	177	203	200	−2,0	−1,5	53	61	60**	2,0	−1,6
España	10	15	15	6,0	0,0	222	201	192	−1,4	−4,5	222	302	288	4,5	−4,6
France	46	47	46	0,3	−2,1	114	106	129	−1,0	21,7	526	497	592	−0,8	19,1
Ireland	1	1	1**	1,3	0,0	140	130	130	−1,1	0,0	14	13	13**	−1,1	0,0
Italia	31	34	33	4,2	−2,9	161	188	180	2,2	−4,3	500	638	594	3,5	−6,9
Nederland	3	4	3	0,0	−25,0	163	165	200	0,2	21,2	49	66	60**	4,3	−9,1
Portugal	1	1**	1**	4,2	0,0	180	200	220	1,5	10,0	18	20**	22**	1,5	10,0
United Kingdom	20	25	25	3,2	0,0	185	164	175	−1,7	6,7	369	410	438	1,5	6,8

C.3 Tomatoes

EUR 12	300	249**	246**	−2,6	−1,2	466	510	514	1,3	0,8	13 985	12 709	12 640**	−1,4	−0,5
Belgique/België	0	1	1	0,0	0,0	1 600	3 300	3 470	10,9	5,2	160	330	347	10,9	5,2
Danmark	0	0	0	x	x	x	x	x	x	x	17	20	20	2,3	0,0
BR Deutschland	0	0	0	−3,2	x	x	x	x	x	x	23	30	24	3,9	−20,0
Elláda	49	39	39**	−1,2	−3,2	462	507	484	1,3	−4,5	2 264	1 979	1 888	−1,9	−4,6
España	61	56	57	−4,9	1,8	398	473	474	2,5	0,2	2 429	2 647	2 699	1,2	2,0
France	17	12	12	0,3	0,0	553	650	660	2,3	1,5	940	780	792	−2,6	1,5
Ireland	0	0	0**	x	x	x	x	x	x	x	14	12	11**	−2,2	−8,3
Italia	143	118	115	−2,7	−2,5	459	465	454	0,2	−2,4	6 563	5 483	5 223	−2,5	−4,7
Nederland	2	2	0**	0,0	−50,0	2 625	3 260	6 060	3,1	85,9	525	652	606	3,1	−7,1
Portugal	26	20**	20**	−3,7	0,0	357	320	445	−1,6	39,1	928	640	890	−5,2	39,1
United Kingdom	1	1	1	0,0	0,0	1 220	1 360	1 400	1,6	2,9	122	136	140	1,6	2,9

C.4 Aubergines

EUR 12	20	18	18**	−1,5	0,0	267	318	324	2,5	1,9	534	573	583**	1,0	1,7
Elláda	3	3	3**	0,0	0,0	237	267	267	1,7	0,0	71	80	80**	1,7	0,0
España	4	4	4**	0,0	0,0	305	340	338	1,6	−0,6	122	136	135**	1,6	−0,7
France	1	1	1	0,0	0,0	290	250	240	−2,1	−4,0	29	25	24	−2,1	−4,0
Italia	12	10	10	−2,6	0,0	248	303	314	2,9	3,6	297	303	314	0,3	3,6
Nederland	0	0	0	x	x	x	x	x	x	x	15	29	30**	9,9	3,4

Source : Eurostat.

4.5.3.1 Intra-EC trade and external trade in fresh fruit and vegetables ([3])

EUR 12

(1 000 t)

1	2	3	1985	1990	1991 ∞	1992 ∞	1993 ∞ (p)	% TAV	
								1992 / 1985	1993 / 1992
1	2	3	4	5	6	7	8	9	10
Intra-EC trade ([1])	Vegetables of which:	Total	3 801	4 639	5 135	5 421	4 413	5,2	− 18,6
		Cauliflowers	190	257	257	302	226	6,8	− 25,2
		Tomatoes	733	891	955	1 032	809	5,0	− 21,6
		Cucumbers	396	483	552	556	428	5,0	− 23,0
	Fruits ([2]) of which:	Total	3 168	3 888	4 284	4 364	2 105	4,7	− 51,8
		Apples	1 163	1 312	1 425	1 330	1 194	1,9	− 10,2
		Pears	276	317	375	391	327	5,1	− 16,4
		Peaches	415	560	538	591	434	5,2	− 26,6
	Citrus fruit of which:	Total	1 872	2 629	2 511	2 683	2 255	5,3	− 16,0
		Oranges	867	1 403	1 288	1 336	1 073	6,4	− 19,7
		Lemons	272	321	303	315	291	2,1	− 7,6
		Clementines	434	616	613	692	609	6,9	− 12,0
Imports	Vegetables of which:	Total	662	886	940	861	822	3,8	− 4,5
		Cauliflowers	1	1	2	1	2	0,0	100,0
		Tomatoes	272	288	350	357	391	4,0	9,5
		Cucumbers	44	50	80	57	38	3,8	− 33,3
	Fruits ([2]) of which:	Total	1 014	1 690	2 237	2 138	1 166	11,2	− 45,5
		Apples	461	643	839	849	562	9,1	− 33,8
		Pears	101	228	259	295	245	16,5	− 16,9
		Peaches	2	11	14	17	9	35,8	− 47,1
	Citrus fruit of which:	Total	1 419	1 527	1 636	1 581	1 465	1,6	− 7,3
		Oranges	862	912	918	877	783	0,2	− 10,7
		Lemons	89	79	113	122	89	4,6	− 27,0
		Clementines	103	74	89	77	98	− 4,1	27,3
Exports	Vegetables of which:	Total	457	634	672	797	984	8,3	23,5
		Cauliflowers	21	34	31	37	38	8,4	2,7
		Tomatoes	72	106	132	147	210	10,7	42,9
		Cucumbers	30	43	52	50	59	7,6	18,0
	Fruits ([2]) of which:	Total	623	589	654	734	511	2,4	− 30,4
		Apples	189	153	144	160	363	− 2,4	126,9
		Pears	58	44	52	59	60	0,2	1,7
		Peaches	98	99	97	122	104	3,2	− 14,8
	Citrus fruit of which:	Total	685	753	864	924	1 324	4,4	43,3
		Oranges	324	416	539	610	817	9,5	33,9
		Lemons	251	203	194	142	257	− 7,8	81,0
		Clementines	45	71	65	82	129	9,0	57,3

Source: Eurostat.

([1]) Based on goods entering.
([2]) Citrus fruit not included.
([3]) For tax reasons, the Canary Islands are still included under non-member countries.

4.5.4.6   **Market balance — processed tomatoes**
— processed peaches   **EUR 12**

	1 000 t			% TAV	
	1985/86	1991/92 ∞	1992/93 ∞	1991/92 / 1985/86	1992/93 / 1991/92
1	2	3	4	5	6
Processed tomatoes					
Usable production	7 155	6 427	6 225	− 1,8	− 3,1
Imports	94	602	341	36,3	− 43,4
Exports	3 065	1 543	1 906	− 10,8	23,5
Intra-EC trade	1 441	2 996	2 849	13,0	− 4,9
Change in stocks	100	:	:	×	×
Internal use	4 083	5 486	4 660	5,0	− 15,1
of which:					
— losses (market)	0	0	0	×	×
— human consumption (¹)	4 083	5 486	4 660	5,0	− 15,1
Human consumption (kg/head)	13	16	14	3,5	− 12,5
Self-sufficiency (%)	175	117	134	− 6,5	14,5
Processed peaches					
Usable production	602	566	647	− 1,0	14,3
Imports	27	145	146	32,3	0,7
Exports	103	252	277	16,1	9,9
Intra-EC trade	134	746	669	33,1	− 10,3
Change in stocks	0	30	0	×	×
Internal use	526	429	516	− 3,3	20,3
of which:					
— losses (market)	0	0	0	×	×
— human consumption (¹)	526	429	516	− 3,3	20,3
Human consumption (kg/head)	2	1,3	1,5	− 6,9	15,4
Self-sufficiency (%)	114	132	125	2,5	− 5,3

Source: Eurostat.
(¹) According to the market balance.

4.5.5.1   Producer prices of certain types of fruit and vegetables

		ECU/100 kg			% TAV	
		1991/92	1992/93	1993/94	1992/93 / 1991/92	1993/94 / 1992/93
1	2	3	4	5	6	7
Apples 'Golden Delicious'	Belgique/België	61,70	17,90	19,97	−71,0	11,6
	Danmark	49,25	17,78	19,18	−63,9	7,9
	BR Deutschland	74,44	25,13	29,76	−66,2	18,4
	Elláda	89,55	35,89	45,28	−59,9	26,2
	España	67,34	14,51	26,26	−78,5	81,0
	France	73,87	21,94	28,38	−70,3	29,4
	Ireland	38,66	:	:	x	x
	Italia	60,76	21,35	26,87	−64,9	25,9
	Nederland	84,21	21,68	25,18	−74,3	16,1
	Portugal	52,97	25,32	29,18	−52,2	15,2
Pears	Belgique/België	70,68	40,96	29,66	−42,0	−27,6
	Danmark	66,43	25,27	19,80	−62,0	−21,6
	BR Deutschland	66,05	27,54	32,56	−58,3	18,2
	Elláda	105,11	53,88	60,02	−48,7	11,4
	España	66,16	27,03	42,08	−59,1	55,7
	France	82,46	32,65	41,06	−60,4	25,8
	Italia	73,65	27,33	43,42	−62,9	58,9
	Nederland	90,24	43,58	36,33	−51,7	−16,6
	Portugal	42,79	28,34	32,45	−33,8	14,5
	United Kingdom	64,66	40,61	30,19	−37,2	−25,7
Peaches	Elláda	54,50	36,40	35,13	−33,2	−3,5
	España	42,62	30,05	31,51	−29,5	4,9
	France	89,25	45,85	61,13	−48,6	33,3
	Italia	78,30	41,41	65,52	−47,1	58,2
	Portugal	:	:	45,26	x	x
Nectarines	España	68,88	43,78	47,48	−36,4	8,5
	France	95,89	54,47	68,06	−43,2	24,9
	Italia	91,87	58,23	75,75	−36,6	30,1
	Portugal	:	:	61,60	x	x
Apricots	Elláda	75,13	52,00	43,16	−30,8	−17,0
	España	55,57	53,44	40,70	−3,8	−23,8
	France	85,11	55,91	94,28	−34,3	68,6
	Italia	74,35	48,50	48,22	−34,8	−0,6
	Portugal	:	37,08	48,62	x	31,1
Table grapes	Elláda	41,32	36,20	52,43	−12,4	44,8
	España	52,40	47,05	37,21	−10,2	−20,9
	France	83,44	55,79	66,68	−33,1	19,5
	Italia	36,25	34,96	26,94	−3,6	−22,9
	Portugal	:	:	39,95	x	x

Citrus fruit:					
Oranges	Elláda	28,08	20,55	29,43	− 26,8	43,2
	España	29,49	19,53	26,98	− 33,8	38,1
	Italia	30,41	21,26	23,41	− 30,1	10,1
	Portugal	32,39	26,05	24,79	− 19,6	− 4,8
Mandarins	Elláda	35,36	26,58	27,86	− 24,8	4,8
	España	71,97	45,78	45,62	− 36,4	− 0,3
	Italia	40,45	29,39	30,64	− 27,3	4,3
	Portugal	33,33	36,31	30,63	8,9	− 15,6
Lemons	Elláda	39,55	33,39	28,90	− 15,6	− 13,4
	España	36,68	33,90	36,54	− 7,6	7,8
	Italia	42,61	38,25	29,66	− 10,2	− 22,5
	Portugal	22,19	22,71	25,25	2,3	11,2
Clementines	Elláda	51,72	32,53	38,35	− 37,1	17,9
	España	61,88	42,07	44,01	− 32,0	4,6
	France	43,81	37,09	33,82	− 15,3	− 8,8
	Italia	97,37	21,21	:	− 78,2	x
	Portugal	43,54	42,26	38,64	− 2,9	− 8,6
Satsumas	España	46,37	27,66	25,37	− 40,3	− 8,3
	Portugal	:	30,62	33,88	x	10,6
Cauliflowers	Belgique/België	60,28	43,92	45,28	− 27,1	3,1
	Danmark	70,09	:	:	x	x
	BR Deutschland	32,38	26,40	24,55	− 18,5	− 7,0
	Elláda	76,73	41,91	35,90	− 45,4	− 14,3
	España	25,47	22,71	20,32	− 10,8	− 10,5
	France	24,84	24,82	18,41	− 0,1	− 25,8
	Italia	37,02	26,72	20,31	− 27,8	− 24,0
	Nederland	57,35	41,18	39,57	− 28,2	− 3,9
	Portugal	19,02	30,37	32,25	59,7	6,2
	United Kingdom	25,12	19,80	18,78	− 21,2	− 5,2
'Round' tomatoes	Belgique/België (¹)	61,76	44,97	43,39	− 27,2	− 3,5
	Danmark (¹)	100,99	68,42	81,42	− 32,3	19,0
	BR Deutschland (²)	54,23	37,18	36,52	− 31,4	− 1,8
	Elláda (²)	35,27	29,46	27,51	− 16,5	− 6,6
	España (²)	42,04	35,07	30,65	− 16,6	− 12,6
	France (²)	54,43	45,53	42,62	− 16,4	− 6,4
	Ireland (¹)	64,63	48,53	50,56	− 24,9	4,2
	Italia (²)	45,14	32,31	23,89	− 28,4	− 26,1
	Nederland (¹)	68,28	43,43	41,67	− 36,4	− 4,1
	Portugal	73,03	19,26	25,25	x	31,1
	United Kingdom (¹)	73,03	53,52	55,68	− 26,7	4,0
Aubergines	España	29,65	40,81	20,92	37,6	− 48,7
	France	78,95	79,55	56,45	0,8	− 29,0
	Italia	41,26	39,06	26,75	− 5,3	− 31,5
	Nederland	99,18	111,45	71,55	12,4	− 35,8
	Portugal	:	:	54,81	x	x

Source: EC Commission, Directorate-General for Agriculture.

(¹) Tomatoes grown under glass.
(²) Open-grown tomatoes.

4.5.6.1   Quantities of fruit and vegetables bought in

1	2	1 000 kg			% of harvested production	
		1991/92	1992/93	1993/94 (p)	1992/93	1993/94
		3	4	5	6	7
Apples	EUR 12	35 737	1 761 123	988 030	16,1	11,6
	Belgique/België	33	100 465	73 713	20,9	15,0
	Danmark	0	0	129	0,0	0,3
	BR Deutschland	0	54 143	46 383	1,8	3,1
	Elláda	26 912	160 149	95 949	41,6	27,2
	España	0	144 273	52 541	13,5	6,4
	France	0	849 317	455 000	35,4	22,4
	Ireland	555	709	440	8,7	5,5
	Italia	7 367	314 429	102 382	13,1	4,8
	Nederland	3	104 188	127 169	18,3	22,3
	Portugal	0	11 231	9 970	4,0	4,1
	United Kingdom	867	22 219	24 354	5,9	6,5
Pears	EUR 12	3 269	196 712	40 338	6,3	1,6
	Belgique/België	342	2 218	4 932	2,0	3,3
	BR Deutschland	0	333	656	0,1	0,2
	Elláda	353	1 235	3 506	1,4	3,7
	España	0	19 594	6 436	3,3	1,4
	France	1	37 938	6 576	9,6	2,8
	Italia	1 954	131 226	9 366	11,5	1,0
	Nederland	268	2 144	6 714	2,1	4,5
	Portugal	268	1 977	1 389	2,0	1,6
	United Kingdom	83	47	763	0,2	1,7
Peaches	EUR 12	430 001	890 311	690 554	24,8	21,7**
	BR Deutschland	0	0	223	0,0	1,1
	Elláda	326 489	597 684	580 788	61,5	60,6
	España	638	38 356	35 043	4,4	5,0
	France	12 611	92 243	25 557	26,5	9,2
	Italia	90 263	159 914	47 950	12,4	4,1
	Portugal	0	2 114	993	2,5	1,6
Nectarines	EUR 12	83 075	340 974	157 091	30,6	17,5
	Elláda	31 138	114 587	88 472	76,6	72,5
	España	353	3 895	3 655	2,5	2,3
	France	6 689	57 529	17 173	31,7	13,7
	Italia	44 895	164 963	47 791	27,4	10,1

Table grapes		EUR 12	291	3 791	18 859	0,1	0,8**
	Elláda		291	0	15 662	0,0	5,2**
	España		0	90	2 940	0,0	0,9
	France		0	3 701	257	4,6	0,2
	Italia		0	0	0	0,0	0,0
Apricots		EUR 12	1 004	20 946	85 773	3,3	16,2
	Elláda		841	5 393	32 218	6,3	37,5
	España		67	710	6	0,4	0,0
	France		96	492	52 838	0,2	26,8
	Italia		0	14 351	711	8,6	0,9
Oranges		EUR 12	164 588	607 454	312 800	10,0	6,3**
	Elláda		156 960	234 434	151 368	26,9	17,0**
	España		0	88 126	92 727	4,2	5,8
	France		7 334	283 825	67 519	9,7	2,9
	Italia		294	626	578	39,0	17,9
	Portugal		0	443	608	0,3	0,4**
Mandarins		EUR 12	2 723	7 266	6 313	1,8	1,6**
	Elláda		2 723	2 722	2 647	5,3	4,8**
	España		0	0	0	0,0	0,0
	Italia		0	4 544	3 666	2,3	2,2
Lemons		EUR 12	13 736	84 578	61 482	5,2**	4,2**
	Elláda		0	1 610	89	1,4	0,1**
	España		13 728	82 205	61 007	11,1	11,0
	France		8	116	58	23,7**	8,5**
	Italia		0	647	328	0,1	0,0
Clementines		EUR 12	1 713	58 164	21 433	4,2	1,5**
	Elláda		41	987	64	4,4	0,3**
	España		68	29 755	12 336	2,9	1,0
	France		1 604	8 891	2 773	44,1	10,7
	Italia		0	18 531	6 260	6,1	3,6
Satsumas		EUR 12	0	4 030	3 578	1,1	0,8**
	Elláda		0	0	7	0,0	0,2**
	España		0	4 030	3 571	1,1	0,8

4.5.6.1   (cont.)

		1 000 kg			% of harvested production	
1	2	1991/92	1992/93	1993/94 (p)	1992/93	1993/94
		3	4	5	6	7
Cauliflowers	EUR 12	75 948	140 808	108 679	6,2**	4,6**
	Belgique/België	833	1 958	626	2,0	0,5
	BR Deutschland	2 200	6 506	2 533	4,3	1,7
	Elláda	6 290	739	1 635	1,2	2,7**
	España	25	2 154	8 815	0,7	3,1
	France	29 376	108 094	65 551	21,7	11,1
	Ireland	851	307	369	2,3	2,8**
	Italia	27 966	2 076	17 770	0,3	3,0
	Portugal	0	261	291	1,3**	1,5**
	United Kingdom	8 407	18 713	11 089	4,6	2,5
Tomatoes	EUR 12	60 273	252 697	50 090	2,0	0,4**
	Belgique/België	312	6 543	5 444	2,0	1,6
	BR Deutschland	46	312	99	1,0	0,4
	Elláda	19 091	3 169	5 006	0,2	0,3
	España	71	49	2 613	0,0	0,1
	France	26 407	25 528	11 964	3,3	1,5
	Ireland	47	120	88	1,0	0,8**
	Italia	9 072	175 792	5 852	3,2	0,1
	Nederland	5 227	41 181	18 894	6,3	3,1
	Portugal	0	0	130	0,0	0,0
	United Kingdom	0	3	0	0,0	0,0
Aubergines	EUR 12	553	148	384	0,0	0,1**
	Elláda	0	0	5	0,0	0,0**
	España	0	0	54	0,0	0,0**
	France	553	148	325	0,6	1,4
	Italia	0	0	0	0,0	0,0

Source : EC Commission, Directorate-General for Agriculture.

4.6.1.1 Area under vines, yield and production of wine and must

1	Area					Yield					Production				
	1 000 ha			% TAV		hl/ha			% TAV		1 000 hl			% TAV	
	1985/86	1991/92	1992/93	1991/92 1985/86	1992/93 1991/92	1985/86	1991/92	1992/93	1991/92 1985/86	1992/93 1991/92	1985/86	1991/92	1992/93	1991/92 1985/86	1992/93 1991/92
	2	3	4	5	6	7	8	9	10	11	12	13	14	15	16
EUR 12	4 026	3 599∞	3 513∞	-1,9	-2,4	46,1	50,5∞	54,4∞	1,5	7,7	185 735	156 315∞	190 977∞	-2,8	22,2
Belgique/België	0	0	0	x	x	x	x	x	x	x	2	1	2	-10,9	100,0
BR Deutschland	93	102∞	106∞	1,6	3,9	65,6	104,9∞	127,2∞	8,1	21,3	6 097	10 699∞	13 482∞	9,8	26,0
Elláda	86	84	65	-0,4	-22,6	55,6	47,9	62,3	-2,5	30,1	4 782	4 021	4 050	-2,8	0,7
España	1 572	1 385	1 317	-2,1	-4,9	22,5	22,2	25,9	-0,2	16,7	33 103	30 796	34 032	-1,2	10,5
France	1 011	889	883	-2,1	-0,7	69,3	46,6	71,6	-6,4	53,6	70 055	41 438	63 256	-8,4	52,7
Italia	993	881	875	-2,0	-0,7	62,1	67,2	77,8	1,3	15,8	61 690	59 238	68 086	-0,7	14,9
Luxembourg	1	1	1	0,0	0,0	92,0	86,0	271	-1,1	215,1	107	86	271	-3,6	215,1
Portugal	373	325	264	-2,3	-18,8	26,5	30,8	29,4	2,5	-4,5	9 893	10 021	7 771	0,2	-22,5
United Kingdom	0	1	1	x	0,0	18,3	15,0	27	-3,3	80,0	6	15	27	16,5	80,0

Source : Eurostat.

4.6.3.1   Trade (¹) in wine and share in world trade

(1 000 hl)

1	Imports 1985	Imports 1992	Imports 1993	% TAV 1992/1985	% TAV 1993/1992	Exports 1985	Exports 1992	Exports 1993	% TAV 1992/1985	% TAV 1993/1992	% of world trade (1992)
	2	3	4	5	6	7	8	9	10	11	12
EUR 12	27 403	30 872∞	26 606∞	1,7	− 13,8	40 548	36 073∞	38 941∞	− 1,7	8,0	74,4∞
BLEU/UEBL	2 012	2 352	2 294	2,3	− 2,5	103	194	159	9,5	− 18,0	2,8
Danmark	1 021	1 169	81	2,0	− 93,1	36	32	11	− 1,7	− 65,6	1,3
BR Deutschland	9 161	10 414∞	8 902∞	1,8	− 14,5	2 897	2 791∞	2 555∞	− 0,5	− 8,5	14,7∞
Elláda	7	83	3	42,4	− 96,4	1 291	606	136	− 10,2	− 77,6	0,8
España	9	129	60	46,3	− 53,5	6 256	6 148	10 293	− 0,2	67,4	7
France	6 859	7 357	5 794	1,0	− 21,2	11 617	11 457	10 626	− 0,2	− 7,3	20,9
Ireland	120	148	12	3,0	− 91,9	2	2	1	0,0	− 50,0	0,2
Italia	689	796	402	2,1	− 49,5	16 694	12 429	12 871	− 4,1	3,6	14,7
Nederland	2 148	2 229	1 916	0,5	− 14,0	94	46	93	− 9,7	102,2	2,5
Portugal	1	30	182	62,6	506,7	1 479	2 309	2 147	6,6	− 7,0	2,6
United Kingdom	5376	6165	6960	2,0	12,9	79	59	49	− 4,1	− 16,9	6,9

Source : Eurostat and FAO.

(¹) Intra and extra.

4.6.4.1   Supply balance — wine

EUR 12

	1 000 hl			% TAV	
	1985/86	1991/92 ∞	1992/93 ∞	1991/92 / 1985/86	1992/93 / 1991/92
1	2	3	4	5	6
1. Total wine :					
Usable production	185 735	160 650	197 676	− 2,4	23,0
Change in stocks	295	12 452	3 681	86,6	− 70,4
Imports	4 614	3 324	3 298	− 5,3	− 0,8
Exports	17 053	8 525	9 592	− 10,9	12,5
Intra-EC trade	20 597	26 335	25 503	4,2	− 3,2
Internal uses	173 001	163 566	181 003	− 0,9	10,7
— losses – production	545	529	944	− 0,5	78,4
– marketing	519	326	566	− 7,5	73,6
— processing	40 578	34 748	50 476	− 2,6	45,3
— human consumption	131 359	127 554	129 016	− 0,5	1,1
Human consumption (l/head)	40,8	37,3	37,7	− 1,5	1,1
Self-sufficiency (%)	126,7	112	133	− 2,0	18,8
2. Quality wines produced in specified regions (Total) :					
Usable production	44 665	49 416	59 099	1,7	19,6
Internal uses	32 583	45 550	49 271	5,7	8,2
3. Table wines (Total) :					
Usable production	120 904	101 205	117 385	− 2,9	16,0
Internal uses of which :	115 410	102 184	108 527	− 2,0	6,2
— human consumption	86 806	73 710	71 443	− 2,7	− 3,1
— Community distillation (¹)	21 929	24 430	32 878	1,8	34,6

Source : Eurostat and EC Commission, Directorate-General for Agriculture.

(¹) Excluding distillation for the production of wine spirits bearing a designation of origin and national distillation operations.

4.6.5.1   Producer prices (¹) for table wines

	ECU				% TAV	
	1985/86	1991/92	1992/93	1993/94	1992/93 ÷ 1985/86	1993/94 ÷ 1992/93
1	2	3	4	5	6	7
Type R I : Red, 10 to 12°, % vol./hl						
Elláda	3,050	3,800		2,130	x	x
Heraklion					x	x
Patras	3,050	3,800		2,130	x	x
España	2,457	2,148	1,723	1,824	−4,9	5,9
Requena	2,257	2,096	1,718	1,831	−3,8	6,6
Reus		2,532	1,642	1,674		1,9
Villafranca del Bierzo	2,713	2,297	1,834	1,686	−5,4	−8,1
France	2,662	3,115	3,005	3,044	1,7	1,3
Bastia	2,450	3,099	2,912	3,158	2,5	8,4
Béziers	2,657	3,111	2,989	3,014	1,7	0,8
Montpellier	2,671	3,117	3,016	3,090	1,8	2,5
Narbonne	2,683	3,161	3,059	3,060	1,9	0,0
Nîmes	2,664	3,116	2,998	3,043	1,7	1,5
Perpignan	2,695	3,086	2,716	3,023	0,1	11,3
Italia	3,037	2,356	2,043	2,299	−5,5	12,5
Asti	3,347	2,797		2,716	x	x
Firenze	2,578	2,157	1,754	1,656	−5,4	−5,6
Lecce					x	x
Pescara	2,691	1,612	2,103	2,031	−3,5	−3,4
Reggio Emilia	3,135	3,375	2,845	2,366	−1,4	−16,8
Treviso	2,762	2,895	2,144	2,104	−3,6	−1,9
Verona (local wines)	2,785	3,009		2,496	x	x
E.C.	2,699	2,601	2,849	2,499	0,8	−12,3
Type R II : Red, 12.5 to 15°, % vol./hl						
Elláda					x	x
Heraklion					x	x
Patras					x	x
España	2,843	2,160	1,962	1,873	−5,2	−4,5
Calatayud					x	x
Falset	2,947	2,956	1,852	1,966	−6,4	6,2
Jumilla	2,861	2,428	2,028	1,927	−4,8	−5,0
Navalcarnero	2,747	2,500	1,840	2,284	−5,6	24,1
Requena		2,143	1,431		x	x
Toro					x	x
Villena	2,867	2,996	2,015	2,385	−4,9	18,4
France	2,464	2,722	2,979	2,848	2,7	−4,4
Bastia	2,464	2,722	2,979	2,848	2,7	−4,4
Brignoles					x	x

Italia	2,676	2,528	2,193	1,889	-2,8	-13,9
Bari	2,676	2,507	2,108	1,861	-3,4	-11,7
Barletta	:	2,549	2,118	1,978	x	-6,6
Cagliari	2,935	3,667	3,092	2,703	0,7	-12,6
Lecce	2,711	:	:	:	x	x
Taranto	2,627	:	:	:	x	x
E.C.	2,603	2,531	2,306	2,113	-1,7	-8,4

Type R III : Red, Portuguese type, hl

BR Deutschland	115,154	47,817	:	51,328	x	x
Rheinpfalz-Rheinhessen (Hügelland)	115,154	47,817	:	:	x	x
E.C.	115,154	47,817	:	:	x	x

Type A I : White, 10 to 13°, % vol./hl

Ellada	3,005	3,800	:	:	x	x
Athens	2,907	:	:	:	x	x
Heraklion	:	:	:	:	x	x
Patras	3,080	3,800	:	:	x	x
España	1,991	1,957	1,379	2,159	-5,1	56,6
Alcázar de San Juan	2,084	1,826	1,366	2,159	-5,9	58,1
Almedralejo	1,947	1,933	1,376	1,631	-4,8	18,5
Medina del Campo	:	:	:	:	x	x
Ribadavia	:	:	:	:	x	x
Villafranca del Penedès	2,186	2,857	2,077	2,496	-0,7	20,2
Villar del Arzobispo	:	2,071	:	:	x	x
Villarrobledo	2,041	1,941	1,469	2,068	-4,6	40,8
France	2,902	4,235	:	:	x	x
Bordeaux	2,949	4,235	:	:	x	x
Nantes	2,840	:	:	:	x	x
Italia	2,621	2,682	1,985	1,865	-3,9	-6,0
Bari	2,456	2,694	1,918	1,902	-3,5	-0,8
Cagliari	2,632	3,208	2,655	2,198	0,1	-17,2
Chieti	2,576	2,782	2,051	1,741	-3,2	-15,1
Ravenna (Lugo, Faenza)	2,773	2,825	2,010	1,973	-4,5	-1,8
Trapani (Alcamo)	2,373	2,252	1,796	1,642	-3,9	-8,6
Treviso	2,991	3,009	2,195	2,129	-4,3	-3,0
E.C.	2,599	2,446	1,783	1,790	-5,2	0,4

Type A II : White, Sylvaner type, hl

BR Deutschland	83,238	57,716	33,543	40,309	-12,2	20,2
Rheinpfalz (Oberhaardt)	82,681	55,848	34,579	39,944	-11,7	15,5
Rheinhessen (Hügelland)	85,186	59,262	32,686	40,548	-12,8	24,1
E.C.	83,238	57,716	33,600	40,309	-12,2	20,0

Type A III : White, Riesling type, hl

BR Deutschland	69,590	77,207	:	40,834	x	x
Mosel/Rheingau	69,590	77,207	:	40,834	x	x
E.C.	69,590	77,207	:	40,834	x	x

Source : EC Commission, Directorate-General for Agriculture.
(1) Weighted average market prices.

4.7.1.1   Area, yield and production of potatoes

	Area					Yield					Production				
	1 000 ha			% TAV		100 kg/ha			% TAV		1 000 t			% TAV	
	1985	1992 ∞	1993 ∞	1992/1985	1993/1992	1985	1992 ∞	1993 ∞	1992/1985	1993/1992	1985	1992 ∞	1993 ∞	1992/1985	1993/1992
1	2	3	4	5	6	7	8	9	10	11	12	13	14	15	16
EUR 12	1 545	1 574	1 366	0,3	−13,2	277	309	333	1,6	7,8	42 832	47 128	45 534	1,4	−3,4
Belgique/België	48	62	50	3,7	−19,4	376	415	419	1,4	1,0	1 805	2 616	2 093	5,4	−20,0
Danmark	30	54	46	8,8	−14,8	358	329	378	−1,2	14,9	1 073	1 775	1 741	7,5	−1,9
BR Deutschland	220	361	312	7,3	−13,6	359	302	393	−2,4	30,1	7 905	10 897	12 260	4,7	12,5
Elláda	49	43	49	−1,8	14,0	206	228	208	1,5	−8,8	1 009	980	1 021	−0,4	4,2
España	331	257	212	−3,6	−17,5	175	202	188	2,1	−6,9	5 781	5 181	3 977	−1,6	−23,2
France	210	184	164	−1,9	−10,9	326	363	371	1,5	2,2	6 856	6 676	5 801	−0,4	−13,1
Ireland	33	22	22	−5,6	0,0	208	292	183	5,0	−37,3	686	642	402	−0,9	−37,4
Italia	136	101	85	−4,2	−15,8	176	247	238	5,0	−3,6	2 397	2 498	2 026	0,6	−18,9
Luxembourg	1	1	1	0,0	0,0	290	270	23	−1,0	−91,5	29	27	23	−1,0	−14,8
Nederland	169	187	166	1,5	−11,2	423	409	464	−0,5	13,4	7 150	7 641	7 699	1,0	0,8
Portugal	126	105	87	−2,6	−17,1	99	143	158	5,4	10,5	1 249	1 500	1 373	2,7	−8,5
United Kingdom	192	180	170	−0,9	−5,6	359	433	419	2,7	−3,2	6 892	7 802	7 117	1,8	−8,8

Source : Eurostat.

4.7.1.2 Area, yield and production of early potatoes

	Area					Yield					Production				
	1000 ha			% TAV		100 kg/ha			% TAV		1000 t			% TAV	
1	1985	1992 ∞	1993 ∞	1992/1985	1993/1992	1985	1992 ∞	1993 ∞	1992/1985	1993/1992	1985	1992 ∞	1993 ∞	1992/1985	1993/1992
	2	3	4	5	6	7	8	9	10	11	12	13	14	15	16
EUR 12	161	155	152	−0,5	−1,9	206	218	210	0,8	−3,7	3 321	3 378	3 195	0,2	−5,4
Belgique/België	5	9	6	8,8	−33,3	212	243	310	2,0	27,6	106	244	186	12,6	−23,8
BR Deutschland	22	25	22	1,8	−12,0	286	258	330	−1,5	27,9	630	650	660	0,4	1,5
Elláda	16	14	13	−1,9	−7,1	219	236	250	1,1	5,9	350	330	325	−0,8	−1,5
España	50	43	48	−2,1	11,6	170	192	138	1,8	−28,1	850	787	663	−1,1	−15,8
France	24	24	23	0,0	−4,2	202	240	212	2,5	−11,7	484	557	487	2,0	−12,6
Italia	28	27	24	−0,5	−11,1	178	166	187	−1,0	12,7	498	495	449	−0,1	−9,3
United Kingdom	16	15	16	−0,9	6,7	252	279	266	1,5	−4,7	403	408	425	0,2	4,2

Source : Eurostat.

4.7.4.1   **Supply balance — potatoes**   **EUR 12**

	1 000 t			% TAV	
	1985/86	1991/92 ∞	1992/93 ∞	1991/92 / 1985/86	1992/93 / 1991/92
1	2	3	4	5	6
Usable production	43 908	43 094	48 059	− 0,3	11,5
Change in stocks	142	− 49	440	×	×
Imports	430	570	738	4,8	29,5
Exports	1 169	843	8 085	− 5,3	859,1
Intra-EC trade	4 960	7 007	6 420	5,9	− 8,4
Internal use	43 040	42 833	46 448	− 0,1	8,4
of which:					
— animal feed	6 117	3 441	4 753	− 9,1	38,1
— seed	3 072	3 288	2 962	1,1	− 9,9
— industrial use	475	550	584	2,5	6,2
— alcohol	475	550	584	2,5	6,2
— processing	5 170	8 270	7 840	8,1	− 5,2
— losses (market)	2 045	1 661	1 712	− 3,4	3,1
— human consumption	26 161	27 103	28 598	0,6	5,5
Human consumption (kg/head/year)	81,2	78,3	82,47	− 0,6	5,3
Self-sufficiency (%)	102	100,6	103,47	− 0,2	2,9

Source: Eurostat.

4.8.1.1   Area, yield and production of leaf tobacco, by groups of varieties

1	2	Area ha 1985	Area ha 1992 ∞	Area ha 1993 ∞	% TAV 1992/1985	% TAV 1993/1992
		3	4	5	6	7
	EUR 12	29 489	72 528	51 847	13,7	− 28,5
I Flue cured	BR Deutschland	424	912	1 041	11,6	14,1
	Elláda	164	28 796	12 362	109,2	− 57,1
	España	5 626	13 800	12 410	13,7	− 10,1
	France	2 647	3 560	3 963	4,3	11,3
	Italia	19 274	23 561	19 984	2,9	− 15,2
	Portugal	1 354	1 899	2 087	5,0	9,9
	EUR 12	34 477	29 624	26 267	− 2,1	− 11,3
II Light air cured	BR Deutschland	1 069	1 717	1 547	7,0	− 9,9
	Elláda	9 926	5 213	3 664	− 8,8	− 29,7
	España	3 248	2 906	1 930	− 1,6	− 33,6
	France	1 815	2 560	2 619	5,0	2,3
	Italia	17 974	16 936	16 175	− 0,8	− 4,5
	Portugal	445	292	332	− 5,8	13,7
	EUR 12	44 253	23 282	21 257	− 8,8	− 8,7
III Dark air cured	Belgique/België	543	417	400	− 3,7	− 4,1
	BR Deutschland	1 495	1 172	1 205	− 3,4	2,8
	España	15 530	3 070	3 252	− 20,7	5,9
	France	10 406	5 598	4 515	− 8,5	− 19,3
	Italia	16 279	13 025	11 885	− 3,1	− 8,8
	EUR 12	6 577	3 463	3 435	− 8,8	− 0,8
IV Fire cured	España	:	6	11	×	83,3
	France	:	1	:	×	×
	Italia	6 577	3 456	3 424	− 8,8	− 0,9
	EUR 12	46 967	27 382	17 239	− 7,4	− 37,0
V Sun cured	Elláda	25 897	19 255	9 304	− 4,1	− 51,7
	Italia	21 070	8 127	7 935	− 12,7	− 2,4
	EUR 12	62 692	49 047	47 169	− 3,4	− 3,8
VI, VII, VIII Special sun cured	Elláda	62 692	49 047	47 169	− 3,4	− 3,8
	EUR 12	224 455	205 326	167 214	− 1,3	− 18,6
Raw tobacco	Belgique/België	543	417	400	− 3,7	− 4,1
	BR Deutschland	2 988	3 801	3 793	3,5	− 0,2
	Elláda	98 679	102 311	72 499	0,5	− 29,1
	España	24 404	19 782	17 603	− 3,0	− 11,0
	France	14 868	11 719	11 097	− 3,3	− 5,3
	Italia	81 174	65 105	59 403	− 3,1	− 8,8
	Portugal	1 799	2 191	2 419	2,9	10,4

NB: Classification of tobacco varieties as set out in the Annex to Regulation (EEC) No 2075/92, 30.6.1992.
Source: EC Commission, Directorate-General for Agriculture.

Yield					Production				
100 kg/ha			% TAV		t			% TAV	
1985	1992 ∞	1993 ∞	$\frac{1992}{1985}$	$\frac{1993}{1992}$	1985	1992 ∞	1993 ∞	$\frac{1992}{1985}$	$\frac{1993}{1992}$
8	9	10	11	12	13	14	15	16	17
21,8	23,0	23,9	0,8	4,0	64 387	167 040	124 129	14,6	− 25,7
17,9	18,0	17,1	0,1	− 5,1	758	1 640	1 777	11,7	8,4
22,9	24,8	30,7	1,2	23,5	375	71 526	37 921	111,7	− 47,0
17,2	21,9	22,2	3,5	1,7	9 677	30 158	27 569	17,6	− 8,6
22,0	18,5	20,1	− 2,4	8,2	5 834	6 602	7 951	1,8	20,4
23,3	22,7	23,6	− 0,3	4,0	44 831	53 517	47 193	2,6	− 11,8
21,5	18,9	8,2	− 1,8	− 56,5	2 912	3 597	1 718	3,1	− 52,2
28,4	27,0	27,1	− 0,7	0,4	97 765	79 859	71 086	− 2,8	− 11,0
27,7	25,0	20,4	− 1,5	− 18,3	2 962	4 290	3 158	5,4	− 26,4
30,6	25,2	31,5	− 2,7	25,0	30 341	13 127	11 530	− 11,3	− 12,2
18,1	23,0	24,3	3,5	5,7	5 879	6 681	4 689	1,8	− 29,8
22,9	20,0	23,4	− 2,0	17,1	4 163	5 113	6 125	3,0	19,8
29,7	29,5	27,7	− 0,1	− 6,0	53 385	49 890	44 775	− 1,0	− 10,3
23,3	26,0	24,4	1,6	− 6,1	1 035	758	809	− 4,4	6,7
18,9	23,0	21,4	2,8	− 6,8	83 530	53 481	45 524	− 6,2	− 14,9
38,6	33,8	37,6	− 1,9	11,2	2 097	1 409	1 503	− 5,5	6,7
29,4	27,0	26,6	− 1,2	− 1,3	4 393	3 160	3 206	− 4,6	1,5
16,7	23,3	26,0	4,9	11,7	25 900	7 156	8 466	− 16,8	18,3
24,5	22,4	26,5	− 1,3	18,2	25 520	12 567	11 980	− 9,6	− 4,7
15,7	22,4	17,1	5,2	− 23,5	25 620	29 189	20 369	1,9	− 30,2
18,3	18,8	19,5	0,4	3,6	12 038	6 513	6 690	− 8,4	2,7
×	18,3	19,1	×	4,1	:	11	21	×	90,9
×	×	×	×	×	:	0	:	×	×
18,3	18,8	19,5	0,4	3,5	12 038	6 502	6 669	− 8,4	2,6
14,9	15,1	18,6	0,2	23,3	70 008	41 263	32 042	− 7,3	− 22,3
15,3	15,2	20,1	− 0,1	32,3	39 730	29 305	18 740	− 4,3	− 36,1
14,4	14,7	16,8	0,3	13,9	30 278	11 958	13 302	− 12,4	11,2
12,4	12,2	13,3	− 0,3	9,1	78 029	59 628	62 563	− 3,8	4,9
12,4	12,2	13,3	− 0,3	9,1	78 029	59 628	62 563	− 3,8	4,9
18,1	19,9	20,5	1,4	3,0	405 757	407 784	342 034	0,1	− 16,1
38,6	33,8	37,6	− 1,9	11,2	2 097	1 409	1 503	− 5,5	6,7
27,2	23,9	21,5	− 1,8	− 10,3	8 113	9 090	8 141	1,6	− 10,4
15,0	17,0	18,0	1,7	6,3	148 475	173 586	130 754	2,3	− 24,7
17,0	22,2	23,1	3,9	4,1	41 456	44 006	40 745	0,9	− 7,4
23,9	20,7	23,5	− 2,0	13,3	35 517	24 282	26 056	− 5,3	7,3
20,5	23,2	22,3	1,8	− 4,0	166 152	151 056	132 308	− 1,4	− 12,4
21,9	19,9	10,4	− 1,4	− 47,4	3 947	4 355	2 527	1,4	− 42,0

4.8.2.1   World production of raw tobacco and production in principal exporting countries

	%			1 000 t			% TAV	
	1985	1992	1993	1985	1992 ∞	1993 ∞	$\frac{1992}{1985}$	$\frac{1993}{1992}$
1	2	3	4	5	6	7	8	9
World	100	100	100	7 000,0	8 324,9	8 352,2	2,5	0,3
of which:								
— EUR 12	5,8	4,9	4,1	405,8	407,8	342,0	0,1	− 16,1
— Turkey	2,4	4,0	3,9	170,0	331,8	326,1	10,0	− 1,7
— USSR/CIS	5,4	1,7	2,9	376,0	143,3	245,1	− 12,9	71,0
— Bulgaria	1,7	0,9	0,6	119,0	71,9	49,7	− 6,9	− 30,9
— Zimbabwe	1,6	2,5	2,8	111,0	211,4	235,3	9,6	11,3
— Malawi	1,0	1,7	1,6	67,0	137,9	133,8	10,9	− 3,0
— India	6,9	7,0	7,0	486,0	584,4	580,6	2,7	− 0,7
— Rep. of Korea	1,1	1,0	1,3	76,0	79,6	106,5	0,7	33,8
— USA	9,8	9,4	8,8	686,0	780,9	731,9	1,9	− 6,3
— Canada	1,3	0,9	1,0	88,0	71,8	83,7	− 2,9	16,6
— Mexico	0,8	0,4	0,9	54,0	29,8	71,4	− 8,1	139,6
— Brazil	5,9	6,9	7,3	411,0	577,0	608,0	5,0	5,4
— Argentina	0,9	1,3	1,3	61,0	108,6	112,3	8,6	3,4
— Peop. Rep. China	34,3	42,0	41,4	2 400,0	3 499,0	3 456,6	5,5	− 1,2

Source : EC Commission, Directorate-General for Agriculture and USDA.

4.8.3.1   EC share of world trade (¹) in raw tobacco

	Provenance or destination %	1 000 t			% TAV	
		1991	1992	1993	$\frac{1992}{1991}$	$\frac{1993}{1992}$
1	2	3	4	5	6	7
Imports	World	1 634,4	1 670,7	1 750,2	2,2	4,8
	EUR 12 ∞	527,7	526,9	417,4	− 0,2	− 20,8
	%	32,3	31,5	23,8	×	×
Exports	World	1 689,3	1 628,9	1 776,6	− 3,6	9,1
	EUR 12 ∞	210,6	199,8	209,7	− 5,1	5,0
	%	12,5	12,3	11,8	×	×

Source : Eurostat and World Tobacco Situation (USDA).

(¹) Excl. intra-EC trade, except for 1993.

4.8.3.2   **EC tobacco exports to third countries**   **EUR 12**

Destination	t			% of 1993 total	% TAV	
	1991 ∞	1992 ∞	1993 ∞		$\frac{1992}{1991}$	$\frac{1993}{1992}$
1	2	3	4	5	6	7
USA	36 796	44 320	33 530	16,0	20,4	− 24,3
Russia	0	7 424	32 832	15,7	×	342,2
Egypt	14 341	23 422	15 632	7,5	63,3	− 33,3
Algeria	11 823	12 924	15 553	7,4	9,3	20,3
Romania	4 799	1 377	14 370	6,9	− 71,3	943,6
Switzerland	9 779	9 688	9 281	4,4	− 0,9	− 4,2
Japan	6 636	11 640	9 082	4,3	75,4	− 22,0
Poland	10 617	7 145	7 439	3,5	− 32,7	4,1
Bulgaria	6 370	8 641	5 520	2,6	35,7	− 36,1
Cuba	6 994	7 691	5 458	2,6	10,0	− 29,0
Hungary	10 738	2 771	4 786	2,3	− 74,2	72,7
Czech Republic	0	0	3 933	1,9	×	×
Mexico	2 124	2 915	3 866	1,8	37,2	32,6
Territory of the former Yugoslav Rep. of Macedonia	0	0	3 246	1,5	×	×
Tunisia	2 673	3 373	2 778	1,3	26,2	− 17,6
Others	86 946	56 446	42 432	20,2	− 35,1	− 24,8
World	210 636	199 777	209 738	100,0	− 5,2	5,0

Source : Eurostat.

4.8.3.3   Imports and exports of raw tobacco

EUR 12

(1000 t)

1	1991 ∞			1992 ∞			1993 ∞		
	Intra	Extra	Total	Intra	Extra	Total	Intra	Extra	Total
1	2	3	4	5	6	7	8	9	10
A. Imports									
Flue cured Virginia	23,4	255,2	278,6	28,2	244,0	272,2	:	192,7	192,7
Light air cured Burley	17,5	61,6	79,1	19,1	69,4	88,5	:	54,9	54,9
Light air cured Maryland	0,1	1,7	1,8	0,0	1,1	1,1	:	1,5	1,5
Fire cured Kentucky	2,5	8,6	11,1	2,3	11,2	13,5	:	8,6	8,6
Other fire cured tobacco	0,9	9,0	9,9	0,7	8,2	8,9	:	8,2	8,2
Light air cured (other)	6,9	5,9	12,8	6,4	5,3	11,7	:	5,0	5
Sun cured	24,9	26,8	51,7	23,1	34,5	57,6	:	27,0	27
Dark air cured	7,8	44,6	52,4	7,5	47,2	54,7	:	35,9	35,9
Flue cured (other)	10,4	22,7	33,1	10,9	21,4	32,3	:	19,6	19,6
Other tobacco	16,4	11,2	27,6	11,5	10,2	21,7	:	5,3	5,3
Tobacco refuse	31,2	80,4	111,6	26,2	74,6	100,8	:	58,6	58,6
Total	142,0	527,7	669,7	135,9	527,1	663,0		417,3	417,3
B. Exports									
Flue cured Virginia	30,0	25,0	55,0	30,7	23,1	53,8	:	17,2	17,2
Light air cured Burley	27,7	33,1	60,8	18,4	27,7	46,1	:	24,3	24,3
Light air cured Maryland	0,2	0,8	1,0	0,2	1,0	1,2	:	1,3	1,3
Fire cured Kentucky	2,6	2,5	5,1	1,9	2,3	4,2	:	2,9	2,9
Other fire cured tobacco	2,3	1,8	4,1	1,3	2,9	4,2	:	1,7	1,7
Light air cured (other)	0,2	0,0	0,2	0,1	0,0	0,1	:	0,0	0,0
Sun cured	23,2	72,7	95,9	22,3	78,6	100,9	:	88,8	88,8
Dark air cured	9,1	54,8	63,9	6,7	34,1	40,8	:	30,0	30,0
Flue cured (other)	7,9	8,5	16,4	7,0	9,5	16,5	:	20,8	20,8
Other tobacco	14,5	7,0	21,5	10,3	7,0	17,3	:	10,8	10,8
Tobacco refuse	23,6	4,5	28,1	20,0	13,3	33,3	:	11,9	11,9
Total	141,3	210,7	352,0	118,9	199,5	318,4	:	209,7	209,7

Source : Eurostat.

4.8.6.1   **Quantities of tobacco bought in**

	t			% of production		
	Harvest					
	1990	1991	1992	1990	1991	1992
1	2	3	4	5	6	7
Paraguay	:	:	:	×	×	×
Bad. Geudertheimer	:	:	:	×	×	×
Bad. Burley	89	:	:	1,9	×	×
Bright	:	463	3 270	×	0,9	6,9
Burley I	:	:	:	×	×	×
Maryland	:	:	:	×	×	×
Kentucky	92	313	:	1,3	4,3	×
F. Havanna	717	158	:	3,4	3,3	×
Xanti-Yaka	:	91	26	×	1,6	0,6
Perustitza	:	:	:	×	×	×
Erzegovina	:	21	:	×	1,0	×
Basmas	1 578	1 635	2 638	8,6	6,7	12,7
Katerini	670	431	454	4,1	2,5	2,9
Kaba Koulak c.	738	647	819	6,0	5,5	6,9
Kaba Koulak n.c.	24	54	73	1,6	3,8	6,3
Myrodata	9	76	70	0,2	1,8	1,5
Zichnomyrodata	18	23	:	5,1	7,6	×
Tsebelia	3 145	340	51	16,4	1,8	0,3
Mavra	1 093	291	:	13,4	3,4	×
Burley EL	:	:	:	×	×	×
Virginia EL	:	:	204	×	×	0,3
Total	8 173	4 543	7 605	2,6	1,6	2,7

Source : EC Commission, Directorate-General for Agriculture.

4.9.1.1 Seed production and related aid (1993)

Product	100 kg					
	EUR 12	Belgique/ België	Danmark	BR Deutschland	Elláda	España
1	2	3	4	5	6	7
1. *Gramineae*						
Festuca pratensis Huds.	21 999	0	8 560	12 840	0	0
Poa pratensis L.	80 648	396	43 170	2 124	0	0
Poa trivialis L.	4 580	0	4 580	0	0	0
Lolium per. L. (high persistence)	379 382	4 704	114 070	35 826	0	0
Lolium per. L. (new. var. & others)	203 160	3 996	99 770	32 888	0	275
Lolium per. L. (low persistence)	40 762	0	2 850	0	0	0
Lolium multiflorum Lam.	302 272	20 622	42 490	138 639	0	7 900
Phleum pratense L.	49 336	0	730	5 688	0	0
Phleum bertolonii (DC)	229	0	50	0	0	0
Festuca rubra L.	249 900	1 472	153 620	16 655	0	0
Dactylis glomerata L.	33 637	0	8 650	1 248	0	275
Agrostis canina L.	0	0	0	0	0	0
Agrostis gigantea Roth.	0	0	0	0	0	0
Agrostis stolonifera L.	196	0	100	0	0	0
Agrostis capillaris L.	905	0	0	42	0	0
Festuca ovina L.	19 259	1 413	5 460	8 282	0	0
Lolium X hybridum Hausskn.	25 373	0	11 610	2 772	0	0
Arrhenatherum elatius L – P	1 047	0	0	1 047	0	0
Festuca arundinacae Schreb.	37 156	0	190	163	0	350
Poa nemoralis L.	1 681	0	0	0	0	0
Festololium	0	0	0	0	0	0
2. *Leguminosae*						
Pisum sativum L. partim	1 334 695	0	0	164 263	420	12 540
Vicia faba L. partim	301 013	0	0	79 091	0	1 175
Vicia sativa L.	257 380	150	0	2 363	45 000	56 400
Vicia villosa roth.	3 133	0	0	868	0	1 700
Trifolium pratense L.	25 293	35	380	3 874	0	0
Trifolium repens L.	9 002	0	8 660	342	0	0
Trifolium repens L. giganteum	976	0	0	472	0	0
Trifolium alexandrinum L.	10 626	0	0	0	0	0
Trifolium hybridum L.	4	0	2	0	0	0
Trifolium incarnatum L.	15 099	0	0	165	0	0
Trifolium resupinatum L.	3 629	0	0	0	0	0
Medicago sativa L. (ecotypes)	79 745	0	0	0	4 500	16 700
Medicago sativa L. (varieties)	66 028	0	60	55	0	4 550
Medicago lupulina L.	2 250	0	2 250	0	0	0
Onobrichis viciifolia scop.	872	0	0	0	0	0
Hedysarium coronarium L.	226	0	0	0	0	0
3. *Ceres*						
Triticum spelta L.	21 619	14 700	0	6 919	0	0
Oryza sativa L.	39 480	0	0	0	16 000	0
— type japonica	187 815	0	0	0	0	87 815
— type indica	550 540	0	0	0	0	150 450
4. *Oleagineae*						
Linum usitatiss. (fibre flax)	105 660	52 560	2 600	0	0	0
Linum usitatiss. (seed flax)	170 202	250	0	26 202	0	0
Cannabis sativa L.	3 780	0	0	0	0	0

Source : EC Commission, Directorate-General for Agriculture.

| | | 100 kg | | | | | ECU/ 100 kg | 1 000 ECU EUR 12 |
France	Ireland	Italia	Luxembourg	Nederland	Portugal	United Kingdom		
8	9	10	11	12	13	14	15	16
20	:	0	0	333	0	246	36,1	794
0	:	0	0	34 958	0	0	31,9	2 573
0	:	0	0	0	0	0	32,2	147
25 000	:	0	694	151 188	0	47 900	28,9	10 964
23 000	:	926	110	:	0	42 195	21,5	4 368
0	:	0	0	37 912	0	0	15,8	648
50 000	:	20 127	3 888	0	0	18 606	17,5	5 290
100	:	85	6	41 639	0	1 088	69,2	3 414
0	:	0	0	179	0	0	42,2	10
22 000	:	0	0	51 831	0	4 322	36,1	9 021
22 000	:	47	0	104	0	1 313	43,7	1 470
0	:	0	0	0	0	0	62,9	0
0	:	0	0	0	0	0	62,9	0
0	:	0	0	96	0	0	62,9	12
0	:	0	0	863	0	0	62,9	57
2 000	:	0	0	2 104	0	0	36,1	695
6 000	:	0	116	727	0	4 148	17,5	444
0	:	0	0	0	0	0	55,6	58
20 000	:	54	0	16 325	0	74	48,8	1 813
0	:	0	0	1 681	0	0	32,2	54
0	:	0	0	0	0	0	26,8	0
950 000	:	0	2 772	33 000	0	171 700	0,0	0
14 000	:	0	547	4 300	0	201 900	0,0	0
110 000	:	37 224	0	0	60	6 183	25,4	6 537
500	:	40	0	25	0	0	19,9	62
20 000	:	917	20	0	0	67	43,4	1 120
0	:	0	0	0	0	0	62,2	560
0	:	135	0	0	0	369	58,6	57
0	:	10 606	0	0	20	0	37,9	403
0	:	2	0	0	0	0	38,0	0
7 000	:	7 934	0	0	0	0	37,9	572
1 500	:	2 129	0	0	0	0	37,9	138
0	:	58 545	0	0	0	0	18,3	1 459
40 000	:	21 363	0	0	0	0	30,3	2 001
0	:	0	0	0	0	0	26,4	59
800	:	72	0	0	0	0	16,6	14
0	:	226	0	0	0	0	30,2	7
0	:	0	0	0	0	0	11,9	257
12 000	:	0	0	0	11 480	0	13,8	545
0	:	100 000	0	0	:	0	12,3	7 873
0	:	400 000	0	0	:	0	14,3	2 310
23 500	:	0	0	26 000	0	1 000	23,5	2 483
30 000	:	0	0	3 400	:	110 350	18,6	3 166
3 780	:	0	0	0	0	0	17,0	64

4.9.1.2   Area under seed (1993/94)

Product	Total	Belgique/ Belgïe	Danmark	BR Deutschland	Elláda
1	2	3	4	5	6
1. Gramineae					
Festuca pratensis Huds.	3 150	0	931	2 140	0
Poa pratensis L.	10 364	36	6 038	531	0
Poa palustris & triviali	594	0	594	0	0
Lolium per. L. (high persistence)	25 134	0	12 831	5 118	0
Lolium per. L. (new. var. & others)	34 005	392	10 315	4 385	0
Lolium per. L. (low persistence)	640	333	307	0	0
Lolium multiflorum Lam.	30 474	1 473	3 891	14 211	0
Phleum pratense L.	2 071	0	203	1 625	0
Phleum bertolonii (DC)	14	0	14	0	0
Festuca rubra L.	27 746	157	15 257	3 331	0
Dactylis glomerata L.	4 929	0	1 158	378	0
Agrostis canina L.	0	0	0	0	0
Agrostis gigantea Roth.	0	0	0	0	0
Agrostis stolonifera L.	37	0	0	15	0
Agrostis capillaris L.	187	0	0	23	0
Festuca ovina L.	3 717	157	713	2 366	0
Lolium X hybridum Hausskn.	2 508	0	1 111	396	0
Arrhenatherum elatius L – P	238	0	0	238	0
Festuca arundinacae Schreb.	4 113	0	45	25	0
Poa nemoralis L.	121	0	0	0	0
Festololium	0	0	0	0	0
2. Leguminosae					
Pisum sativum L. partim	56 504	0	15 366	8 511	42
Vicia faba L. partim	11 366	0	134	2 845	0
Vicia sativa L.	15 705	5	1	303	4 000
Vicia villosa roth.	528	0	0	248	0
Trifolium pratense L.	6 902	7	85	1 937	0
Trifolium repens L.	2 863	0	2 665	190	0
Trifolium repens L. giganteum	646	0	0	295	0
Trifolium alexandrinum L.	1 857	0	0	0	0
Trifolium hybridum L.	16	0	10	0	0
Trifolium incarnatum L.	3 041	0	0	33	0
Trifolium resupinatum L.	576	0	0	0	0
Medicago sativa L. (ecotypes)	16 049	0	0	0	0
Medicago sativa L. (varieties)	20 462	0	10	55	2 000
Medicago lupulina L.	216	0	215	0	0
Onobrichis viciifolia scop.	179	0	0	0	0
Hedysarium coronarium L.	0	0	0	0	0
3. Ceres					
Triticum spelta L.	608	420	0	188	0
Oryza sativa L.	4 466	0	0	0	4 000
— type japonica	11 935	0	0	0	:
— type indica	3 638	0	0	0	:
4. Oleagineae					
Linum usitatiss. (fibre flax)	14 025	4 380	0	0	0
Linum usitatiss. (seed flax)	16 459	10	235	2 382	0
Cannabis sativa L.	418	0	0	0	0

Source : EC Commission, Directorate-General for Agriculture.

(ha)

España	France	Ireland	Italia	Luxembourg	Nederland	Portugal	United Kingdom
7	8	9	10	11	12	13	14
0	8	0	0	0	53	0	18
0	9	0	0	0	3 750	0	0
0	0	0	0	0	0	0	0
0	2 399	0	0	53	0	0	4 733
45	2 229	260	93	90	12 816	0	3 461
0	0	0	0	389	0	0	0
800	3 987	34	1 406	1	2 688	0	1 595
0	21	0	8	0	17	0	196
0	0	0	0	0	0	0	0
0	2 982	0	0	0	5 067	0	952
55	3 107	0	47	0	15	0	169
0	0	0	0	0	0	0	0
0	0	0	0	0	0	0	0
0	0	0	0	0	22	0	0
0	2	0	0	0	162	0	0
0	219	0	0	0	253	0	9
0	570	0	0	15	60	0	356
0	0	0	0	0	0	0	0
50	2 805	0	34	0	1 121	0	33
0	0	0	0	0	121	0	0
0	0	0	0	0	0	0	0
418	22 598	0	0	79	838	0	8 652
47	455	910	0	14	85	0	6 876
4 700	4 394	0	1 520	0	0	0	737
180	90	0	1	0	9	0	0
0	4 640	0	191	5	0	0	37
0	8	0	0	0	0	0	0
0	6	0	65	0	0	0	280
0	2	0	1 842	0	0	13	0
0	5	0	1	0	0	0	0
0	1 386	0	1 622	0	0	0	0
0	319	0	257	0	0	0	0
4 175	0	0	11 874	0	0	0	0
1 250	13 090	0	4 057	0	0	0	0
0	0	0	1	0	0	0	0
0	167	0	12	0	0	0	0
0	0	0	0	0	0	0	0
0	0	0	0	0	0	0	0
:	293	0	:	0	0	173	0
1 351	:	:	10 584	:	:	:	:
2 316	:	:	1 322	:	:	:	:
0	6 784	0	0	0	2 743	0	118
0	1 895	73	0	0	156	0	11 708
0	418	0	0	0	0	0	0

4.10.1.1　Area, yield and production of hops

	Area					Yield					Production				
	ha			% TAV		100 kg/ha			% TAV		t			% TAV	
	1985	1992 ∞	1993 ∞	1992/1985	1993/1992	1985	1992 ∞	1993 ∞	1992/1985	1993/1992	1985	1992 ∞	1993 ∞	1992/1985	1993/1992
1	2	3	4	5	6	7	8	9	10	11	12	13	14	15	16
EUR 12	25 748	28 554	28 675	1,5	0,4	17,3	12,7	18,0	−4,3	41,7	44 644	36 367	51 695	−2,9	42,1
Belgique/België	701	394	409	−7,9	3,8	16,3	15,5	14,3	−0,7	−7,7	1 141	609	585	−8,6	−3,9
BR Deutschland	19 598	22 938	23 015	2,3	0,3	18,2	12,5	18,4	−5,2	47,2	35 697	28 725	42 428	−3,1	47,7
España	:	1 148	1 142	x	−0,5	:	11,1	18,3	x	64,9	:	1 277	2 093	x	63,9
France	655	639	670	−0,4	4,9	19,2	14,3	16,0	−4,1	11,9	1 257	914	1 071	−4,5	17,2
Ireland	45	12	13	−17,2	8,3	11,6	17,3	14,5	5,9	−16,2	52	21	19	−12,1	−9,5
Portugal (¹)	:	10	96	x	860,0	:	2,1	0,4	x	−81,0	:	2	39	x	1 850,0
United Kingdom	4 749	3 413	3 329	−4,6	−2,5	13,7	14,1	16,4	0,4	16,3	6 497	4 818	5 460	−4,2	13,3

Source : EC Commission, Directorate-General for Agriculture.

(¹) New varieties since 1993.

4.10.4.1 Market balance — hops

		Unit	EUR 12			% TAV		World			% TAV	
			1985	1992 ∞	1993 ∞	$\frac{1992}{1985}$	$\frac{1993}{1992}$	1985	1992 ∞	1993 ∞	$\frac{1992}{1985}$	$\frac{1993}{1992}$
1	2	3	4	5	6	7	8	9	10	11	12	13
Hops												
A	Area	1 000 ha	25,75	28,56	28,68	1,5	0,4	85,70	92,35	92,00	1,1	– 0,4
B	Yield	t/ha	1,73	1,27	1,80	– 4,3	41,7	1,40	1,30	1,47	– 1,1	13,1
C = A × B	Production: hops	1 000 t	44,55	36,27	51,62	– 2,9	42,3	119,98	120,06	135,24	– 0,2	12,6
D	of which — alpha acid	%	5,69	5,28	6,47	– 1,1	22,5	5,78	6,54	6,89	1,8	5,4
E = C × D/100	— alpha acid	t	2 534,75	1 915,12	3 339,49	– 3,9	74,4	6 934,84	7 851,60	9 318,04	1,6	18,7
Beer												
F	Beer production (¹)	Mio hl	255,00	297,15	286,57	2,6	– 3,6	997,00	1 140,00	1 188,62	2,3	4,3
G	of which — alpha acid	grams/hl	8,00	6,70	6,47	– 2,9	– 3,4	7,10	6,70	7,30	– 1,0	9,0
H = F × G × 1 000	— alpha acid	t	2 040,00	1 990,91	1 854,04	– 0,4	– 6,9	7 078,70	7 638,00	8 676,93	1,3	13,6

Source: EC Commission, Directorate-General for Agriculture.
(¹) Following year.

4.10.5.1   **Market price for hops**

1		Zentner = 50 kg			% TAV	
		1991/92 ∞	1992/93 ∞	1993/94 ∞	1992/93 / 1991/92	1993/94 / 1992/93
		2	3	4	5	6
EUR 12 (no contract)	ECU	165	168	66	1,8	− 61,0
EUR 12 (under contract)	ECU	167	160	158	− 4,2	− 1,1
	Total ECU	165	161	117	− 2,4	− 27,6
Belgique/België	BFR	7 584	9 769	4 685	28,8	− 52,0
BR Deutschland	DM	380	370	263	− 2,6	− 28,9
España	PTA	23 619	21 703	23 356	− 8,1	7,6
France	FF	1 539	1 569	1 549	1,9	− 1,3
Ireland	IRL	222	233	225	5,0	− 3,4
Portugal	ESC	10 039	33 727	15 000	236,0	− 55,5
United Kingdom	UKL	160	172	128	7,5	− 25,6

Source : EC Commission, Directorate-General for Agriculture.

4.11.1.1 Area, production and yield of cotton (unginned and ginned)

1	Elláda			% TAV		España			% TAV		EUR 12			% TAV	
	1985	1992	1993	1992/1985	1993/1992	1985	1992	1993	1992/1985	1993/1992	1985	1992	1993	1992/1985	1993/1992
	2	3	4	5	6	7	8	9	10	11	12	13	14	15	16
Area (1 000 ha)	209	321,2	351,6	6,3	9,5	60,3	76,0	31,8	3,4	− 58,2	269,3	397,2	383,4	5,7	− 3,5
Production (t):															
unginned cotton	526 145	750 440	979 192	5,2	30,5	194 166	214 039	94 964	1,4	− 55,6	720 211	964 479	1 074 156	4,3	11,4
ginned cotton	163 277	242 359	314 598	5,8	29,8	62 133	70 007	30 836	1,7	− 56,0	225 410	312 366	345 434	4,8	10,6
cotton seed	281 406	404 458	525 826	5,3	30,0	104 860	112 353	49 612	1,0	− 55,8	386 256	516 811	575 438	4,2	11,3
Yield (kg/ha):															
unginned cotton	2 517	2 336	2 785	− 1,1	19,2	3 220	2 815	2 989	− 1,9	6,2	2 674	2 428	2 802	− 1,4	15,4
ginned cotton	781	755	895	− 0,5	18,5	1 030	921	971	− 1,6	5,4	837	786	901	− 0,9	14,6
cotton seed	1 346	1 259	1 496	− 1,0	18,8	1 739	1 478	1 562	− 2,3	5,7	1 434	1 301	1 501	− 1,4	15,4

Source : EC Commission, Directorate-General for Agriculture.

4.11.1.2   Area, yield and production of fibre flax

	Area					Yield					Production				
	1 000 ha			% TAV		100 kg/ha			% TAV		1 000 t			% TAV	
	1985	1992 ∞	1993 ∞	1992/1985	1993/1992	1985	1992 ∞	1993 ∞	1992/1985	1993/1992	1985	1992 ∞	1993 ∞	1992/1985	1993/1992
1	2	3	4	5	6	7	8	9	10	11	12	13	14	15	16
Flax straw															
EUR 12	75,7	44,2	52,0	-7,4	17,6	85	66,6	68,8	-3,4	3,3	650,0	294,9	357,9	-10,7	21,4
Belgique/België	10,5	6,8	8,5	-6,0	25,0	65,0	75,0	60,0	2,1	-20,0	68,3	50,9	51,0	-4,1	0,2
Danmark	0,3	0,5	0,2	7,6	-60,0	85,0	66,2	60,0	-3,5	-9,4	2,6	3,0	1,2	2,1	-60,0
BR Deutschland	0,0	1,0	1,0	×	0,0	×	66,2	60,0	×	-9,4	0,0	5,5	6,0	×	9,1
France	59,5	33,1	36,7	-8,0	10,9	90,0	65,0	76,0	-4,5	16,9	535,5	215,2	278,9	-12,2	29,6
Nederland	4,7	2,9	3,3	-6,7	13,8	80,0	65,0	63,0	-2,9	-3,1	37,6	18,7	20,8	-9,5	11,2
United Kingdom	0,7	0,1	2,2	-24,3	2 100,0	85,0	66,2	:	-3,5	×	6,0	0,9	:	-23,7	×
Flax fibre															
EUR 12	75,7	44,2	52,0	-7,4	17,6	16,4	10,0	17,1	-6,8	71,0	123,8	44,0	88,8	-13,7	101,8
Belgique/België	10,5	6,8	8,5	-6,0	25,0	14,0	10,8	13,5	-3,6	25,0	14,7	7,3	11,5	-9,5	57,5
Danmark	0,3	0,5	0,2	7,6	-60,0	16,0	10,0	13,7	-6,5	37,0	0,5	0,5	0,3	0,0	-40,0
BR Deutschland	0,0	1,0	1,0	×	0,0	15,0	5,8	14,0	-12,7	141,4	0,0	0,5	1,4	×	180,0
France	59,5	33,1	36,7	-8,0	10,9	16,8	9,5	18,8	-7,8	97,9	100,0	31,6	69,0	-15,2	118,4
Nederland	4,7	2,9	3,3	-6,7	13,8	16,0	8,3	11,8	-9,0	42,2	7,5	2,4	3,9	-15,0	62,5
United Kingdom	0,7	0,1	2,2	-24,3	2 100,0	16,0	10,0	12,5	-6,5	25,0	1,1	0,1	2,7	-29,0	2 600,0

Source: EC Commission, Directorate-General for Agriculture.

4.11.1.3   **Output of silkworm cocoons and number of boxes of silkworm eggs used**

		Quantity			% TAV	
		1985	1992	1993	$\frac{1992}{1985}$	$\frac{1993}{1992}$
1	2	3	4	5	6	7
Silkworm cocoons (kg)	EUR 12	160 528	39 689	52 245	− 18,1	31,6
	Elláda	50 660	25 055	29 000	− 9,6	15,7
	France	3 699	1 187	1 504	− 15,0	26,7
	Italia	106 229	13 447	21 741	− 25,6	61,7
Boxes of silkworm eggs	EUR 12	6 582	3 394	3 299	− 9,0	− 2,8
	Elláda	2 300	1 158	1 255	− 9,3	8,4
	France	177	57	92	− 14,9	61,4
	Italia	4 105	2 152	1 877	− 8,8	− 12,8

Source: EC Commission, Directorate-General for Agriculture.

4.11.3.1   **Imports of flax straw into Belgium**

Exporting Member State	t			% TAV	
	1985	1992	1993	$\frac{1992}{1985}$	$\frac{1993}{1992}$
1	2	3	4	5	6
EUR 12	65 605	36 175	30 844	− 8,2	− 14,7
France	43 990	28 897	28 858	− 5,8	− 0,1
Nederland	21 615	7 278	1 986	− 14,4	− 72,7

Source: EC Commission, Directorate-General for Agriculture.

4.11.3.2   Intra-EC trade and external trade in cotton fibre (¹)

EUR 12

(1 000 t)

		1985	1990	1991	1992	1993	% TAV	
							1992 / 1985	1993 / 1992
1	2	3	4	5	6	7	8	9
Intra-EC trade (²)	EUR 12	40	63	68∞	56∞	71∞	4,9	26,8
	BLEU/UEBL	5	6	8	9	9	8,8	0,0
	BR Deutschland	6	10	20∞	13∞	7∞	11,7	− 46,2
	Elláda	1	1	0	0	1	×	×
	España	0	1	2	1	5	×	400,0
	France	7	15	14	11	18	6,7	63,6
	Ireland	0	3	3	2	1	×	− 50,0
	Italia	10	20	13	10	24	0,0	140,0
	Nederland	2	1	1	1	2	− 9,4	100,0
	Portugal	1	5	4	6	3	29,2	− 50,0
	United Kingdom	7	1	1	1	1	− 24,3	0,0
Imports	EUR 12	1 050	1 004	943∞	937∞	873∞	− 1,6	− 6,8
	BLEU/UEBL	39	32	32	32	29	− 2,8	− 9,4
	Danmark	2	3	4	4	4	10,4	0,0
	BR Deutschland	234	198	191∞	189∞	158∞	− 3,0	− 16,4
	Elláda	50	35	26	20	7	− 12,3	− 65,0
	España	88	98	85	82	79	− 1,0	− 3,7
	France	152	100	101	114	110	− 4,0	− 3,5
	Ireland	17	17	21	14	15	− 2,7	7,1
	Italia	247	306	309	304	302	3,0	− 0,7
	Nederland	8	4	2	3	2	− 13,1	− 33,3
	Portugal	166	182	152	158	154	− 0,7	− 2,5
	United Kingdom	47	30	21	17	13	− 13,5	− 23,5
Exports	EUR 12	53	73	95∞	78∞	141∞	5,7	80,8
	BLEU/UEBL	0	1	1	0	1	×	×
	BR Deutschland	5	11	14∞	17∞	8∞	19,1	− 52,9
	Elláda	38	50	65	44	106	2,1	140,9
	España	8	8	13	12	16	6,0	33,3
	France	2	2	2	4	6	10,4	50,0
	Italia	0	1	0	1	3	×	200,0
	United Kingdom	0	0	0	0	0	×	×

Source : Eurostat.

(¹) Cotton, other than rendered absorbent or bleached.
(²) Based on entries.

4.11.5.1 Producer prices for flax seed

1	ECU/t (¹)			% TAV	
	1985/86	1992/93	1993/94	$\frac{1992/93}{1985/86}$	$\frac{1993/94}{1992/93}$
1	2	3	4	5	6
Belgique/België	242,5	118,46	126,30	− 9,7	6,6
Nederland	296,2	148,65	162,69	− 9,4	9,4

Source: EC Commission, Directorate-General for Agriculture.

(¹) Calculated on the basis of prices in national currencies.

4.11.5.2 Flax tow prices

1	ECU/t (¹)			% TAV	
	1985/86	1992/93	1993/94	$\frac{1992/93}{1985/86}$	$\frac{1993/94}{1992/93}$
1	2	3	4	5	6
Belgique/België — water-retted					
Scutched flax:					
— average – low	1 468,6	745,2	1 065,6	− 9,2	43,0
– normal	1 688,9	863,3	1 166,8	− 9,1	35,2
– good	1 903,1	1 181,1	1 393,6	− 6,6	18,0

Source: EC Commission, Directorate-General for Agriculture.

(¹) Calculated on the basis of prices in national currencies.

4.11.5.3   Ginned cotton, world prices ([1])

(ECU/100 kg)

1	1985	1990	1991	1992	1993	1994
1	2	3	4	5	6	7
I	211,5	130,2	121,2	89,5	82,5	101,0
II	208,5	119,1	118,1	86,4	88,9	126,1
III	218,6	124,4	119,8	87,4	95,6	134,8
IV	207,4	127,1	128,7	89,4	95,1	131,6
V	189,0	120,4	127,4	99,9	90,9	138,2
VI	185,6	123,4	130,9	92,8	90,2	136,3
VII	181,5	129,9	135,7	94,3	87,4	127,2
VIII	161,9	127,2	123,9	83,8	92,1	117,7
IX	147,0	117,5	118,6	76,6	88,9	113,6
X	137,8	117,9	109,6	76,0	84,8	108,7
XI	120,8	117,4	106,0	75,8	84,5	109,7
XII	117,3	114,8	93,2	79,9	91,5	:
Ø	173,9	122,4	119,4	86,0	89,4	:
% TAV in relation to preceding year	− 23,0	− 8,4	− 2,4	− 28,0	4,0	:

([1]) 'Mid. 1-3/32' in force the first day of each month.

4.13.7.3 Industrial production of compound feedingstuffs, by species and by Member State

(1 000 t)

1	2	EUR 12 (¹)	Belgique/België	Danmark	BR Deutschland	Elláda	España	France	Ireland	Italia	Nederland	Portugal	United Kingdom
		3	4	5	6	7	8	9	10	11	12	13	14
Cattle (²)	1991	36 375	1 405	1 580	8 149	1 460**	3 157	4 664	1 383	4 590	4 784	1 145	4 058
	1992	35 355	1 302	1 649	7 708	1 476	3 134	4 347	1 460	4 348	4 739	1 119	4 073
	1993 (p)	35 775	1 293	1 706	8 063	1 502	2 620	4 423	1 655	4 300	4 841	1 016	4 356
% TAV 1992/1991		-2,8	-7,3	4,4	-5,4	1,1	-0,7	-6,8	5,6	-5,3	-0,9	-2,3	0,4
% TAV 1993/1992		1,2	-0,7	3,5	4,6	1,8	-16,4	1,7	13,4	-1,1	2,2	-9,2	6,9
Pigs	1991	38 677	2 813	2 799	7 476	600**	5 431	5 493	550	2 429	7 311	1 422	2 353
	1992	39 359	2 919	3 295	6 737	610	5 391	5 904	589	2 446	7 678	1 356	2 434
	1993 (p)	40 910	3 112	3 621	6 947	620	5 120	6 373	625	2 450	8 007	1 534	2 501
% TAV 1992/1991		1,8	3,8	17,7	-9,9	1,7	-0,7	7,5	7,1	0,7	5,0	-4,6	3,4
% TAV 1993/1992		3,9	6,6	9,9	3,1	1,6	-5,0	7,9	6,1	0,2	4,3	13,1	2,8
Poultry	1991	31 377	939	564	4 195	620**	4 159	7 398	408	4 514	3 482	1 232	3 866
	1992	31 456	939	619	4 157	635	4 129	7 675	396	4 306	3 536	1 241	3 823
	1993 (p)	31 693	993	622	4 075	650	4 050	7 999	425	4 350	3 465	1 245	3 819
% TAV 1992/1991		0,3	0,0	9,8	-0,9	2,4	-0,7	3,7	-2,9	-4,6	1,6	0,7	-1,1
% TAV 1993/1992		0,8	5,8	0,5	-2,0	2,4	-1,9	4,2	7,3	1,0	-2,0	0,3	-0,1
Other	1991	6 307	109	124	562	35**	1 153	1 464	236	1 121	520	150	833
	1992	6 494	130	129	536	40	1 146	1 497	254	1 180	518	190	874
	1993 (p)	6 633	151	116	566	50	1 150	1 495	270	1 200	517	204	914
% TAV 1992/1991		3,0	19,3	4,0	-4,6	14,3	-0,6	2,3	7,6	5,3	-0,4	26,7	4,9
% TAV 1993/1992		2,1	16,2	-10,1	5,6	25,0	0,3	-0,1	6,3	1,7	-0,2	7,4	4,6
Total	1991	112 737	5 267	5 067	20 382	2 715**	13 900	19 019	2 577	12 654	16 097	3 949	11 110
	1992	112 664	5 290	5 692	19 138	2 761	13 800	19 423	2 699	12 280	16 471	3 906	11 204
	1993 (p)	115 011	5 549	6 065	19 651	2 822	12 940	20 290	2 975	12 300	16 830	3 999	11 590
% TAV 1992/1991		-0,1	0,4	12,3	-6,1	1,7	-0,7	2,1	4,7	-3,0	2,3	-1,1	0,8
% TAV 1993/1992		2,1	4,9	6,6	2,7	2,2	-6,2	4,5	10,2	0,2	2,2	2,4	3,4

Source : Fefac.
(¹) Luxembourg is not included.
(²) Including milk-replacer feed.

4.13.7.5   Use of cereals by the compound feedingstuffs industry

1	% of production of compound feedingstuffs			1 000 t			% TAV	
	1991 ∞	1992 ∞	1993 ∞	1991 ∞	1992 ∞	1993 ∞	$\frac{1992}{1991}$	$\frac{1993}{1992}$
1	2	3	4	5	6	7	8	9
EUR 12 (^{1})	29,3	28,1	29,8	32 320	31 710	33 420	– 1,9	5,4
Belgique/België	11,3	13,2	14,5	640	670	780	4,7	16,4
Danmark	25,4	26,3	27,9	1 290	1 510	1 730	17,1	14,6
BR Deutschland	26,0	25,7	25,9	5 340	4 860	5 090	– 9,0	4,7
España	41,0	39,9	42,6	5 680	5 250	5 450	– 7,6	3,8
France	31,1	30,9	33,0	5 880	5 960	6 720	1,4	12,8
Ireland	26,9	25,9	23,3	670	720	750	7,5	4,2
Italia	46,5	46,3	44,7	5 880	5 960	6 720	13,6	12,8
Nederland	13,6	14,5	15,5	2 270	2 420	2 610	6,6	7,9
Portugal	23,1	28,2	27,5	920	990	1 080	7,6	9,1
United Kingdom	32,5	32,1	31,9	3 780	3 630	3 710	– 4,0	2,2

Source: Fefac.
(^{1}) Greece and Luxembourg not included.

4.13.7.7   **Production of dehydrated fodder** (excl. potatoes)

	1 000 t			% TAV	
	1991/92	1992/93	1993/94 (p)	1992/93 / 1991/92	1993/94 / 1992/93
1	2	3	4	5	6
EUR 12	3 687	4 220	4 529	14,5	7,3
BLEU/UEBL	7	8	8	14,3	0,0
Danmark	335	307	360	− 8,4	17,3
BR Deutschland	370	405	436	9,5	7,7
Elláda	20	17	30	− 15,0	76,5
España	794	1 081	1 367	36,1	26,5
France	1 360	1 494	1 415	9,9	− 5,3
Ireland	4	4	5	0,0	25,0
Italia	463	528	517	14,0	− 2,1
Nederland	245	267	291	9,0	9,0
Portugal	0	0	5	×	×
United Kingdom	89	109	95	22,5	− 12,8

Source : EC Commission, Directorate-General for Agriculture.

4.13.7.8   **Community supplies of dehydrated and dried fodder**   **EUR 12**

	1 000 t			% TAV	
	1991	1992	1993	1992 / 1991	1993 / 1992
1	2	3	4	5	6
Production	4 118	4 660	4 914	13,2	5,5
Imports	80	95	43	18,8	− 54,7
Exports	69	120	95	73,9	− 20,8
Availabilities	4 129	4 635	4 862	12,3	4,9

Source : Eurostat and EC Commission, Directorate-General for Agriculture.

4.13.7.9   Area, yield and production of dry pulses, feed peas and field beans

		Area				
		1 000 ha			% TAV	
		1985	1992 ∞	1993 ∞	$\frac{1992}{1985}$	$\frac{1993}{1992}$
1	2	3	4	5	6	7
Dried pulses, total	EUR 12	827	1 257	1 345	6,2	7,0
	Belgique/België	1	6	9	29,2	50,0
	Danmark	124	118	121	− 0,7	2,5
	BR Deutschland	28	41	89	5,6	117,1
	Elláda	27	5	5	− 21,4	0,0
	España	75	30	32	− 12,3	6,7
	France	242	714	753	16,7	5,5
	Ireland	2	2	2	0,0	0,0
	Italia	170	103	93	− 6,9	− 9,7
	Nederland	20	7	4	− 13,9	− 42,9
	Portugal	1	17	19	49,9	11,8
	United Kingdom	137	208	214	6,1	2,9
Feed peas	EUR 12	444	950	1 023	11,5	7,7
	Belgique/België	1	5	7	25,8	40,0
	Danmark	123	117	119	− 0,7	1,7
	BR Deutschland	14	29	59	11,0	103,4
	Elláda	0	1	1	×	0,0
	España	4	8	8	10,4	0,0
	France	189	695	737	20,4	6,0
	Ireland	2	1	1	− 9,4	0,0
	Italia	2	11	9	27,6	− 18,2
	Nederland	17	5	3	− 16,0	− 40,0
	Portugal	1	0	0	×	×
	United Kingdom	91	79	79	− 2,0	12,7
Field beans	EUR 12	383	298	310	− 3,5	4,0
	Belgique/België	0	1	2	×	100,0
	Danmark	1	2	2	10,4	0,0
	BR Deutschland	14	18	30	3,7	66,7
	Elláda	27	4	4	− 23,9	0,0
	España	71	22	21	− 15,4	− 4,5
	France	53	14	13	− 17,3	− 7,1
	Ireland	0	1	5	×	400,0
	Italia	168	89	81	− 8,7	− 9,0
	Nederland	3	1	1	− 14,5	0,0
	Portugal	0	17	17	×	0,0
	United Kingdom	46	129	135	15,9	4,7

Source: EC Commission, Directorate-General for Agriculture.

Yield					Production				
100 kg/ha			% TAV		1 000 t			% TAV	
1985	1992 ∞	1993 ∞	$\frac{1992}{1985}$	$\frac{1993}{1992}$	1985	1992 ∞	1993 ∞	$\frac{1992}{1985}$	$\frac{1993}{1992}$
8	9	10	11	12	13	14	15	16	17
32,2	**37,8**	**42,3**	**2,3**	**11,9**	**2 670**	**4 746**	**5 684**	**8,6**	**19,8**
40,0	44,0	44,9	1,4	2,0	4	26	38	30,7	46,2
41,0	25,8	37,6	− 6,4	45,7	508	305	456	− 7,0	49,5
36,4	29,8	32,6	− 2,8	9,4	112	137	289	2,9	110,9
14,4	22,7	21,2	6,7	− 6,6	39	11	11	− 16,5	0,0
8,5	12,7	12,2	5,9	− 3,9	64	38	39	− 7,2	2,6
45,8	46,1	50,7	0,1	10,0	1 113	3 293	3 811	16,8	15,7
35,0	42,0	47,8	2,6	13,8	7	8	29	1,9	262,5
13,9	17,9	15,8	3,7	− 11,7	237	183	147	− 3,6	− 21,3
35,0	42,1	47,1	2,7	11,9	70	30	19	− 11,4	− 36,7
8,0	4,1	8,0	− 9,1	95,1	1	7	16	32,0	128,6
35,4	33,9	38,9	− 0,6	14,7	515	706	830	4,6	17,6
43,1	**41,7**	**46,9**	**− 0,5**	**12,5**	**1 909**	**3 967**	**4 799**	**11,0**	**21,0**
30,3	44,7	46,0	5,7	2,9	3	22	32	32,9	45,5
41,1	25,8	37,6	− 6,4	45,7	506	302	448	− 7,1	48,7
33,6	26,9	30,2	− 3,1	12,3	47	78	179	7,5	129,5
:	33,3	20,0	×	×	0	3	2	×	− 33,3
10,0	13,8	13,8	4,7	0,0	4	11	11	15,5	0,0
50,3	46,5	51,0	− 1,1	9,7	950	3 230	3 758	19,1	16,3
35,0	36,0	37,0	0,4	2,8	7	4	4	− 7,7	0,0
10,0	32,5	30,7	18,3	− 5,5	2	35	29	50,5	− 17,1
35,3	42,0	44,5	2,5	6,0	60	21	13	− 13,9	− 38,1
10,0	:	:	×	×	0	0	0	×	×
36,3	33,0	41,0	− 1,4	24,2	330	261	324	− 3,3	35,6
20,4	**25,6**	**27,9**	**3,3**	**9,0**	**755**	**763**	**867**	**0,2**	**13,6**
:	40,0	40,0	×	0,0	1	4	6	21,9	50,0
20,0	25,8	37,6	3,7	45,7	2	3	8	6,0	166,7
39,3	33,3	37,3	− 2,3	12,0	65	60	111	− 1,1	85,0
14,4	20,0	21,4	4,8	7,0	39	8	9	− 20,3	12,5
8,5	12,3	10,4	5,4	− 15,4	60	27	23	− 10,8	− 14,8
29,8	37,1	36,2	3,2	− 2,4	158	52	47	− 14,7	− 9,6
:	48,0	50,0	×	4,2	0	4	25	×	525,0
14,0	16,2	14,0	2,1	− 13,6	235	144	112	− 6,8	− 22,2
33,3	42,5	55,0	3,5	29,4	10	9	7	− 1,5	− 22,2
8,0	4,1	8,2	− 9,1	100,0	0	7	14	×	100,0
33,7	34,5	37,5	0,3	8,7	185	445	506	13,4	13,7

4.13.7.12 Cif offer price (Rotterdam) for soya cake

(ECU/100 kg)

	1985	1986	1987	1988	1989	1990	1991	1992	1993	1994
1	2	3	4	5	6	7	8	9	10	11
I	24,60	21,84	17,86	19,82	26,10	18,66	15,33	17,07	18,85	20,24
II	23,97	21,44	18,29	19,62	25,08	17,64	15,47	17,65	18,62	19,90
III	24,81	21,72	17,62	20,04	26,24	17,88	16,21	17,56	17,92	19,49
IV	24,21	21,57	17,83	20,57	25,73	17,85	17,38	17,53	17,59	20,02
V	22,53	20,68	17,95	21,80	24,55	17,75	18,32	17,33	18,10	:
VI	21,22	19,88	19,85	28,18	24,93	16,70	19,31	17,14	18,41	:
VII	19,78	19,27	19,55	27,30	23,79	16,52	17,37	16,46	21,92	:
VIII	19,13	19,62	18,54	26,29	20,73	15,84	16,99	16,00	21,28	:
IX	20,81	19,98	18,93	27,32	21,38	16,32	17,79	15,85	19,61	:
X	21,18	19,41	19,53	26,05	21,37	16,27	17,86	16,51	18,81	:
XI	21,47	18,59	20,20	24,64	20,91	15,25	17,25	17,31	19,91	:
XII	21,13	17,48	20,63	24,60	19,75	15,36	16,77	18,39	19,74	:
Average 12 months	22,06	20,16	18,94	23,86	23,38	16,82	17,18	17,06	19,22	:
% TAV compared with previous year	− 13,7	− 8,6	− 6,1	26,0	− 2,0	− 28,1	2,1	− 0,7	12,7	:

Source : Eurostat.

4.14.1.1   **Gross internal production and consumption of meat** (¹)   **EUR 12**

	Relative share %			1 000 t			% TAV	
	1991	1992	1993	1991	1992	1993	$\frac{1992}{1991}$	$\frac{1993}{1992}$
1	2	3	4	5	6	7	8	9
Gross internal production								
— pigmeat	42,4	42,6	44,9	14 339	14 387	15 277	0,3	6,2
— beef/veal	25,7	24,8	22,8	8 705	8 380	7 750	− 3,7	− 7,5
— poultrymeat	20,0	20,5	20,6	6 755	6 930	7 009	2,6	1,1
— sheepmeat and goatmeat	3,6	3,5	3,4	1 221	1 177	1 148	− 3,6	− 2,5
— equine meat	0,1	0,1	0,1	46	49	42	5,7	− 12,8
— other	2,3	2,4	2,4	767	817	819	6,5	0,3
Total	94,0	94,1	94,2	31 833	31 739	32 046	− 0,3	1,0
Edible offals	6,0	5,9	5,8	2 020	2 000	1 960	− 1,0	− 2,0
Total	100	100	100	33 853	33 739	34 006	− 0,3	0,8
				Kg/head				
Meat consumption								
— pigmeat	42,6	43,0	43,9	39,8	40,2	42,0	1,0	4,5
— beef/veal	23,8	23,0	23,1	22,2	21,5	22,1	− 3,2	2,8
— poultrymeat	19,9	20,2	19,7	18,6	18,9	18,9	1,8	0,0
— sheepmeat and goatmeat	4,5	4,4	4,2	4,2	4,1	4,0	− 2,4	− 2,4
— equine meat	0,5	0,5	0,5	0,5	0,5	0,5	2,8	− 6,3
— other	2,6	2,7	2,7	2,4	2,5	2,5	5,1	1,0
Total	93,8	93,8	94,0	87,7	87,7	90,0	0,1	2,6
Edible offals	6,2	6,2	6,0	5,8	5,8	5,7	0,5	− 2,2
Total	100	100	100	93,5	93,6	95,7	0,1	2,3

Source : EC Commission, Directorate-General for Agriculture.

(¹) Carcass weight for meat.

4.14.3.1   Net balance of external trade (^{1}) in meat (^{2}) and self-sufficiency

EUR 12	Net balance (^{1})			Self-sufficiency (%)		
	1 000 t					
	1991	1992	1993	1991	1992	1993
1	2	3	4	5	6	7
Meat (^{2})						
— pigmeat	567	477	687	104,5	103,4	104,6
— beef/veal	790	751	729	113,5	112,9	100,7
— poultrymeat	326	361	504	105,4	105,7	107,1
— sheepmeat and goatmeat	− 239	− 256	− 240	83,6	82,7	82,7
— equine meat	− 141	− 129	− 125	24,6	27,3	25,3
— other	− 64	− 57	− 67	92,3	93,5	92,4
Total	1 239	1 147	1 488	105,2	104,5	102,5
Edible offals	13	− 18	− 14	100,6	99,1	99,3
Total	1 252	1 128	1 473	104,0	104,2	102,3

Source : EC Commission, Directorate-General for Agriculture.
(^{1}) Exports minus imports.
(^{2}) Including live animals, carcass weight equivalent.

4.15.0.1   Cattle numbers (December of previous year)

1	1 000 head			% of EUR 12	% TAV	
	» 1987 «	1993 ∞	1994 ∞	1994	» 1992 « / » 1987 «	1994 / 1993
1	2	3	4	5	6	7
EUR 12	80 351	79 321	78 642	100,0	0,4	− 0,9
Belgique/België	2 961	3 100	3 127	4,0	1,1	0,9
Danmark	2 348	2 180	2 115	2,7	− 1,1	− 3,0
BR Deutschland	14 950	16 207	15 897	20,2	3,6	− 1,9
Elláda	733	629	608	0,8	− 2,3	− 3,3
España	5 018	4 962	5 002	6,4	0,1	0,8
France	21 521	20 328	20 099	25,6	− 0,6	− 1,1
Ireland	5 614	6 265	، 6 308	8,0	2,0	0,7
Italia	8 887	7 704	7 621	9,7	− 2,0	− 1,1
Luxembourg	210	202	205	0,3	− 0,3	1,5
Nederland	4 692	4 794	4 629	5,9	0,6	− 3,4
Portugal	1 340	1 345	1 322	1,7	0,2	− 1,7
United Kingdom	12 077	11 605	11 709	14,9	− 0,6	0,9

Source : Eurostat and EC Commission, Directorate-General for Agriculture.

4.15.1.1   Slaughterings of adult bovine animals and calves (¹)

		1000 head			% TAV		Average carcass weight in kg			% TAV	
		» 1987 «	» 1992 «	1993 ∞	» 1992 « / » 1987 «	1993 / 1992	» 1987 «	1992 ∞	1993 ∞	» 1992 « / » 1987 «	1993 / 1992
1	2	3	4	5	6	7	8	9	10	11	12
Adult bovine animals	EUR 12	23 903	24 553	22 279	0,3	− 9,3	295,8	307,8	311,3	0,8	1,1
	Belgique/België	687	706	722	1,2	2,3	392,3	415,0	423,2	1,3	2,0
	Danmark	914	830	773	− 2,4	− 6,9	251,2	260,2	261,0	0,7	0,3
	BR Deutschland	5 088	5 607	4 801	2,7	− 14,4	310,0	314,1	320,3	0,2	2,0
	Elláda	330	279	261	− 2,9	− 6,5	233,0	247,3	246,6	1,1	− 0,3
	España	1 814	2 102	1 923	2,1	− 8,5	235,5	249,3	247,4	1,1	− 0,7
	France	4 665	4 595	4 057	− 1,1	− 11,7	332,8	345,6	352,9	0,8	2,1
	Ireland	1 576	1 707	1 606	1,3	− 5,9	305,7	330,4	328,6	1,4	− 0,5
	Italia	3 464	3 553	3 444	0,0	− 3,1	278,8	284,3	288,6	0,5	1,5
	Luxembourg	30	24	21	− 4,7	− 12,5	300,1	291,7	336,7	1,4	15,4
	Nederland	1 192	1 399	1 313	3,1	− 6,1	293,2	322,4	323,4	1,9	0,3
	Portugal	417	444	422	1,2	− 5,0	242,8	256,8	252,6	1,0	− 1,6
	United Kingdom	3 726	3 307	2 936	− 2,4	− 11,2	276,0	289,7	291,4	0,9	0,6
Calves	EUR 12	7 056	6 327	5 980	− 2,3	− 5,5	126,4	132,7	135,6	1,1	2,2
	Belgique/België	308	376	379	4,1	0,8	135,4	156,9	160,8	3,4	2,5
	Danmark	40	27	28	− 6,5	3,7	47,0	37,0	42,9	− 3,2	15,8
	BR Deutschland	692	556	526	− 3,3	− 5,4	122,4	122,3	125,5	− 0,1	2,6
	Elláda	54	78	82	7,7	5,1	113,7	141,0	145,1	4,1	2,9
	España	119	85	72	− 6,3	− 15,3	118,8	129,4	123,0	1,2	− 4,9
	France	2 964	2 376	2 205	− 4,3	− 7,2	116,8	121,6	123,4	0,9	1,5
	Ireland	3	1	1	− 8,9	0,0	137,1	132,1	132,1	− 0,7	0,0
	Italia	1 547	1 514	1 419	− 0,7	− 6,3	133,0	136,7	136,6	0,7	− 0,1
	Luxembourg	0	1	1	0,0	0,0	108,6	99,8	127,9	0,7	28,2
	Nederland	1 207	1 197	1 174	− 0,5	− 1,9	150,7	153,7	159,0	0,1	3,4
	Portugal	62	84	74	5,1	− 11,9	97,3	107,1	110,3	0,6	3,0
	United Kingdom	60	32	19	− 8,9	− 40,6	51,2	62,5	63,5	1,6	1,6

Source : Eurostat.

(¹) Total slaughterings of animals of domestic and foreign origin.

4.15.1.2 Net production of beef/veal (adult bovine animals and calves) ([1])

		1 000 t ([2])			% TAV	
		»1987«	1992 ∞	1993 ∞	»1992«/»1987«	1993/1992
1	2	3	4	5	6	7
Adult bovine animals	EUR 12	7 069	7 555	6 935	1,1	− 8,2
	Belgique/België	272	293	306	2,4	4,4
	Danmark	230	216	202	− 1,8	− 6,5
	BR Deutschland	1 577	1 761	1 538	2,8	− 12,7
	Elláda	77	69	64	− 2,1	− 7,2
	España	427	524	476	3,3	− 9,2
	France	1 553	1 588	1 432	− 0,3	− 9,8
	Ireland	481	564	528	2,8	− 6,4
	Italia	966	1 010	994	0,5	− 1,6
	Luxembourg	9	7	7	− 3,7	0,0
	Nederland	349	451	425	5,3	− 5,8
	Portugal	101	114	107	2,4	− 6,1
	United Kingdom	1 027	958	856	− 1,6	− 10,6
Calves	EUR 12	891	841	811	− 1,3	− 3,6
	Belgique/België	42	59	61	8,1	3,4
	Danmark	2	1	1	− 10,0	0,0
	BR Deutschland	84	68	66	− 3,3	− 2,9
	Elláda	6	11	12	14,4	9,1
	España	14	11	9	− 5,2	− 18,2
	France	346	289	272	− 3,6	− 5,9
	Ireland	0	0	0	0,0	×
	Italia	206	207	194	− 0,1	− 6,3
	Luxembourg	0	0	0	0,0	×
	Nederland	182	184	187	0,0	1,6
	Portugal	6	9	8	6,7	− 11,1
	United Kingdom	3	2	1	− 8,9	− 50,0
Beef/veal	EUR 12	7 961	8 397	7 746	0,8	− 7,8
	Belgique/België	314	352	367	3,2	4,3
	Danmark	232	217	203	− 1,8	− 6,5
	BR Deutschland	1 662	1 829	1 604	2,5	− 12,3
	Elláda	83	80	76	− 0,9	− 5,0
	España	442	535	485	3,0	− 9,3
	France	1 899	1 877	1 704	− 0,9	− 9,2
	Ireland	481	564	528	2,8	− 6,4
	Italia	1 171	1 218	1 188	0,4	− 2,5
	Luxembourg	9	7	7	− 3,7	0,0
	Nederland	531	635	612	3,5	− 3,6
	Portugal	107	123	115	2,7	− 6,5
	United Kingdom	1 030	960	857	− 1,6	− 10,7

NB : These figures do not correspond to gross domestic production; for this see Table 4.14.1.1.

Source : Eurostat.

([1]) Total slaughterings of animals including those of foreign origin.
([2]) Carcass weight.

4.15.2.1   **World production and production of principal beef/veal producing/exporting countries** (^{1})

1	%			1 000 t			% TAV	
	» 1985 «	1992 ∞	1993 ∞	» 1985 «	1992 ∞	1993 ∞	» 1992 « / » 1985 «	1993 / 1992
1	2	3	4	5	6	7	8	9
World	100,0	100,0	100,0	48 977	53 356	52 969	1,5	− 0,7
— EUR 12	17,2	15,7	14,6	8 424	8 396	7 744	− 0,3	− 7,8
— USA	22,6	19,9	20,0	11 058	10 611	10 584	0,0	− 0,3
— CIS	16,0	13,5	13,0	7 843	7 222	6 886	− 0,8	− 4,7
— Brazil (^{2})	4,2	8,6	8,9	2 075	4 590	4 719	3,0	2,8
— Argentina	5,7	4,7	4,7	2 772	2 487	2 508	− 1,5	0,8
— Uruguay	0,7	0,7	0,6	323	360	330	1,2	− 8,3
— Australia	3,0	3,4	3,4	1 459	1 838	1 796	3,5	− 2,3
— New Zealand	1,0	1,0	1,1	501	525	582	1,4	10,9
— Peop. Rep. China	1,3	3,3	3,8	616	1 755	2 005	×	14,2
— Canada	2,0	1,7	1,6	1 003	898	871	− 2,2	− 3,0
— Mexico	2,3	3,1	3,2	1 137	1 660	1 720	6,4	3,6
— Colombia	1,2	1,4	1,4	588	734	768	4,7	4,6
— Poland	1,4	1,0	1,1	692	544	587	− 2,4	7,9
— India	4,1	4,5	4,6	2 005	2 398	2 458	3,1	2,5
— Japan	1,1	1,1	1,1	560	592	593	0,8	0,2
— South Africa	1,3	1,3	1,3	626	715	693	2,0	− 3,1
— Austria	0,5	0,5	0,5	235	244	246	0,4	0,8
— Switzerland	0,3	0,3	0,3	171	167	152	− 0,8	− 9,0
— Sweden	0,3	0,2	0,3	146	130	141	− 1,2	8,5
— Hungary	0,3	0,2	0,2	132	110	97	− 2,8	− 11,8
— Finland	0,3	0,2	0,2	125	117	106	− 1,4	− 9,4
— Norway	0,2	0,2	0,2	74	84	83	1,7	− 1,2

Source : FAO and other international organizations (GATT).

(^{1}) Net production.
(^{2}) New series as from 1991.

EUR 12

4.15.3.1 Beef/veal — EC trade by species

Denomination	Imports						Exports					
	1992			1993			1992			1993		
	Extra-EC	Intra-EC	World	Extra-EC	Intra-EC	World	Extra-EC	Intra-EC	World	Extra-EC	Intra-EC	World
1	2	3	4	5	6	7	8	9	10	11	12	13
1. *Live animals in number (per 1 000 head):*												
— Calves	247,5	1 605,9	1 853,4	395,4	1 757,2	2 152,6	3,2	1 549,7	1 552,9	3,1	1 712,0	1 715,1
— Adult bovine animals	384,7	1 560,4	1 958,1	88,9	861,9	950,8	207,5	1 566,3	1 773,8	400,4	1 024,7	1 425,1
— Pure-bred breeding animals	83,8	43,1	126,9	114,3	134,3	248,6	115,8	59,4	175,2	78,9	117,2	196,1
Total live animals	716,0	3 209,4	3 938,4	598,6	2 753,5	3 352,1	326,5	3 175,4	3 501,9	482,5	2 853,9	3 336,4
2. *Live animals converted to meat weight (per 1 000 t carcass weight)*	92,5	391,8	486,9	80,1	308,3	388,4	84,7	387,7	472,5	143,2	336,5	479,7
3. *Meat (1 000 t carcass weight)*												
— Fresh or chilled from:												
Adult bovine animals	163,3	1 340,4	1 503,8	144,7	1 160,3	1 305,0	102,7	1 315,7	1 418,4	104,2	1 113,3	1 217,5
— Frozen	102,0	241,0	342,9	93,2	170,3	263,5	1 013,7	276,9	1 293,6	887,2	185,0	1 072,1
— Salted or in brine, dried or smoked	0,3	2,9	3,2	0,4	5,8	6,2	1,8	3,7	5,5	2,0	3,8	5,8
— Prepared and preserved (cooked or uncooked)	213,8	53,3	267,1	180,8	59,3	240,1	118,1	43,1	161,3	91,9	19,0	110,9
Total beef/veal (2 + 3)	572,4	2 109,8	2 685,0	499,3	1 704,0	2 203,3	1 323,4	2 119,5	3 446,1	1 228,5	1 657,6	2 886,1

Source : EC Commission, Directorate-General for Agriculture and Eurostat — Comext.

Coefficients : – Live animals : Carcass weight = live weight × 0,50.
  – Boneless meat } Product weight × 1,3 = carcass weight.
  – Prepared and preserved meat

4.15.3.2 Beef/veal — trade with non-member countries

(1 000 tonnes carcass weight)

Reporting countries	1990	%	1991	%	1992	%	1993	%
1	2	3	4	5	6	7	8	9
A. Exports								
EUR 12	816,0	100,0	1 324,9∞	100,0∞	1 323,4∞	100,0∞	1 228,5∞	100,0∞
BLEU/UEBL	3,8	0,5	18,0	1,4	33,5	2,5	40,9	3,3
Danmark	25,7	3,2	41,0	3,1	46,4	3,5	47,3	3,8
BR Deutschland	298,9	36,6	576,9∞	43,5∞	393,8∞	29,8∞	347,9∞	28,3∞
Elláda	1,1	0,1	0,9	0,1	1,0	0,1	2,2	0,2
España	8,4	1,0	21,1	1,6	33,3	2,5	22,3	1,8
France	171,6	21,0	240,1	18,1	283,3	21,4	257,0	20,9
Ireland	187,4	23,0	182,2	13,8	213,4	16,1	257,1	20,9
Italia	50,8	6,2	120,8	9,1	173,2	13,1	101,7	8,3
Nederland	45,9	5,6	91,4	6,9	120,1	9,1	106,3	8,7
Portugal	0,5	0,1	0,5	0,0	0,0	0,0	0,3	0,0
United Kingdom	21,9	2,7	32,0	2,4	25,5	1,9	45,5	3,7
B. Imports								
EUR 12	500,6	100,0	534,5∞	100,0∞	572,4∞	100,0∞	499,2∞	100,0∞
BLEU/UEBL	7,3	1,5	4,0	0,7	3,0	0,5	2,7	0,5
Danmark	0,9	0,2	0,5	0,1	0,5	0,1	0,6	0,1
BR Deutschland	136,2	27,2	169,0∞	31,6∞	176,8∞	30,9∞	158,6∞	31,8∞
Elláda	21,1	4,2	21,7	4,1	14,1	2,5	11,7	2,3
España	4,7	0,9	3,4	0,7	3,1	0,5	2,3	0,5
France	15,0	3,0	14,7	2,7	15,2	2,6	13,9	2,8
Ireland	0,2	0,0	0,0	0,0	0,0	0,0	0,0	0,0
Italia	149,1	29,8	146,5	27,4	143,5	25,1	112,8	22,6
Nederland	35,9	7,2	31,1	5,8	47,3	8,3	47,0	9,4
Portugal	2,3	0,5	3,6	0,7	10,1	1,8	9,7	1,9
United Kingdom	127,9	25,5	140,0	26,2	158,8	27,7	139,9	28,0

Source : Eurostat – Comext.

Coefficients : – Live animals : Carcass weight = live weight × 0,50.
– Boneless meat } Product weight × 1,3 = carcass weight.
– Prepared and preserved meat

4.15.4.1   Supply balance — beef/veal

EUR 12

1	1 000 t (^{3})			% TAV	
	»1987«	1992 ∞	1993 ∞	$\dfrac{\text{»1992«}}{\text{»1987«}}$	$\dfrac{1993}{1992}$
1	2	3	4	5	6
Gross internal production	7 889	8 380	7 750	1,0	− 7,5
Net production	7 962	8 396	7 746	0,8	− 7,7
Changes in stocks	×	212	− 615	×	×
Imports (^{1})	412	480	419	1,8	− 12,7
Exports (^{1})	912	1 239	1 085	6,1	− 12,4
Intra-Community trade (^{2})	1 640	2 110	1 704	3,8	− 19,2
Internal use (total)	7 557	7 425	7 695	0,1	3,6
Gross consumption (kg/head/year)	23,4	21,5	22,1	− 1,3	2,8
Self-sufficiency (%)	104,4	112,9	100,7	0,9	− 10,8

Source : Eurostat and EC Commission, Directorate-General for Agriculture.

(^{1}) Total trade, with the exception of live animals.
(^{2}) All trade, including live animals (figures based on imports).
(^{3}) Carcass weight.

4.15.5.1 Market prices (¹) for beef/veal

(ECU/100 kg) (²)

		Belgique/België	Danmark	BR Deutschland	Elláda	España	France	Ireland	Italia	Luxembourg	Nederland	Portugal	United Kingdom	EUR 12 (³)
1	2	3	4	5	6	7	8	9	10	11	12	13	14	15
Young male bovines R3 (⁴)	1991	255,30	274,26	253,15	356,48	281,08	273,21	253,16	277,69	272,59	261,68	328,61	253,44	265,91
	1992	254,41	264,26	258,34	351,73	265,33	270,50	249,42	285,44	268,86	267,05	319,14	251,18	265,80
	1993	254,42	251,79	255,95	340,21	280,66	269,79	240,33	282,53	270,20	269,07	303,65	253,37	263,66
% TAV	1992/1991	-0,3	-3,6	2,1	-1,3	-5,6	-1,0	-1,5	2,8	-1,4	2,0	-2,9	-0,9	0,0
% TAV	1993/1992	0,0	-4,7	-0,9	-3,3	5,8	-0,3	-3,6	-1,0	0,5	0,8	-4,9	0,9	-0,8
Heifers R3 (⁴)	1991	224,03	240,34	216,56	337,65	294,42	276,02	231,44	333,12	277,94	215,81	305,89	245,96	256,36
	1992	222,34	246,36	229,91	344,37	272,58	278,65	235,49	323,53	274,59	231,16	299,22	248,63	261,16
	1993	236,82	241,38	237,23	332,82	286,40	289,64	239,35	311,46	279,56	236,25	282,57	248,36	264,31
% TAV	1992/1991	0,8	2,5	6,2	2,0	-7,4	1,0	1,7	-2,9	-1,2	7,1	-2,2	1,1	1,9
% TAV	1993/1992	6,5	-2,0	3,2	-3,4	5,1	3,9	1,6	-3,7	1,8	2,2	-5,6	-0,1	1,2
Cows O3 (⁵)	1991	205,99	204,05	168,11	245,55	137,82	211,29	182,63	172,56	196,17	189,67	206,63	215,31	188,10
	1992	215,23	209,37	196,36	222,95	150,65	230,00	201,50	198,26	209,19	216,63	208,80	205,88	207,16
	1993	219,00	207,87	207,16	242,04	164,71	240,33	212,06	209,13	224,10	224,65	215,24	204,88	215,26
% TAV	1992/1991	-4,5	2,6	16,8	-9,2	9,3	8,9	10,3	14,9	6,6	14,2	1,1	-4,4	10,1
% TAV	1993/1992	1,8	-0,7	5,5	8,6	9,3	4,5	5,2	5,5	7,1	3,7	3,1	-0,5	3,9

Source: EC Commission, Directorate-General for Agriculture.

(¹) Country Ø.
(²) Slaughter weight.
(³) Weighted Ø ECU/100 kg.
(⁴) Good conformation and average fat cover.
(⁵) Fairly good conformation and average fat cover.

4.16.0.1   **Pig numbers** (December of previous year)

	1 000 head			% of EUR 12	% TAV	
	1987	1993 ∞	1994 ∞	1994 ∞	$\frac{1993}{1987}$	$\frac{1994}{1993}$
1	2	3	4	5	6	7
EUR 12	102 071	109 338	110 089	100,0	1,2	0,7
Belgique/België	5 989	6 903	7 069	6,4	2,4	2,4
Danmark	9 192	10 345	10 870	9,9	2,0	5,1
BR Deutschland	23 480	26 514	26 075	23,7	2,0	− 1,7
Elláda	1 165	1 099	1 144	1,0	− 1,0	4,1
España	16 507	18 219	17 929	16,3	1,7	− 1,6
France	11 895	12 564	12 868	11,7	0,9	2,4
Ireland	967	1 423	1 487	1,4	6,7	4,5
Italia	9 340	8 244	8 050	7,3	− 2,1	− 2,4
Luxembourg	74	66	72	0,1	− 2,0	9,1
Nederland	14 036	13 709	13 991	12,7	− 0,4	2,1
Portugal	2 392	2 547	2 665	2,4	1,1	4,6
United Kingdom	7 832	7 704	7 869	7,1	− 0,3	2,1

Source : Eurostat.

4.16.1.1   **Number of pigs slaughtered** ([1])

	1 000 head			% TAV		Average carcass weight in kg			% TAV	
	1987	1992 ∞	1993 ∞	$\frac{1992}{1987}$	$\frac{1993}{1992}$	1987	1992 ∞	1993 ∞	$\frac{1992}{1987}$	$\frac{1993}{1992}$
1	2	3	4	5	6	7	8	9	10	11
EUR 12	158 878	171 042	179 626	1,5	5,0	80,6	84,1	84,9	0,9	1,0
Belgique/België	8 864	10 428	11 087	3,3	6,3	87,3	90,4	89,5	0,7	− 1,0
Danmark	16 124	18 236	19 723	2,5	8,2	71,6	75,3	76,4	1,0	1,5
BR Deutschland	39 294	40 771	40 999	0,7	0,6	85,1	90,4	91,3	1,2	1,0
Elláda	2 371	2 403	2 330	0,3	− 3,0	67,0	64,0	63,0	− 0,9	− 1,6
España	20 555	24 901	26 811	3,9	7,7	74,3	76,8	77,6	0,7	1,0
France	20 526	22 458	24 112	1,8	7,4	85,4	88,8	89,2	0,8	0,5
Ireland	2 240	2 964	3 087	5,8	4,1	63,9	68,2	69,0	1,3	1,2
Italia	11 374	12 259	12 241	1,5	− 0,1	107,6	108,3	112,0	0,1	3,4
Luxembourg	132	95	108	− 6,4	13,7	87,3	90,4	89,5	0,7	− 1,0
Nederland	19 001	18 789	20 345	− 0,2	8,3	80,7	84,4	85,9	0,9	1,8
Portugal	2 624	3 451	4 068	5,6	17,9	69,0	68,8	70,7	− 0,1	2,8
United Kingdom	15 773	14 286	14 713	− 2,0	3,0	63,9	67,7	68,4	1,2	1,0

Source : Eurostat.

([1]) Animals of domestic and foreign origin.

4.16.1.2   Net pigmeat production (¹)

	1 000 t			% TAV	
	1987	1992 ∞	1993 ∞	$\dfrac{1992}{1987}$	$\dfrac{1993}{1992}$
1	2	3	4	5	6
EUR 12	12 812	14 387	15 255	2,3	6,0
BLEU/UEBL	785	951	1 002	3,9	5,4
Danmark	1 154	1 372	1 507	3,5	9,8
BR Deutschland	3 343	3 684	3 742	2,0	1,6
Elláda	159	153	147	− 0,8	− 3,9
España	1 529	1 912	2 081	4,6	8,8
France	1 753	1 993	2 151	2,6	7,9
Ireland	143	202	213	7,2	5,4
Italia	1 224	1 328	1 370	1,6	3,2
Nederland	1 534	1 585	1 747	0,7	10,2
Portugal	181	237	288	5,5	21,5
United Kingdom	1 008	967	1 006	− 0,8	4,0

Source: Eurostat.

(¹) Animals of domestic and foreign origin.

4.16.2.1   World production and gross domestic production of principal pigmeat-producing or exporting countries

	%			1 000 t			% TAV	
	1985	1991 ∞	1992 ∞	1985	1991 ∞	1992 ∞	$\dfrac{1991}{1985}$	$\dfrac{1992}{1991}$
1	2	3	4	5	6	7	8	9
World	100,0	100,0	100,0	59 228	70 654	71 041	3,0	0,5
EUR 12	20,4	20,3	20,3	12 105	14 339	14 388	2,9	0,3
Peop. Rep. China	29,5	36,3	38,7	17 492	25 646	27 460	6,6	7,1
USA	11,1	10,3	11,0	6 594	7 258	7 817	1,6	7,7
USSR	9,9	8,5	7,3	5 853	6 012	5 185	0,4	− 13,8
Poland	2,5	2,8	2,8	1 475	2 013	1 998	5,3	− 0,7
Japan	2,6	2,1	2,0	1 531	1 483	1 446	− 0,5	− 2,5
Brazil	1,3	1,6	1,6	770	1 160	1 160	7,1	0,0
Canada	1,6	1,6	1,7	972	1 129	1 205	2,5	6,7
Yugoslavia	1,5	1,0	0,9	874	715	620	− 3,3	− 13,3
Romania	1,6	1,2	1,2	966	850	846	− 2,1	− 0,5
Hungary	1,8	1,4	1,1	1 059	965	815	− 1,5	− 15,5
Czechoslovakia	1,4	1,1	1,1	820	771	750	− 1,0	− 2,7

Source: FAO.

4.16.4.1   **Supply balance — pigmeat**   **EUR 12**

	1 000 t ([1])			% TAV	
	1987	1992 ∞	1993 ∞	1992 / 1987	1993 / 1992
1	2	3	4	5	6
Gross internal production	12 869	14 388	15 277	2,3	6,2
Imports — Live animals	33,6	1,6	0,1	− 45,6	− 93,8
Exports — Live animals	1,0	1,3	0,9	5,4	− 30,8
Intra-Community trade	441	489	:	2,1	×
Net production	12 836	14 388	15 277	2,3	6,2
Changes in stocks	10	0	− 10	×	×
Imports	84	59	26	− 6,8	− 55,9
Exports	440	536	714	4,0	33,2
Intra-Community trade	2 101	2 793	:	5,9	×
Internal use	12 510	13 911	14 599	2,1	4,9
Gross consumption in kg/head/year	38,7	40,2	42,0	0,8	4,5
Self-sufficiency (%)	102,6	103,4	104,6	0,2	1,2

Source: Eurostat.

([1]) Carcass weight.

4.16.5.1   **Market prices for pigmeat** ([1])

	ECU/100 kg ([2])			% TAV ([3])	
	1991	1992	1993	1992 / 1991	1993 / 1992
1	2	3	4	5	6
EUR 12 ([4])	137,459	142,355	104,528	3,6	− 26,6
Belgique/België	146,093	149,562	112,983	2,4	− 24,2
Danmark	119,861	127,179	96,861	6,1	− 22,7
BR Deutschland	139,251	140,384	103,872	3,0	− 26,0
Elláda	142,459	161,595	119,699	25,3	− 15,9
España	143,227	147,837	108,086	6,2	− 13,6
France	138,637	141,465	103,248	2,0	− 26,7
Ireland	123,017	131,890	104,606	7,2	− 13,4
Italia	163,424	173,732	121,330	9,8	− 15,9
Luxembourg	147,881	163,273	124,037	10,4	− 23,8
Nederland	128,133	126,856	89,808	− 1,0	− 29,2
Portugal	153,524	167,957	115,003	7,4	− 24,0
United Kingdom	123,632	133,284	105,380	12,5	− 26,6

Source: EC Commission, Directorate-General for Agriculture.

([1]) Representative markets.

([2]) Slaughter weight — Class I.

([3]) Calculated on the basis of prices in national currencies.

([4]) Weighted ∅ ECU/100 kg.

4.17.0.1 **Sheep and goat numbers** (preceding December)

1	1 000 head			% of EUR 12	% TAV	
	»1987«	1992 ∞	1993 ∞	1993	$\frac{»1992«}{»1987«}$	$\frac{1993}{1992}$
1	2	3	4	5	6	7
Sheep						
EUR 12	89 023	98 365	98 620	100,0	2,2	0,3
Belgique/België	128	129	129	0,1	0,5	− 0,5
Danmark	65	122	102	0,1	11,4	− 16,4
BR Deutschland	1 364	2 488	2 386	2,4	14,7	− 4,1
Elláda	10 612	9 837	10 108	10,2	− 1,1	2,8
España	19 267	24 608	24 575	24,9	4,8	− 0,1
France	12 140	10 640	10 380	10,5	− 2,5	− 2,4
Ireland	3 759	5 988	6 125	6,2	9,8	2,3
Italia	11 400	10 435	10 403	10,5	− 1,5	− 0,3
Luxembourg	5	6	7	0,0	5,9	16,7
Nederland	1 140	1 800	1 580	1,6	9,0	− 12,2
Portugal	3 012	3 380	3 348	3,4	2,2	− 0,9
United Kingdom	26 130	28 932	29 477	29,9	2,5	1,9
Goats						
EUR 12	11 858	12 202	12 028	100,0	0,9	− 1,4
Belgique/België	7	9	9	0,1	3,0	− 4,4
Danmark	0	0	0	0,0	×	×
BR Deutschland	46	83	88	0,7	13,6	6,0
Elláda	6 057	5 832	5 850	48,6	− 0,6	0,3
España	2 656	2 799	2 664	22,1	2,4	− 4,8
France	996	1 121	1 071	8,9	2,3	− 4,5
Ireland	0	0	0	0,0	×	×
Italia	1 192	1 314	1 321	11,0	1,9	0,5
Luxembourg	1	1	1	0,0	0,0	0,0
Nederland	45	77	73	0,6	10,8	− 5,2
Portugal	803	862	858	7,1	1,4	− 0,5
United Kingdom	55	105	94	0,8	13,7	− 10,5

Source: Eurostat.

4.17.1.1   Sheep and goats slaughtered

	1 000 head			% TAV		Average carcass weight in kg			% TAV	
	»1987«	1992 ∞	1993 ∞	»1992«/»1987«	1993/1992	»1987«	1992 ∞	1993 ∞	»1992«/»1987«	1993/1992
1	2	3	4	5	6	7	8	9	10	11
EUR 12	70 665	83 359	81 120	3,3	− 2,7	14,1	14,1	13,9	− 0,0	− 1,4
BLEU/UEBL	331	316	238	− 2,0	− 24,7	28,7	20,1	20,0	− 7,0	− 0,5
Danmark	42	90	93	16,5	3,5	23,3	36,7	23,7	3,2	− 35,4
BR Deutschland	1 381	2 274	2 206	10,7	− 3,0	20,3	19,3	18,5	− 0,8	− 4,1
Elláda	11 297	12 492	12 459	1,8	− 0,3	10,7	11,0	10,0	0,0	− 9,1
España	19 673	22 845	22 305	2,7	− 2,4	11,2	10,8	10,7	− 0,8	− 0,9
France	9 749	10 005	9 857	1,0	− 1,5	17,2	17,2	16,6	− 0,3	− 3,5
Ireland	2 070	4 357	4 643	16,3	6,6	23,0	21,7	20,7	− 1,4	− 4,6
Italia	8 163	9 755	8 901	2,9	− 8,8	8,6	8,8	9,1	0,7	3,4
Nederland	507	723	789	7,8	9,1	23,7	23,7	23,2	− 0,1	− 2,1
Portugal	1 355	1 372	1 302	0,6	− 5,1	10,2	10,2	10,5	0,6	2,9
United Kingdom	16 098	19 130	18 328	3,9	− 4,2	19,0	18,6	18,5	− 0,5	− 0,5

Source: Eurostat.

4.17.1.2   Gross internal sheepmeat and goatmeat production

	1 000 t			% TAV	
	»1987«	1992	1993	»1992«/»1987«	1993/1992
1	2	3	4	5	6
EUR 12	993	1 177 ∞	1 148 ∞	3,5	− 2,5
BLEU/UEBL	4	3	3**	− 3,6	0,0
Danmark	1	2	3**	18,5	50,0
BR Deutschland	25	45 ∞	40 ∞	14,9	− 11,1
Elláda	119	129	126**	1,2	− 2,3
España	222	235	228	0,8	− 3,0
France	157	156	156	0,4	0,0
Ireland	49	91	94**	13,6	3,3
Italia	49	59	59**	3,7	0,0
Nederland	20	28	30**	8,6	7,1
Portugal	27	27	30**	1,7	11,1
United Kingdom	321	402	379	4,5	− 5,7

Source: Eurostat.

4.17.3.1   Sheepmeat and goatmeat — EC trade, by species

Description	Imports						Exports					
	1992 ∞			1993 ∞			1992 ∞			1993 ∞		
	Extra EC	Intra EC	World	Extra EC	Intra EC	World	Extra EC	Intra EC	World	Extra EC	Intra EC	World
1	2	3	4	5	6	7	8	9	10	11	12	13
1. Live animals, in number (per 1 000 head)												
— *Pure-bred breeding animals*												
Sheep	10,7	14,9	25,5	1,3	..	..	0,6	22,7	23,3	1,2	..	..
Goats	0,0	1,0	1,0	0,1	..	..	1,1	0,6	1,6	0,8	..	..
— *Other live animals*												
Sheep	2 329,8	4 739,7	7 069,5	1 572,7	..	..	52,6	4 664,1	4 716,7	46,4	..	..
Goats	1,4	76,2	77,5	2,1	..	..	2,1	58,5	60,5	2,4	..	..
Total live animals	2 341,9	4 831,8	7 173,7	1 576,2	..	..	56,4	4 745,9	4 802,3	50,8	..	..
2. Live animals converted to meat weight (1 000 tonnes carcass weight)	19,3	70,1	89,4	15,3	..	..	0,8	69,9	70,6	0,9	..	..
3. Meat (1 000 tonnes carcass weight)												
— Fresh or chilled	16,3	177,0	193,4	15,2	..	..	3,1	173,2	176,3	3,6	..	..
— Frozen	226,1	16,9	243,0	215,5	..	..	1,4	18,2	19,6	1,5	..	..
— Salted or in brine, dried or smoked	0,0	0,0	0,0	0,0	..	..	0,0	0,1	0,1	0,0	..	..
— Prepared and preserved	0,0	1,3	1,3	0,0	..	..	0,1	1,0	1,1	0,1	..	..
Total sheepmeat and goatmeat (2 + 3)	261,8	265,3	527,1	246,0	..	..	5,4	262,4	267,8	6,2	..	..

Source: EC Commission, Directorate-General for Agriculture and Eurostat — Comext.
Coefficients: Live animals: Carcass weight = live weight × 0,47
    — Boneless meat   } Product weight × 1,7 = carcass weight.
    — Prepared and preserved meat

4.17.3.2   Sheepmeat and goatmeat — trade with non-member countries

(1 000 t carcass weight)

Reporting countries	1990	%	1991 ∞	%	1992 ∞	%	1993 ∞	%
1	2	3	4	5	6	7	8	9
A. Exports								
EUR 12	5,9	100,0	18,9	100,0	5,5	100,0	6,2	100,0
BLEU/UEBL	0,1	1,7	0,1	0,5	0,0	0,0	0,1	1,5
Danmark	0,1	1,7	0,1	0,5	0,1	1,8	0,1	1,6
BR Deutschland	0,9	15,5	13,3	70,7	0,3	5,5	0,3	4,9
Elláda	0,1	1,7	0,3	1,6	0,2	3,6	0,2	3,7
España	1,1	19,0	1,3	6,9	1,5	27,3	1,9	29,8
France	1,0	17,2	1,5	8,0	0,9	16,4	0,9	14,9
Ireland	0,0	0,0	0,0	0,0	0,1	1,8	0,0	0,1
Italia	0,1	1,7	0,0	0,0	0,3	5,5	0,0	0,1
Nederland	0,0	0,4	0,0	0,0	0,0	0,0	0,2	3,0
Portugal	0,0	0,0	0,0	0,0	0,0	0,0	0,1	0,9
United Kingdom	2,4	41,4	2,2	11,7	2,1	38,2	2,5	39,5
B. Imports								
EUR 12	283,4	100,0	257,9	100,0	261,8	100,0	246,0	100,0
BLEU/UEBL	9,3	3,3	7,1	2,8	8,6	3,3	11,0	4,5
Danmark	3,4	1,2	3,1	1,2	3,4	1,3	3,9	1,6
BR Deutschland	37,2	13,1	36,9	14,3	40,0	15,3	40,7	16,5
Elláda	17,5	6,2	16,6	6,4	21,8	8,3	17,0	6,9
España	13,6	4,8	11,3	4,4	10,6	4,0	5,0	2,0
France	10,9	3,8	22,9	8,9	21,7	8,3	22,5	9,1
Ireland	0,1	0,0	0,0	0,0	0,1	0,0	0,0	0,0
Italia	29,7	10,5	27,2	10,5	25,1	9,6	17,6	7,2
Nederland	3,9	1,4	5,9	2,3	4,7	1,8	6,4	2,6
Portugal	8,1	2,9	8,1	3,1	7,8	3,0	3,6	1,5
United Kingdom	149,6	52,8	118,9	46,1	118,1	45,1	118,4	48,1

Source : EC Commission, Directorate-General for Agriculture, and Eurostat — Comext.

Coefficients : – Live animals : Carcass weight = live weight × 0,47.
 – Boneless meat
 – Prepared and preserved meat } Product weight × 1,7 = carcass weight.

4.17.3.3 **Imports of sheepmeat** (¹)

EUR 12	t (²)			% TAV	
	1991 ∞	1992 ∞	1993 ∞	1992/1991	1993/1992
1	2	3	4	5	6
Total imports					
— New Zealand	198 872	209 374	202 906	5,3	− 3,1
— Argentina	8 823	4 580	2 883	− 48,1	− 37,1
— Australia	16 837	16 829	15 939	0,0	− 5,3
— Hungary	14 131	10 823	8 466	− 23,4	− 21,8
— Bulgaria	3 638	2 141	2 503	− 41,1	16,9
— Poland	5 265	6 511	4 736	23,7	− 27,3
— ex-Yugoslavia	2 786	3 189	87	14,5	− 97,3
— Uruguay	4 296	5 025	5 508	17,0	9,6
— Romania	477	507	405	6,3	− 20,1
— Other countries	2 792	2 856	2 661	2,3	− 6,8
Grand total	257 917	261 835	246 094	1,5	− 6,0

Source : EC Commission, Directorate-General for Agriculture and Eurostat — Comext.

(¹) Incl. live animals.
(²) Tonnes carcass weight.

4.17.4.1 **Supply balance — sheepmeat and goatmeat**   **EUR 12**

1	1 000 t			% TAV	
	» 1987 «	1992 ∞	1993 ∞	» 1992 «/» 1987 «	1993/1992
1	2	3	4	5	6
Gross internal production	993	1 177	1 148	3,5	− 2,5
Imports — live animals (¹)	21	19	15	− 2,2	− 20,7
Exports — live animals (¹)	1	1	1	20,4	12,5
Intra-Community trade (¹)	29	70	:	×	×
Net production	1 007	1 195	1 162	3,5	− 2,8
Changes in stocks	− 9	− 10	:	×	×
Imports (²)	229	243	231	0,6	− 4,9
Exports (²)	5	5	5	8,2	0,0
Intra-Community trade (³)	114	195	:	×	×
Internal use	1 222	1 423	1 388	3,1	− 2,4
Gross consumption (kg/head/year)	3,8	4,1	4,0	1,7	− 2,4
Self-sufficiency (%)	81,2	82,7	82,7	0,4	− 0,0

Source : Eurostat and EC Commission, Directorate-General for Agriculture.

(¹) Carcass weight.
(²) Carcass weight — All trade with the exception of live animals.
(³) All trade in carcass weight, with the exception of live animals (figures based on imports).

4.17.5.1 Market prices for sheepmeat ([1])

	ECU/kg ([2])			% TAV ([3])	
	1991	1992	1993	$\frac{1992}{1991}$	$\frac{1993}{1992}$
1	2	3	4	5	6
Belgique/België	3,398	3,402	2,867	0,1	− 15,5
Danmark	2,388	2,196	2,044	− 8,1	− 5,6
BR Deutschland	2,598	2,729	2,591	5,1	− 5,1
Elláda	3,872	3,727	3,077	6,5	2,4
España	3,053	3,292	2,847	7,5	2,5
France	2,937	2,849	2,758	− 3,0	− 2,8
Ireland	2,269	2,047	2,074	− 9,7	10,7
Italia	3,831	3,399	2,927	− 9,6	5,2
Nederland	2,977	2,907	2,768	− 2,4	− 4,8
Portugal	3,060	3,251	3,033	6,3	0,7
United Kingdom	1,927	2,245	2,295	21,0	19,8

Source: EC Commission, Directorate-General for Agriculture.

([1]) Belgique/België: Average price:
    1) moutons extra (carcass weight 30 kg) — schapen extra (30 kg per stuk).
    2) agneaux extra (carcass weight 16 kg) — lammeren extra (16 kg per stuk).
    Danmark: country ∅: lambs 1st quality.
    Deutschland: country ∅: lambs carcasses of 'L'-Mastlämmer quality.
    Ellada: country ∅: 76% amnos galaktos,
                 24% amnos.
    España:
    France: country ∅ for 'carcasses d'agneaux de boucherie'.
    Ireland: country ∅: 70% prime quality.
               30% second quality.
    Italia: average price:
    1) agnelloni (± 20 kg carcass weight) = 36% (country ∅).
    2) agnelli (± 10 kg carcass weight) = 64% (markets: Cagliari, Roma, Napoli, Firenze - L'Aquila).
    Nederland: country ∅ 'Vette Lammeren'.
    Portugal:
    United Kingdom: ∅ market prices for sheep qualifying for guaranteed prices (pence/kg net on the hoof).
([2]) Slaughter weight.
([3]) Calculated on the basis of prices in national currency.

4.18.0.1   Number of utility chicks of table strains hatched

	1 000 head				% TAV	
	»1987«	1991	1992	1993	$\frac{»1992«}{»1987«}$	$\frac{1993}{1992}$
1	2	3	4	5	6	7
EUR 12	3 054 795	3 584 300∞	3 627 137∞	3 667 836**∞	5,9	1,1
BLEU/UEBL	91 869	114 593	125 634	134 966	11,0	7,4
Danmark	87 458	104 808	114 689	119 850**	9,5	4,5
BR Deutschland	218 993	286 378∞	318 351∞	324 400**∞	13,3	1,9
Elláda	68 700	80 315	83 954	85 752	6,9	2,1
España	498 383	543 669	529 396	513 076	2,0	− 3,1
France	658 344	824 012	828 216**	855 260	8,0	3,3
Ireland	38 326	48 770	50 025	52 954	9,3	5,9
Italia	373 922	413 211	417 739	403 470	3,8	− 3,4
Nederland	300 876	340 826	336 406	330 067	3,8	− 1,9
Portugal	118 109	178 340	171 808	177 595	13,3	3,4
United Kingdom	599 815	649 378	650 919	670 447**	2,8	3,0

Source: Eurostat.

4.18.1.1   Gross internal production of poultrymeat

	1 000 t				% TAV	
	»1987«	1991	1992	1993	$\frac{»1992«}{»1987«}$	$\frac{1993}{1992}$
1	2	3	4	5	6	7
EUR 12	5 784	6 756∞	6 922∞	6 930**∞	6,2	0,1
BLEU/UEBL	141	178	194	198**	11,2	2,1
Danmark	113	142	160	172**	12,3	7,5
BR Deutschland	390	574∞	604∞	615∞	15,7	1,8
Elláda	149	160	175	173**	5,5	− 1,2
España	786	880	858	821	3,0	− 4,3
France	1 408	1 780	1 865	1 875	9,8	0,5
Ireland	67	90	90	92**	10,3	2,2
Italia	1 046	1 114	1 095	1 099**	1,5	0,4
Nederland	484	548	574	585**	5,8	1,9
Portugal	171	190	207	210**	6,6	1,4
United Kingdom	1 029	1 099	1 100	1 090	2,2	− 0,9

Source: Eurostat.

4.18.3.1   Trade in poultrymeat with non-member countries (¹)

(t carcass weight)

Reporting country	1990 ∞	%	1991 ∞	%	1992 ∞	%	1993 ∞	%
1	2	3	4	5	6	7	8	9
A. Exports								
EUR 12	428 134	100,0	477 707	100,0	519 306	100,0	659 472	100,0
BLEU/UEBL	30 463	7,1	33 185	6,9	36 082	6,9	42 540	6,5
Danmark	47 823	11,2	51 475	10,8	53 612	10,3	66 546	10,1
BR Deutschland	11 298	2,6	18 675	3,9	28 288	5,4	38 226	5,8
Elláda	582	0,1	605	0,1	1 043	0,2	7 695	1,2
España	6 178	1,4	7 035	1,5	6 571	1,3	6 562	1,0
France	275 359	64,3	292 708	61,3	304 639	58,7	359 119	54,5
Ireland	52	0,0	188	0,0	277	0,1	439	0,1
Italia	5 297	1,2	3 289	0,7	3 769	0,7	5 571	0,8
Nederland	34 489	8,1	42 735	8,9	54 457	10,5	98 040	14,9
Portugal	969	0,2	6 243	1,3	12 994	2,5	10 166	1,5
United Kingdom	15 625	3,6	21 567	4,5	17 574	3,4	24 547	3,7
B. Imports								
EUR 12	137 970	100,0	151 511	100,0	158 117	100,0	154 920	100,0
BLEU/UEBL	3 425	2,5	4 515	3,0	3 843	2,4	1 304	0,8
Danmark	77	0,1	51	0,0	29	0,0	51	0,0
BR Deutschland	74 538	54,0	92 155	60,8	90 797	57,4	90 215	58,2
Elláda	4 099	3,0	2 217	1,5	1 561	1,0	694	0,4
España	1 325	1,0	799	0,5	1 446	0,9	2 434	1,6
France	10 555	7,7	12 265	8,1	11 823	7,5	13 881	9,0
Ireland	0	0,0	0	0,0	2	0,0	12	0,0
Italia	31 020	22,5	28 358	18,7	29 006	18,3	28 152	18,2
Nederland	10 236	7,4	8 885	5,9	11 364	7,2	11 155	7,2
Portugal	508	0,4	2	0,0	0	0,0	29	0,0
United Kingdom	2 186	1,6	2 263	1,5	8 246	5,2	7 004	4,5

Source : Eurostat — Comext.

(¹) Live animals, expressed as carcass weight (live weight × 0,7), and poultrymeat, including offals, livers and fats of Chapter 02 of CN.

4.18.4.1   **Supply balance — poultrymeat**   **EUR 12**

	1 000 t (¹)				% TAV	
	»1987«	1991 ∞	1992 ∞	1993 ∞	»1992«/»1987«	1993/1992
1	2	3	4	5	6	7
Gross internal production	5 784	6 755	6 930	7 008	− 5,8	1,1
Imports — live birds	3	3	2	2	14,5	0,0
Exports — live birds	7	2	1	1	91,3	0,0
Intra-Community trade	83	135	150	:	− 17,9	×
Net production	5 780	6 756	6 930	7 009	− 5,9	1,1
Changes in stocks	41	23	13	− 40	×	×
Imports	81	152	158	155	− 20,0	− 1,9
Exports	367	478	519	659	− 10,9	27,0
Intra-Community trade	467	791	905	:	− 19,8	×
Internal use (total)	5 452	6 406	6 555	6 545	− 6,0	− 0,2
Human consumption (kg/head/year)	16,8	18,6	18,9	18,9	− 3,9	0,0
Self-sufficiency (%)	106,1	105,4	105,7	107,1	0,1	1,3

Source : Eurostat.
(¹) Carcass weight.

4.18.5.1 Market prices for chickens ([1])

	ECU/Kg ([2])			% TAV ([3])	
	1991	1992	1993	$\dfrac{1992}{1991}$	$\dfrac{1993}{1992}$
1	2	3	4	5	6
Belgique/België	1,269	1,259	1,242	– 0,8	– 1,0
Danmark	1,358	1,343	1,299	– 1,1	– 1,8
BR Deutschland	1,387	1,344	1,288	– 3,1	– 4,4
Elláda	2,016	1,853	1,644	6,2	5,6
España	1,044	1,092	0,983	4,2	6,1
France	0,932	0,928	0,953	– 0,5	3,3
Ireland	1,971	1,776	1,465	26,8	– 10,3
Italia	1,494	1,316	1,232	– 10,4	14,5
Luxembourg	:	:	:	×	×
Nederland	1,246	1,208	1,158	– 2,7	– 4,4
Portugal	:	:	1,244	×	×
United Kingdom	1,364	1,323	1,189	1,3	4,7

Source: EC Commission, Directorate-General for Agriculture.

([1]) Belgique/België   : Poulets à 70%, prix de gros à la vente. Kuikens 70%, groothandelsverkoopprijs. A partir de juillet 1982 prix franco frontière. Vanaf juli 1982 prijs
franco grens.
Danmark   : Kyllinger, 70%, slagterie til detailhandel.
BR Deutschland : Schlachterei – Abgabepreis frei Empfänger, 70% gefroren.
Ellada   : Chondriki timi 70% (prix de gros).
España   : Precio de mercado.
France   : Paris-Rungis: poulets, classe A (moyens), 83%, prix de gros à la vente.
Ireland   : Chickens, 70%, wholesale price.
Italia   : Milano: prezzi d'acquisto all'ingrosso, 83%.
Nederland   : LEI: Kuikens 70% – Groothandelsverkoopprijs.
Portugal   : Preço à produção.
United Kingdom : London: Chickens, 83%, wholesale price.
([2]) Slaughter weight.
([3]) Calculated on the basis of prices in national currencies.

4.19.0.1   Laying hens, numbers

	1 000 head				% TAV	
	»1987«	1991	1992	1993	»1992«/»1987«	1993/1992
1	2	3	4	5	6	7
EUR 12	288 870 (¹)	341 509∞	338 449∞	333 125∞	5,4	− 1,6
BLEU/UEBL	10 652	11 445**	11 823**	12 532**	3,5	6,0
Danmark	3 828	3 854	3 866	3 723**	0,3	− 3,7
BR Deutschland	47 100	58 400∞	56 600∞	55 468**∞	6,3	− 2,0
Elláda	17 177	16 329	16 072	14 465**	− 2,2	− 10,0
España	:	44 795**	41 615**	36 205**	×	− 13,0
France	62 421	64 700	64 400	67 700	1,0	5,1
Ireland	3 462	3 459	3 597**	3 752**	1,3	4,3
Italia	51 950	50 997	51 503	51 761**	− 0,3	0,5
Nederland	38 246	37 085**	38 605**	37 563**	0,3	− 2,7
Portugal	:	6 405**	7 116**	6 867**	×	− 3,5
United Kingdom	54 034	44 040	43 252	43 090	− 7,2	− 0,4

Source: Eurostat.

4.19.0.2   Number of utility chicks hatched from laying hens

	1 000 head				% TAV	
	»1987«	1991	1992	1993	»1992«/»1987«	1993/1992
1	2	3	4	5	6	7
EUR 12	234 021	241 972∞	235 419∞	241 016	0,2	2,4
BLEU/UEBL	8 172	10 148	10 576	13 345	9,0	26,2
Danmark	4 091	4 242	4 946	5 000**	6,5	1,1
BR Deutschland	33 850	36 868∞	41 862**∞	41 778**	7,3	− 0,2
Elláda	2 781	4 706	3 526	3 657	8,2	3,7
España	37 954	35 773	29 673	35 893	− 7,9	21,0
France	40 776	46 084	45 938	47 188	4,1	2,7
Ireland	1 597	1 437	1 474	1 283	− 2,6	− 13,0
Italia	31 733	31 050	27 658	23 137	− 4,5	− 16,3
Nederland	31 372	32 891	32 669	32 016**	1,4	− 2,0
Portugal	4 777	6 072	5 322	6 611	3,7	24,2
United Kingdom	36 918	32 701	31 775	31 108**	− 4,9	− 2,1

Source: Eurostat.

4.19.1.1 **Usable production of eggs** (total eggs)

	1 000 t				% TAV	
	»1987«	1991	1992	1993	$\dfrac{»1992«}{»1987«}$	$\dfrac{1993}{1992}$
1	2	3	4	5	6	7
EUR 12	4 884	4 981∞	4 925∞	4 870	0,3	− 1,1
BLEU/UEBL	182	200	201	217**	3,4	8,0
Danmark	77	84	88	87**	4,6	− 1,1
BR Deutschland	721	922∞	902∞	824**	7,8	− 8,6
Elláda	128	123	124	122**	− 1,1	− 1,6
España	671	641	602	612**	− 3,6	1,7
France	891	928	932	938	1,5	0,6
Ireland	39	39	39**	38**	0,0	− 2,6
Italia	631	648	665	672**	1,8	1,1
Nederland	654	652	630**	623**	− 1,2	− 1,1
Portugal	88	98	103	102**	5,4	− 1,0
United Kingdom	802	646	639	635	− 7,3	− 0,6

Source : Eurostat.

(¹)

4.19.3.1 Trade in eggs with non-member countries (¹)

Reporting country	1990	%	1991 ∞	%	1992 ∞	%	1993 ∞	%
1	2	3	4	5	6	7	8	9
A. Exports								
EUR 12	80 809	100,0	90 036	100,0	94 925	100,0	96 795	100,0
BLEU/UEBL	2 028	2,5	2 573	2,9	2 571	2,7	3 871	4,0
Danmark	1 626	2,0	1 456	1,6	1 089	1,1	925	1,0
BR Deutschland	14 029	17,4	17 220	19,1	14 111	14,9	18 569	19,2
Elláda	277	0,3	690	0,8	448	0,5	651	0,7
España	6 182	7,7	6 302	7,0	6 693	7,1	7 580	7,8
France	8 024	9,9	11 089	12,3	10 136	10,7	9 668	10,0
Ireland	0	0,0	0	0,0	0	0,0	15	0,0
Italia	193	0,2	140	0,2	264	0,3	862	0,9
Nederland	48 220	59,7	50 345	55,9	59 144	62,3	54 213	56,0
Portugal	23	0,0	62	0,1	226	0,2	101	0,1
United Kingdom	207	0,3	159	0,2	243	0,3	340	0,4
B. Imports								
EUR 12	21 866	100,0	21 512	100,0	17 491	100,0	15 816	100,0
BLEU/UEBL	3 704	16,9	1 596	7,4	581	3,3	44	0,3
Danmark	3 403	15,6	3 214	14,9	3 959	22,6	3 690	23,3
BR Deutschland	3 641	16,7	3 261	15,2	4 320	24,7	2 809	17,8
Elláda	177	0,8	66	0,3	21	0,1	17	0,1
España	260	1,2	414	1,9	304	1,7	474	3,0
France	361	1,6	312	1,5	293	1,7	189	1,2
Ireland	0	0,0	0	0,0	0	0,0	0	0,0
Italia	2 833	13,0	4 735	22,0	2 844	16,3	4 017	25,4
Nederland	7 278	33,2	7 832	36,4	5 130	29,3	4 563	28,9
Portugal	142	0,6	73	0,3	19	0,1	11	0,1
United Kingdom	65	0,3	9	0,0	20	0,1	2	0,0

Source : Eurostat — Comext.

(¹) Eggs in the shell — Code CN 04070030.

4.19.4.1 **Supply balance — eggs** (total eggs)   **EUR 12**

	1 000 t				% TAV	
1	»1987«	1991 ∞	1992 ∞	1993 ∞	$\dfrac{»1992«}{»1987«}$	$\dfrac{1993}{1992}$
	2	3	4	5	6	7
Usable production	4 884	4 981	4 905	4 871	0,1	− 0,7
Change in stocks	3	− 8	0	0	×	×
Imports	46	36	28	28	− 15,3	0,0
Exports	115	155	157	183	10,9	16,6
Intra-Community trade	650	669	658	:	0,4	×
Internal use of which :	4 812	4 872	4 785	4 713	− 0,2	− 1,5
— eggs for hatching	289	332	333	340	4,8	2,1
— industrial use	15	20	21	22	11,9	4,8
— losses (market)	19	19	19	18	0,0	− 5,3
— human consumption	4 489	4 501	4 412	4 334	− 0,6	− 1,8
Human consumption (kg/head/year)	13,9	13,0	12,8	12,4	− 2,7	− 3,1
Self-sufficiency (%)	101,5	102,3	102,5	103,3	0,3	0,8

Source : Eurostat.

4.19.5.1   Market prices for eggs (^{1})

	ECU/100 pieces			% TAV (^{2})	
	1991	1992	1993	$\frac{1992}{1991}$	$\frac{1993}{1992}$
1	2	3	4	5	6
Belgique/België	4,889	4,244	4,294	− 16,6	1,9
Danmark	6,685	6,595	6,909	− 1,3	6,4
BR Deutschland	5,672	5,056	5,314	− 10,8	5,3
Elláda	9,784	8,528	7,427	1,4	3,5
España	6,280	6,157	6,311	− 2,3	20,1
France	5,411	4,775	4,558	− 11,7	− 4,1
Ireland	5,883	5,465	4,921	− 7,1	− 1,6
Italia	6,015	5,430	4,979	− 8,5	12,4
Luxembourg	6,411	5,856	6,048	− 8,7	3,5
Nederland	4,811	4,179	4,284	− 13,1	2,5
Portugal	:	:	6,099	×	×
United Kingdom	5,063	4,880	4,840	1,2	22,1

Source: EC Commission, Directorate-General for Agriculture.

(^{1}) Eggs: Class IV − weight 55-60 gr. :
Belgique/België	: Kruishoutem: prix de gros à l'achat, franco marché
	groothandelsaankoopprijs, franco markt.
Danmark	: engrospris.
BR Deutschland	: Packstellenabgabepreis, frei Empfänger.
Ellada	: Wholesale prices.
España	: Precio de mercado.
France	: Prix de vente, sortie station.
Ireland	: Dublin: wholesale selling price.
Italia	: Milano: prezzo d'acquisto del commercio all'ingrosso, franco mercato.
Luxembourg	: Prix de gros à la vente, franco détaillant.
Nederland	: Groothandelsverkoopprijs.
Portugal	: Preços de ovos.
United Kingdom	: Eggs Authority: packer to producer price.

(^{2}) Calculated on the basis of prices in national currency.

4.20.0.1   Dairy herds and yield

Dairy cows in December	1 000 head			% TAV	
	1985	1992	1993	$\frac{1992}{1985}$	$\frac{1993}{1992}$
1	2	3	4	5	6
EUR 12	26 240**	21 687*∞	21 388**∞	− 2,7	− 1,4
Belgique/België	951	741	705*	− 3,5	− 4,9
Danmark	913	708	711	− 3,6	0,4
BR Deutschland	5 452	5 365∞	5 301∞	− 0,2	− 1,2
Elláda	219	205*	219*	− 0,9	6,8
España (²)	1 880	1 447	1 371	− 3,7	− 5,3
France	6 506	4 642	4 615	− 4,7	− 0,6
Ireland	1 495	1 262	1 274	− 2,4	1,0
Italia	2 804	2 317	2 220**	− 2,7	− 4,2
Luxembourg	70	51*	51*	− 4,4	0,0
Nederland	2 333	1 821	1 760**	− 3,5	− 3,3
Portugal	360**	381	375*	0,8	− 1,6
United Kingdom	3 257	2 747	2 786	− 2,4	1,4

Dairy cows yield (¹)	kg/head			% TAV	
	1985	1992	1993	$\frac{1992}{1985}$	$\frac{1993}{1992}$
EUR 12	4 255**	4 873**∞	5 132**∞	2,0	5,3
Belgique/België	3 864	4 410	4 493	1,9	1,9
Danmark	5 379	6 173	6 583	2,0	6,6
BR Deutschland	4 599	4 970**∞	5 237∞	1,1	5,4
Elláda	2 959	3 416	3 668**	2,1	7,4
España	3 335	4 051**	4 167	2,8	2,9
France	4 109	5 096**	5 396	3,1	5,9
Ireland	3 822	4 159**	4 208**	1,2	1,2
Italia	3 388	4 068	4 489**	2,6	10,3
Luxembourg	4 237	4 971	5 255*	2,3	5,7
Nederland	5 150	5 795	6 014	1,7	3,8
Portugal	3 138**	4 383	4 344**	4,9	− 0,9
United Kingdom	4 867	5 137	5 383**	0,8	4,8

Source: Eurostat.

(¹) Production of the year divided by the herd in December of previous year.
(²) 1985: in September.

4.20.1.1   Production of milk from dairy herds and delivery of milk to dairies

Production of milk from dairy cows (¹)	1 000 t			% TAV	
	1985	1992	1993	$\frac{1992}{1985}$	$\frac{1993}{1992}$
1	2	3	4	5	6
EUR 12	115 938**	111 157**∞	111 952**	− 0,6	0,1
Belgique/België	3 796	3 514	3 460**	− 1,1	− 5,3
Danmark	5 099	4 605	4 661	− 1,4	1,2
BR Deutschland	25 674	27 991∞	28 600**	1,2	0,4
Elláda	663	731	752**	1,4	2,9
España	6 258**	6 143	6 030	− 0,3	− 1,8
France	27 790	25 315	25 300**	− 1,3	− 1,1
Ireland	5 823	5 378	5 311**	− 1,1	− 1,2
Italia	10 753	10 315	10 300**	− 0,6	− 0,8
Luxembourg	301	260	268**	− 2,1	3,1
Nederland	12 550	10 901	10 969**	− 2,0	0,5
Portugal	1 114**	1 727**	1 655**	6,5	− 4,2
United Kingdom	16 117	14 277	14 646	− 1,7	3,6

Deliveries of cows' milk (²)	1 000 t			% TAV	
	1985	1992	1993	$\frac{1992}{1985}$	$\frac{1993}{1992}$
EUR 12	105 845**	103 449**∞	103 250**	− 0,3	− 0,2
Belgique/België	3 162	2 914	2 878**	− 1,2	− 1,2
Danmark	4 899	4 405	4 461	− 1,5	1,3
BR Deutschland	23 637	25 612∞	26 017	1,2	1,6
Elláda	461	549	590**	2,5	7,5
España	4 761**	5 438	5 352	1,9	− 1,6
France	25 476	23 059**	22 854	− 1,4	− 0,9
Ireland	5 682	5 261**	5 202**	− 1,1	− 1,1
Italia	8 596	9 962**	9 553	2,1	− 4,1
Luxembourg	294	249	258	− 2,3	3,6
Nederland	12 233	10 431	10 500	− 2,3	0,7
Portugal	1 057**	1 541	1 477	5,5	− 4,2
United Kingdom	15 587	14 028	14 108**	− 1,5	0,6

Source : Eurostat.

(¹) Excl. milk for suckling.
(²) Incl. deliveries of cream (milk equivalent).

4.20.1.2   Deliveries of cows' milk to dairies, as a proportion of cows' milk production (¹)

(%)

	1988	1989	1990	1991	1992	1993
1	2	3	4	5	6	7
EUR 12	90,8*	91,4*	91,9	91,7	93,0	92,3
Belgique/België	84,4	85,4	82,8	83,6	83,1	83,3
Danmark	95,8	95,8	95,8	95,7	95,7	95,7
BR Deutschland	90,3	90,5	90,7	90,9	91,5	91,0
Elláda	74,6*	76,1*	78,2	75,1	75,1	78,5
España	76,3	78,1	78,3**	84,5	88,5	88,8
France	90,7	91,5	91,8**	90,8	91,0	91,2
Ireland	97,6	97,6	97,7	97,6	97,8	97,9
Italia	82,3	82,4	90,2	90,1	96,6	91,9
Luxembourg	94,3	93,6	93,6	95,4	95,8	96,3
Nederland	96,7	96,0	95,5	95,7	95,7	95,9
Portugal	97,6	97,3	98,8	91,2	89,2	89,2
United Kingdom	96,4	98,2	96,1	95,7	98,3	96,5

Source : Eurostat.
(¹) Incl. deliveries of cream (milk equivalent).

EUR 12

4.20.1.3   Production of fresh milk and fresh milk products by the dairy industry

	1987	1988	1989	1990	1991 ∞	1992 ∞	% TAV 1991/1987	% TAV 1992/1991
1	2	3	4	5	6	7	8	9
1. Drinking milk	23 443	23 582	23 797	24 082	26 166**	26 356	1,2	0,7
of which: — whole milk	15 078	14 656	14 374	13 831	14 545**	14 052	0,7	− 3,4
— semi-skimmed milk	6 950	7 449	7 791	8 590	9 666**	10 356	1,7	7,1
— skimmed milk	1 308	1 370	1 539	1 630	1 919**	1 932	2,4	0,7
— untreated milk	107	107	87	31	36	16	2,2	− 55,6
2. Buttermilk	368	384	419	418	455**	469	1,2	3,1
3. Cream	960	1 000	1 000	1 061	1 134**	1 122	1,0	− 1,1
4. Acidified milk	2 728	2 980	3 121	3 346	3 686**	3 840	1,4	4,2
5. Milk-based drinks	595	601	620	659	707**	690	1,0	− 2,4
6. Other fresh products	914	1 003	1 002	1 097	1 236**	1 296	1,7	4,9
7. Subtotal (2-6)	5 565	5 968	6 162	6 581	7 218**	7 417	1,3	2,8
Total	29 008	29 550	29 959	30 663	33 384**	33 773	1,2	1,2

1 000 t

Source: Eurostat.

4.20.1.4   Production in dairies of butter and cheese

Butter (¹)	1 000 t			% TAV	
	1985	1992	1993	$\frac{1992}{1985}$	$\frac{1993}{1992}$
1	2	3	4	5	6
EUR 12	2 032**	1 686**∞	1 676**∞	− 2,6	− 0,6
Belgique/België	83	63	61	− 3,9	− 3,2
Danmark	110	78	77	− 4,8	− 1,3
BR Deutschland	515	474∞	482∞	− 1,2	1,7
Elláda	2	2	1	0,0	− 50,0
España	16**	29	25	8,9	− 13,8
France	586	460	445**	− 3,4	− 3,3
Ireland	164	134	128	− 2,8	− 5,2
Italia	76	100	93	4,0	− 7,0
Luxembourg	8	3	3	− 13,1	0,0
Nederland	263	199	194	− 3,9	− 2,5
Portugal	7	17	17	13,5	0,0
United Kingdom	202	127**	151**	− 6,4	18,9

Cheese (²)	1 000 t			% TAV	
	1985	1992	1993	$\frac{1992}{1985}$	$\frac{1993}{1992}$
EUR 12	4 286**	5 390∞	5 481**∞	3,3	1,7
Belgique/België	51	69	69	4,4	0,0
Danmark	256	292	322	1,9	10,3
BR Deutschland	913	1 293∞	1 336∞	5,1	3,3
Elláda	125	98	98**	− 3,4	0,0
España	139**	227	227	7,3	0,0
France	1 283	1 485	1 520	2,1	2,4
Ireland	79	93	93	2,4	0,0
Italia	626	836	816	4,2	− 2,4
Luxembourg	3	4	3	4,2	− 25,0
Nederland	525	640	640	2,9	0,0
Portugal	30**	48	57	6,9	18,8
United Kingdom	256	305	300	2,5	− 1,6

Source : Eurostat.

(¹) Incl. butteroil manufactured from cream (butter equivalent).
(²) Processed cheese excluded.

4.20.1.5 Production in dairies of milk powder ([1])

Skimmed-milk powder	1 000 t			% TAV	
	1985	1992	1993	$\frac{1992}{1985}$	$\frac{1993}{1992}$
1	2	3	4	5	6
EUR 12	1 948**	1 162∞	1 220∞	− 7,1	5,0
Belgique/België	112	52	59	− 10,4	13,5
Danmark	25	13	20	− 8,9	53,8
BR Deutschland	549	400∞	434∞	− 4,4	8,5
Elláda	0	0	0	×	×
España	27**	23	15	− 2,3	− 34,8
France	653	373	360	− 7,7	− 3,5
Ireland	161	126	132	− 3,4	4,8
Italia	0	0	0	×	×
Luxembourg	13	0	0	×	×
Nederland	163	61	68	− 13,1	11,5
Portugal	4**	12	10	17,0	− 16,7
United Kingdom	241	109**	122	− 11,6	19,6

Other milk powder ([2])	1 000 t			% TAV	
	1985	1992	1993	$\frac{1992}{1985}$	$\frac{1993}{1992}$
EUR 12	838**	983∞	977∞	2,3	− 0,6
Belgique/België	39	43	52	1,4	20,9
Danmark	95	106	106	1,6	0,0
BR Deutschland	125	228∞	211∞	9,0	− 7,5
Elláda	0	0	0	×	×
España	5**	13	12	14,6	− 7,7
France	228	304	301	4,2	− 1,0
Ireland	29	31	31	1,0	0,0
Italia	3	3	8	0,0	×
Luxembourg	1	0	0*	×	×
Nederland	245	163	178	− 5,7	9,2
Portugal	7**	8	7	1,9	− 12,5
United Kingdom	61	84	71	4,7	− 15,5

Source : Eurostat.

([1]) Product weight.
([2]) Whole-milk powder, partly-skimmed-milk powder, cream-milk powder and buttermilk powder included.

4.20.1.6   Production in dairies of concentrated milk and casein ([1])

Concentrated milk (a) ([2])	1 000 t			% TAV	
	1985	1992	1993	$\frac{1992}{1985}$	$\frac{1993}{1992}$
1	2	3	4	5	6
EUR 12	1 511**	1 340**∞	1 312∞	− 1,7	− 2,1
Belgique/België	11	22	12	10,4	− 45,5
Danmark	8	11	13	4,7	18,2
BR Deutschland	527	503∞	514∞	− 0,7	2,2
España	66**	42	48	− 6,3	14,3
France	125	77	64	− 6,7	− 16,9
Ireland	54**	91**	91**	7,7	0,0
Italia	3	1**	1	− 14,5	0,0
Nederland	535	386	374	− 4,6	− 3,1
Portugal	1**	1	1	0,0	0,0
United Kingdom	181	206	191	1,9	− 5,8

Casein (b) ([3])	1 000 t			% TAV	
	1985	1992	1993	$\frac{1992}{1985}$	$\frac{1993}{1992}$
EUR 12	144**	145**∞	109∞	0,1	− 24,8
Danmark	15	15	13	0,0	− 13,3
BR Deutschland	21	20∞	13∞	− 0,7	− 35,0
France	45	39	26	− 2,0	− 33,3
Ireland	33	40**	32	2,8	− 20,0
Nederland	27**	30**	25	1,5	− 16,7
United Kingdom	3	1	0	− 14,5	×

Source : (a) Eurostat.
        (b) EC Commission, Directorate-General for Agriculture.

([1]) Product weight.
([2]) Including that of 'chocolate crumb'.
([3]) Excl. caseinates produced from casein.

4.20.2.1   World exports and production (¹) of — butter (²)
— cheese
— casein

	Production				% TAV		Exports				% TAV	
	1 000 t			%			1 000 t			%		
	1985	1991	1992	1992	1991/1985	1992/1991	1985	1991	1992	1992	1991/1985	1992/1991
1	2	3	4	5	6	7	8	9	10	11	12	13
Butter (²)												
World:	7 620	7 243	7 045	100,0	– 0,7	– 0,4	841	876	744	100,0	0,6	– 2,3
— EUR 12	2 033	1 816∞	1 686∞	23,9	– 1,6	– 1,1	387	322∞	242∞	32,5	– 2,6	– 4,0
— Australia	111	112	135	1,9	0,1	2,7	43	75	65	8,7	8,3	– 2,0
— New Zealand	243	269	268	3,8	1,5	– 0,1	217	251	183	24,6	2,1	– 4,4
— USA	566	606	619	8,8	1,0	0,3	44	32	96	12,9	– 4,4	17,0
— Canada	95	97	86	1,2	0,3	– 1,7	1	12	14	1,9	42,6	2,2
— USSR	1 605	1 512	1 432	20,3	– 0,8	– 0,8	17	5	3	0,4	– 16,0	– 7,0
— Czechoslovakia	152	133	118	1,7	– 1,9	– 1,7	21	51	34	4,6	13,5	– 5,6
— Poland	275	220	180	2,6	– 3,1	– 2,8	–	14	4	0,5	x	– 16,4
— India	700	1 020	106	1,5	5,5	– 27,6	–	–	–	0,0	x	x
— Others	1 840	1 458	2 415	34,3	– 3,3	7,5	111	114	103	13,8	0,4	– 1,4
Cheese												
World:	13 100	14 408	14 693	100,0	1,4	0,3	886	931	910	100,0	0,7	– 0,3
— EUR 12	4 428	5 414∞	5 589∞	38,0	2,9	0,5	404	484∞	465∞	51,1	2,6	– 0,6
— Australia	160	179	197	1,3	1,6	1,4	68	64	66	7,3	– 0,9	0,4
— New Zealand	118	125	142	1,0	0,8	1,8	86	109	115	12,6	3,4	0,8
— USA	2 305	2 730	2 943	20,0	2,4	1,1	16	12	15	1,6	– 4,0	3,2
— Canada	213	262	262	1,8	3,0	0,0	11	11	11	1,2	0,0	0,0
— USSR	814	763	592	4,0	– 0,9	– 3,6	5	2	1	0,1	– 12,3	– 9,4
— Switzerland	126	134	132	0,9	0,9	– 0,2	66	61	65	7,1	– 1,1	0,9
— Argentina	210	324	343	2,3	6,4	0,8	6	8	7	0,8	4,2	– 1,9
— Austria	83	80	81	0,6	– 0,5	0,2	43	30	27	3,0	– 5,0	– 1,5
— Others	4 643	4 397	4 412	30,0	– 0,8	0,0	181	150	138	15,2	– 2,6	– 1,2
Casein												
World:	:	:	:	x	x	x	:	:	:	x	x	x
— EUR 12	143	116∞	144∞	x	– 2,9	3,1	80	58∞	69∞	x	– 4,5	2,5
— Australia	8	2	3	x	– 18,0	6,0	8	2	3	x	– 18,0	6,0
— New Zealand	64	64	70	x	0,0	1,3	77	77	67	x	0,0	– 2,0
— Poland	33	21	12	x	– 6,3	– 7,7	22	12	14	x	– 8,3	2,2

Source: EC Commission, Directorate-General for Agriculture.

(¹) Product weight.
(²) Production in dairies, including butteroil made from cream (butter equivalent).

4.20.2.2   World exports and production of (¹) — whole-milk powder and skimmed-milk powder — concentrated milk

	Production						Exports					
	1000 t			%	% TAV		1000 t			%	% TAV	
	1985	1991	1992	1992	1991/1985	1992/1991	1985	1991	1992	1992	1991/1985	1992/1991
1	2	3	4	5	6	7	8	9	10	11	12	13
Skimmed-milk powder												
World:	4 657	3 852	3 660	100,0	−2,7	−5,0	1 140	918	1 072	100,0	−3,0	16,8
— EUR 12	1 948	1 478∞	1 162∞	31,7	−3,9	−21,4	307	253∞	390∞	36,4	−2,7	54,2
— Australia	134	149	163	4,5	−1,5	9,4	90	126	121	11,3	4,9	−4,0
— New Zealand	242	158	155	4,2	−5,9	−1,9	173	175	164	15,3	0,2	−6,3
— USA	630	398	396	10,8	−6,4	−0,5	344	62	100	9,3	−21,7	61,3
— Canada	99	77	55	1,5	−3,5	−28,6	61	51	34	3,2	−2,5	−33,3
— USSR	260	274	263	7,2	0,8	−4,0	–	–	–	0,0	x	x
— Poland	158	147	150	4,1	−1,0	2,0	41	85	123	11,5	11,0	44,7
— Sweden	58	32	30	0,8	−8,1	−6,3	29	17	4	0,4	−7,3	−76,5
— Others	1 128	1 139	1 286	35,1	0,1	12,9	95	149	136	12,7	6,6	−8,7
Whole-milk powder (²)												
World:	1 912	2 222	2 235	100,0	2,2	0,6	811	1 082	1 039	100,0	4,2	−4,0
— EUR 12	837	1 080∞	983∞	44,0	3,7	−9,0	484	618∞	581∞	55,9	3,6	−6,0
— Australia	45	60	67	3,0	4,2	11,7	32	50	52	5,0	6,6	4,0
— New Zealand	138	274	300	13,4	10,3	9,5	135	252	257	24,7	9,3	2,0
— USA	54	48	67	3,0	−1,7	39,6	1	9	28	2,7	36,9	x
— Canada	10	9	10	0,4	−1,5	11,1	26	5	9	0,9	−21,0	80,0
— USSR	257	250	248	11,1	−0,4	−0,8	–	–	–	0,0	x	x
— Argentina	85	69	72	3,2	−2,9	4,3	–	11	4	0,4	x	−63,6
— Finland	34	11	3	0,1	−14,9	−72,7	33	10	3	0,3	−15,7	−70,0
— Others	452	421	485	21,7	−1,0	15,2	100	127	105	10,1	3,5	−17,3
Concentrated milk												
World:	4 735	4 543	4 585	100,0	−0,6	0,9	760	422	439	100,0	−8,1	4,0
— EUR 12	1 511	1 310∞	1 340∞	29,2	−2,0	2,3	545	316∞	343∞	78,1	−7,5	8,5
— Australia	66	95	90	2,0	5,3	−5,3	4	3	3	0,7	−4,0	0,0
— New Zealand	3	1	1	0,2	−14,5	0,0	1	7	7	1,6	32,0	0,0
— USA	931	942	966	21,1	0,2	2,5	13	3	9	2,1	−18,9	x
— Canada	184	74	65	1,4	−12,2	−12,2	109	14	15	3,4	−25,4	7,1
— USSR	564	635	647	14,1	1,7	1,9	26	15	3	0,7	−7,6	−80,0
— Brazil	36	36	36	0,8	0,0	0,0	–	–	–	0,0	x	x
— India	363	395	400	8,7	1,2	1,3	–	–	–	0,0	x	x
— Others	1 077	1 055	1 040	22,7	−0,3	−1,4	62	64	59	13,4	0,5	−7,8

Source: EC Commission, Directorate-General for Agriculture.

(¹) Product weight.
(²) Whole-milk powder, partly-skimmed-milk powder, cream-milk powder and buttermilk powder included.

4.20.3.1   World trade in certain milk products — EC share

EUR 12

(1 000 t)

A. Exports	1987	1988	1989	1990	1991	1992	1993
1	2	3	4	5	6	7	8
Butter/butteroil (¹)							
— World	1 045	1 057	846	810	876	744	824
— EC	621	645	395	260	322∞	242∞	202∞
— Others	424	412	451	550	554	502	622
— EC share	59,4%	61,0%	46,7%	32,1%	36,8%	32,5%	24,5%
Skimmed-milk powder (¹)							
— World	1 127	1 304	991	928	918	1 072	1 030
— EC	390	617	410	330	253∞	390∞	284∞
— Others	737	687	581	590	665	682	746
— EC share	34,6%	47,3%	41,4%	35,6%	27,6%	36,4%	27,6%
Cheese							
— World	884	906	930	887	931	910	974
— EC	406	404	445	451	484∞	465∞	524∞
— Others	478	502	485	436	447	445	450
— EC share	45,9%	44,6%	47,8%	50,8%	52,0%	51,1%	53,8%
Whole-milk powder							
— World	892	935	916	886	1 082	1 039	1 057
— EC	573	602	573	510	618∞	581∞	588∞
— Others	319	333	343	376	464	458	469
— EC share	64,2%	64,4%	62,6%	57,6%	57,1%	55,9%	55,6%
Condensed milk							
— World	564	522	564	440	422	439	449
— EC	387	383	449	335	316∞	343∞	351∞
— Others	177	139	115	105	106	96	98
— EC share	68,6%	73,4%	79,6%	76,1%	74,9%	78,1%	78,2%
Casein and caseinates							
— EC	99**	80**	80**	61**	58**	69**	57

4.20.3.1 *(cont.)*

B. Imports	1987	1988	1989	1990	1991	1992	1993
1	2	3	4	5	6	7	8
Butter/butteroil ([1])							
— World	1 045	1 057	846	810	876	744	824
— EC	79	76	71	89	68∞	48∞	65∞
— Others	966	981	775	721	808	696	759
— EC share	7,6%	7,2%	8,4%	11,0%	7,8%	6,5%	7,9%
Cheese							
— World	884	906	930	887	931	910	974
— EC	109	115	119	113	109∞	110∞	109∞
— Others	775	791	811	774	822	800	865
— EC share	12,3%	12,7%	12,8%	12,7%	11,7%	12,1%	11,2%
Casein and caseinates							
— EC	26	29	36	63	58	54	59

Source : EC Commission, Directorate-General for Agriculture, GATT and FAO.

([1]) Food aid included.

4.20.4.2   **Detailed supply balance** (a) — **skimmed-milk powder**   **EUR 12**

(1 000 t)

1	1987	1988	1989	1990	1991	1992	1993
1	2	3	4	5	6	7	8
Opening stocks							
— private	:	:	:	:	:	:	:
— public (intervention)	772	473	10	5	333	414	47
Production							
— skimmed-milk powder (b) (¹)	1 628	1 313	1 421	1 624	1 478	1 162	1 220
— buttermilk powder	39	39	38	46	38	37	37
Imports (b)	2	5	53	14	5	3	19
Total availability	:	:	:	:	:	:	:
Consumption at full market prices	300	300	300	300	350	350	350
Subsidized consumption							
— animal feed (calves)	1 103	975	754	767	856	809	661
Special measures							
— pigs and poultry	:	–	–	–	–	–	–
Total consumption	1 403	1 275	1 054	1 067	1 206	1 159	1 011
Exports at world market prices	280	500	323	262	192	293	232
Food aid	110	117	87	68	61	99	52
Total exports	390	617	410	330	253	392	284
Closing stocks							
— private	:	:	:	:	:	:	:
— public (intervention)	473	10	5	333	414	47	37
Total	:	:	:	:	:	:	:

Source: (a) EC Commission, Directorate-General for Agriculture.
            (b) Eurostat.
(¹) Including buttermilk powder incorporated directly in animal feed, milk powder for babies.

4.20.4.3   Detailed supply balance (a) — butter (¹)

EUR 12

(1 000 t)

1	1987	1988	1989	1990	1991 ∞	1992 ∞	1993 ** ∞
	2	3	4	5	6	7	8
Opening stocks							
— private, aided by EC	83	98	100	104	84	41	68
— public (intervention)	1 283	860	102	20	251	261	172
Production							
— dairy (b)	1 887	1 682	1 705	1 771	1 816	1 686	1 676
— farm (b)	27	26	26	25	23	21*	21
Imports	79	76	71	89	68	48	65
Total availability	3 360	2 743	2 004	2 008	2 242	2 057	2 002
Consumption							
— at normal prices	1 235	1 262	1 137	1 090	1 186	1 138	1 125
— at reduced prices (²)	0	0	0	0	0	:	:
Special schemes (³)	361	443	348	366	432	437	467
Reg. No 2409/86	184	190	:	:	:	:	:
Total apparent consumption	1 780	1 895	1 485	1 456	1 618	1 575	1 592
Exports at world market prices	272	275	377	200	216	197	194
Food aid	30	47	18	13	10	5	1
Exports at special prices	319	323	:	5	96	45	7
Total exports (b)	621	645	395	218 (⁴)	322	242	202
Closing stocks							
— private, aided by EC	98	100	104	84	41	68	47
— public (intervention)	860	102	20	251	261	172	161
Total closing stocks	958	202	124	335	302	240	208

Source: (a) EC Commission, Directorate-General for Agriculture (including butteroil, butter equivalent).
　　　　 (b) Eurostat.

(¹) Product weight. Includes butteroil made from cream (butter equivalent).
(²) 1977 : Reg. No 2370/77 (Christmas butter),
　　 1978 : Reg. No 1901/78,
　　 1979 : Reg. No 1269/79.
(³) Comprising (1 000 t) :

	1987	1988	1989	1990	1991	1992	1993
— Welfare schemes	25	13	16	19	22	16	15
— Armed forces and non-profit organizations	44	59	42	39	39	36	38
— Butter concentrate	34	43	16	16	19	19	22
— Sales to food processors	258	328	274	292	352	364	392

(⁴) Not including 42 000 t physically exported from the former GDR to the Soviet Union.

4.20.6.1   **Intervention measures for butter and skimmed-milk powder (1993)**

(t)

Butter (¹)	Taken into storage	Public storage					Private storage
		Release from storage					Quantity subject to storage contracts
		On the Community market (⁴)	For export (⁵)	For food aid (³)	Total		
1	2	3	4	5	6		7
EUR 12	32 169	37 478	6 578	0	44 056		162 056
Belgique/België	0	587	0	0	587		19 972
Danmark	0	1 789	0	0	1 789		2 660
BR Deutschland	6 335	10 357	500	0	10 857		18 650
Elláda	0	0	0	0	0		0
España	10 444	5 131	0	0	5 131		0
France	0	2 399	0	0	2 399		36 572
Ireland	9 026	8 235	6 078	0	14 313		12 999
Italia	1 386	1 300	0	0	1 300		1 001
Luxembourg	0	0	0	0	0		0
Nederland	122	1 605	0	0	1 605		64 584
Portugal	744	1 258	0	0	1 258		0
United Kingdom	4 112	4 817	0	0	4 817		5 618

Skimmed-milk powder (²)	Taken into storage	Release from storage				Private storage
		To the Community market (⁴)	For export (⁵)	For food aid	Total	Quantity subject to storage contracts
EUR 12	10 709	20 667	206	0	20 873	0
Belgique/België	0	12	0	0	12	0
Danmark	0	0	0	0	0	0
BR Deutschland	4 574	8 559	206	0	8 765	0
Elláda	0	0	0	0	0	0
España	0	505	0	0	505	0
France	0	0	0	0	0	0
Ireland	4 409	11 591	0	0	11 591	0
Italia	0	0	0	0	0	0
Luxembourg	0	0	0	0	0	0
Nederland	0	0	0	0	0	0
Portugal	0	0	0	0	0	0
United Kingdom	1 726	0	0	0	0	0

Source : EC Commission, Directorate-General for Agriculture.

(¹) In accordance with Regulation (EEC) No 804/68, Article 6.
(²) In accordance with Regulation (EEC) No 804/68, Article 7.
(³) Including quantities removed under Regulation (EEC) No 2315/76 (Regulation (EEC) No 2200/87).
(⁴) Including quantities refused.
(⁵) Including emergency aid delivered to East European countries.

4.20.6.2   **Application of the quota system**

(1 000 t)

1	1993/1994					1994/1995
	Overall guaranteed quantity (¹)	Deliveries (p) (²)	Adjustment of oil and fat (³)	Transfers (⁴)	Difference after the adjustment	Overall guaranteed quantity
1	2	3	4	5	6 = 3 + 4 − 2 − 5	7
EUR 12	106 498 294	102 629 611	2 612 087	3 324 704	− 4 581 300	106 498 294
Belgique/België	3 066 337	2 978 144	179 149	80 894	10 062	3 066 337
Danmark	4 454 459	4 433 000	31 661	0	10 202	4 454 459
BR Deutschland	27 764 778	25 321 154	1 179 268	1 255 057	− 2 519 413	27 764 778
Elláda	625 985	584 718	17 261	4 832	− 28 838	625 985
España	5 200 000	5 252 028	12 978	172 684	− 107 678	5 200 000
France	23 637 283	23 011 880	484 262	0	− 141 141	23 637 283
Ireland	5 233 805	5 211 740	− 15 958	326 389	− 364 412	5 233 805
Italia	9 212 190	9 461 136	:	:	248 946	9 212 190
Luxembourg	268 098	253 111	13 661	0	− 1 326	268 098
Nederland	10 983 195	10 578 741	431 800	472 165	− 444 819	10 983 195
Portugal	1 804 881	1 480 384	7 994	316	− 316 819	1 804 881
United Kingdom	14 247 283	14 063 575	270 011	1 012 367	− 926 064	14 247 283

Source : EC Commission, Direcotrate-General for Agriculture.

(¹) Former columns 2, 3 and 4 are aggregated in a new column 2, pursuant to Article 3 of Regulation (EEC) No 3950/92.
(²) Replies from Member States.
(³) Article 2 (2) of Regulation (EEC) No 536/93.
(⁴) Article 4 (2) of Regulation (EEC) No 3950/92.

4.20.6.3   **Community butter and skimmed-milk powder stocks** (^{1}) **on 1 April**

	t				
	1990	1991	1992	1993	1994
1	2	3	4	5	6
Butter (^{2})					
EUR 12	81 988	323 515	273 773	195 969	161 755
Belgique/België	2 055	8 072	4 281	3 161	554
Danmark	3	6 872	7 048	2 046	1 052
BR Deutschland	17 660	40 238	41 366	39 044	18 544
Elláda	–	–	–	–	–
España	8 825	32 643	34 756	34 613	40 306
France	8 120	24 294	16 806	12 516	12 882
Ireland	9 996	94 214	102 308	58 090	55 798
Italia	1 883	8 377	10 663	7 278	7 508
Luxembourg	–	–	–	–	–
Nederland	27 502	83 077	35 682	26 976	18 609
Portugal	–	–	848	1 994	1 332
United Kingdom	5 944	25 729	20 015	10 251	5 170
Skimmed-milk powder (^{2})					
EUR 12	21 278	354 191	335 719	30 663	40 847
Belgique/België	–	17 031	8 638	487	691
Danmark	–	–	–	–	–
BR Deutschland	15 904	191 891	144 138	8 122	5 987
Elláda	–	–	–	–	–
España	5 314	19 731	10 356	799	294
France	–	21 243	16 671	37	37
Ireland	–	94 261	147 161	20 738	31 632
Italia	–	–	–	–	–
Luxembourg	–	1 200	925	–	–
Nederland	–	2 624	2 088	–	–
Portugal	–	–	107	–	–
United Kingdom	–	6 810	5 635	480	2 206

Source: EC Commission, Directorate-General for Agriculture.

(^{1}) Stocks referred to in Article 6 of Regulation (EEC) No 804/68 (butter, public and private storage; skimmed-milk powder, public storage).
(^{2}) Product weight.

4.20.6.4 **Quantities of skimmed milk and skimmed-milk powder intended for animal feed and of skimmed milk processed into casein and caseinates, for which aids have been granted**

	Skimmed milk (1)(2)					Skimmed-milk powder (2)					Skimmed milk for casein (2)				
	1 000 t			% TAV		1 000 t			% TAV		1 000 t			% TAV	
	1985	1992 ∞	1993 ∞	1992/1985	1993/1992	1985	1992 ∞	1993 ∞	1992/1985	1993/1992	1985	1992 ∞	1993 ∞	1992/1985	1993/1992
1	2	3	4	5	6	7	8	9	10	11	12	13	14	15	16
EUR 12	4 234	581	424	− 24,7	− 27,0	1 126	800	675	− 4,8	− 15,6	5 447	5 037	3 939	− 1,1	− 21,8
Belgique/België	314	90	78	− 16,3	− 14,2	26	23	21	− 1,8	− 10,9	0	0	0	x	x
Danmark	546	23	16	− 36,6	− 29,4	14	2	1	− 25,3	− 22,3	592	526	459	− 1,7	− 12,7
BR Deutschland	2 064	220	143	− 27,4	− 34,9	175	76	62	− 11,3	− 18,2	749	704	451	− 0,9	− 36,0
France	0	0	0	x	x	0	7	2	x	− 71,3	0	0	0	x	x
Ireland	147	43	29	− 16,1	− 33,4	488	330	282	− 5,4	− 14,4	1 709	1 350	924	− 3,3	− 31,6
Italia	391	7	15	− 43,4	x	21	6	5	− 16,0	− 14,4	1 213	1 366	1 204	1,7	− 11,8
Luxembourg	119	166	123	4,9	− 25,7	148	129	96	− 1,9	− 25,4	0	0	0	x	x
Nederland	68	10	4	− 24,3	− 53,5	232	212	194	− 1,3	− 8,6	1 081	1 073	900	− 0,1	− 16,2
United Kingdom	584	22	16	− 37,2	− 30,1	23	16	12	− 4,8	− 23,8	104	19	1	− 21,6	− 93,1

Source: EC Commission, Directorate-General for Agriculture.
(1) Normal aid + special aid.
(2) Product weight.

4.21.4.1. Supply balance — honey

1	EUR 12	BLEU/UEBL	Danmark	BR Deutsch-land	Elláda	España	France	Ireland	Italia	Nederland	Portugal	United Kingdom
	2	3	4	5	6	7	8	9	10	11	12	13
1991/92												
Usable production (1 000 t)	127	0	3	25	14	25	36**	0	12**	1	6**	5**
Imports	135	7	4	90	2	7	9	1	13	9	0	17
Exports	9	2	2	12	0	7	5	0	0	1	0	1
Intra-Community trade	24	2	1	3	2	1	7	1	2	5	0	1
Internal use : — human consumption	252	6	5	103	16	24	39	1	25	9	6	21
Human consumption (kg/head/year)	0,7	0,6	1,0	1,3	1,6	0,6	0,7	0,4	0,4	0,6	0,6	0,4
Self-sufficiency (%)	50,3	3,3	58,8	24,3	87,5	102,5	91,4	7,1	48,8	8,9	93,3	23,7
1992/93												
Usable production (1 000 t)	118	1*	2	25*	14*	25*	31*	0*	10	1*	4	5
Imports	155	6	4	78	2	14	7	1	11	7	0	23
Exports	10	2	1	12	0	8	4	0	1	1	0	1
Intra-Community trade	21	1	1	4	1	1	5	1	2	4	0	1
Internal use : — human consumption	262	5	5	91	16	32	34	1	20	7	4	27
Human consumption (kg/head/year)	0,8	0,5	1,0	1,1	1,6	0,8	0,6	0,4	0,4	0,5	0,4	0,5
Self-sufficiency (%)	40,0	19,2	37,0	27,4	86,5	79,2	91,2	7,0	49,4	11,0	93,2	18,8

Source: Eurostat and EC Commission, Directorate-General for Agriculture.

4.22.1.1 Community forestry statistics

	Belgique/België	Danmark	BR Deutschland	Elláda	España	France	Ireland	Italia	Luxembourg	Nederland	Portugal	United Kingdom	EUR 12
1	2	3	4	5	6	7	8	9	10	11	12	13	14
Total area (1 000 ha) [1]	3 051,805	4 309,245	24 869,340	13 195,7	50 476,55	54 908,70	7 028,336	30 127,680	258,636	4 041,960	9 207,095	24 413,88	225 888,90
Wooded area (1 000 ha) [1]	617,000	493,294	7 360,031	5 755,0	12 511,00	14 688,33	327,000	6 410,066	88,620	330,175	2 986,300	2 297,00	53 845,81
Wooded area/total area (%)	20%	11%	30%	44%	25%	27%	5%	21%	34%	8%	32%	9%	24%
Breakdown of wooded area by species [2]													
— conifers (%)	47%	63%	69%	19%	48%	30%	90%	25%	34%	65%	46%	73%	42%
— deciduous (%)	53%	37%	31%	81%	52%	70%	10%	75%	66%	35%	54%	27%	58%
Breakdown of wooded area by ownership [2]													
— State forests (%)	11%	30%	31%	73%	6%	10%	79%	6%	8%	30%	3%	43%	27%
— private forests (%)	53%	66%	44%	15%	65%	72%	20%	60%	54%	53%	83%	57%	58%
— other forests under public law (%)	36%	4%	25%	12%	29%	18%	1%	34%	38%	17%	14%	0%	15%
Production of timber in the rough (without bark) (1 000 m³) [3]	5 082 [4]	2 300	44 874	2 345	17 272	44 752	1 677	8 393	–	1 351	11 181	6 409	145 636
Timber consumption in round wood equivalent (1 000 m³) [3]	7 112 [4]	2 072	37 940	2 884	20 007	41 544	1 515	15 772	–	1 539	11 405	6 371	148 161
Deficit or surplus in timber production (1 000 m³) [3]	– 2 030 [4]	228	6 934	– 499	– 2 735	– 3 208	162	– 7 379	–	– 188	– 224	38	– 2 525

Source : EC Commission, Directorate-General for Agriculture.

[1] 1990.
[2] Ø 1976-86.
[3] 1992.
[4] BLEU/UEBL.

EUR 12

4.22.3.1   EC external trade in forest products

		1 000 t			Mio ECU		
1	2	Export 3	Import 4	Balance 5	Export 6	Import 7	Balance 8
Timber in the rough, other small round wood, cut or as particles	1991	16 064,0	15 913,1	150,9	436,2	1 091,4	−655,2
	1992	5 120,1	6 207,7	−1 087,6	340,2	1 048,8	−708,6
	1993	4 211,8	5 383,5	−1 171,7	287,1	921,0	−633,9
Sawn timber	1991	3 313,4	17 880,2	−14 566,8	322,2	5 980,9	−5 658,8
	1992	513,4	15 701,3	−15 187,9	288,8	5 947,0	−5 658,1
	1993	605,7	15 390,9	−14 785,2	342,8	5 811,3	−5 468,6
Panels and sheets	1991	4 452,1	7 856,1	−3 804,0	469,1	2 062,1	−1 592,9
	1992	658,2	4 060,1	−3 401,9	471,3	2 098,7	−1 627,4
	1993	1 112,5	3 339,7	−2 827,2	612,4	2 135,5	−1 523,1
Wooden articles	1991	1 274,9	2 373,4	−1 098,5	665,1	1 330,7	−665,6
	1992	269,3	1 674,4	−1 405,1	677,4	1 546,6	−869,2
	1993	287,3	1 770,6	−1 483,3	647,9	1 828,4	−1 180,5
Paper stock/pulp wood	1991	7 242,4	16 446,5	−9 204,1	358,0	4 510,8	−4 152,7
	1992	2 211,4	11 029,5	−8 818,1	398,9	4 361,0	−3 962,1
	1993	2 177,8	10 684,2	−8 506,3	316,8	3 670,3	−3 353,6
Paper and board	1991	16 775,1	31 411,6	−14 636,5	5 164,2	13 498,0	−8 333,8
	1992	4 416,7	18 858,6	−14 441,9	5 477,8	12 787,6	−7 309,9
	1993	4 919,2	18 926,6	−14 007,5	6 045,9	11 738,2	−5 692,3
Cork and cork articles	1991	165,8	129,2	36,6	226,2	29,0	197,2
	1992	50,1	12,6	37,6	235,7	30,5	205,3
	1993	49,8	16,8	33,0	239,8	33,4	206,4
Total	1991	49 287,7	92 009,9	−42 722,2	7 641,0	28 502,8	−20 861,8
	1992	13 239,2	57 544,1	−44 304,9	7 890,1	27 820,2	−19 930,2
	1993	13 364,1	56 112,2	−42 748,2	8 492,6	26 138,2	−17 645,6

Source : Eurostat and EC Commission, Directorate-General for Agriculture.

4.22.3.2 EC external and intra-Community trade in timber and timber products by Member State

1993 (1 000 t)

	BLEU/UEBL	Danmark	BR Deutschland ∞	Elláda	España	France	Ireland	Italia	Nederland	Portugal	United Kingdom	EUR 12 ∞
1	2	3	4	5	6	7	8	9	10	11	12	13
A — Imports												
Intra: Timber in the rough (round, cut or as particles)	..	..	..	..	..	..	..	..	..	..	..	..
Sawn timber	..	..	..	..	..	..	..	..	..	..	..	..
Panels and sheets	..	..	..	..	..	..	..	..	..	..	..	..
Wooden articles	..	..	..	..	..	..	..	..	..	..	..	..
Paper stock/pulpwood	..	..	..	..	..	..	..	..	..	..	..	..
Paper and board	..	..	..	..	..	..	..	..	..	..	..	..
Cork and cork articles	..	..	..	..	..	..	..	..	..	..	..	..
Extra: Timber in the rough (round, cut or as particles)	74,0	192,0	722,4	230,6	284,5	613,7	19,6	2 696,8	124,6	280,0	145,3	5 383,5
Sawn timber	725,0	986,9	3 304,5	259,2	586,7	930,1	173,0	3 163,0	1 681,9	70,4	3 510,1	15 390,8
Panels and sheets	281,6	211,9	1 332,3	56,7	30,8	201,8	36,9	475,4	353,2	3,7	955,3	3 939,7
Wooden articles	63,3	54,2	1 073,9	17,9	26,8	75,5	8,9	116,9	166,7	7,1	159,5	1 770,6
Paper stock/pulpwood	349,4	67,2	3 254,1	82,3	504,8	1 827,4	13,8	2 371,3	559,2	56,9	1 597,7	10 684,2
Paper and board	869,5	876,5	5 108,9	280,7	1 151,9	1 957,6	181,4	1 991,7	1 665,3	135,7	4 707,4	18 926,6
Cork and cork articles	0,0	0,0	1,4	0,0	0,7	1,9	0,0	7,1	0,0	4,5	1,1	16,8
B — Exports												
Intra: Timber in the rough (round, cut or as particles)	..	..	..	..	..	..	..	..	..	..	..	..
Sawn timber	..	..	..	..	..	..	..	..	..	..	..	..
Panels and sheets	..	..	..	..	..	..	..	..	..	..	..	..
Wooden articles	..	..	..	..	..	..	..	..	..	..	..	..
Paper stock/pulpwood	..	..	..	..	..	..	..	..	..	..	..	..
Paper and board	..	..	..	..	..	..	..	..	..	..	..	..
Cork and cork articles	..	..	..	..	..	..	..	..	..	..	..	..
Extra: Timber in the rough (round, cut or as particles)	60,5	174,4	3 581,0	6,2	23,7	276,6	16,0	7,0	37,4	20,8	8,3	4 211,8
Sawn timber	24,1	30,7	213,2	57,4	48,8	109,7	0,4	49,1	11,9	56,8	3,5	605,7
Panels and sheets	102,5	65,1	286,7	36,0	182,4	94,3	14,8	230,9	7,1	68,9	23,9	1 112,5
Wooden articles	17,1	24,9	118,2	1,9	17,8	38,5	1,2	35,8	14,7	2,7	14,4	287,3
Paper stock/pulpwood	155,6	222,1	782,0	9,5	68,4	143,7	0,0	63,7	456,5	215,8	60,6	2 177,8
Paper and board	217,8	118,5	1 994,2	33,0	282,3	811,9	0,9	474,1	395,9	80,6	509,9	4 919,2
Cork and cork articles	0,0	0,1	1,2	0,0	5,1	0,5	0,0	1,2	0,1	40,7	0,9	49,8

Source: Eurostat and EC Commission, Directorate-General for Agriculture.

European Commission

The Agricultural Situation in the European Union — 1994 Report

Luxembourg: Office for Official Publications of the European Communities
1995 – 432 pp., 20 figs – 16.2 × 22.9 cm

ISBN 92-826-8676-0

Price (excluding VAT) in Luxembourg: ECU 35

This report is the 20th published version of the annual report on the agricultural situation in the European Union. It contains analyses and statistics on the general situation (economic environment and world market), the factors of production, the structures and situation of the markets in the various agricultural products, the obstacles to the common agricultural market, the position of consumers and producers, and the financial aspects. The general prospects and the market outlook for agricultural products are also dealt with.